Introduction to Food and Agribusiness Management

Gregory A. Baker
Santa Clara University

Orlen Grunewald
Kansas State University

William D. Gorman
New Mexico State University

Upper Saddle River, New Jersey 07458

Library of Congress Cataloging-in-Publication Data
Baker, Gregory A.
Introduction to food and agribusiness management / Gregory A. Baker, Orlen Grunewald, William D. Gorman.
p. cm.
Includes index.
ISBN 0-13-014577-7
1. Agricultural industries—Management. 2. Food industry and trade—Management. 3. Food—Marketing—Case studies. 4. Corporations—Finance—Case studies. 5. Industries—Environmental aspects—Case studies. 6. Corporation law. I. Grunewald, Orlen. II. Gorman, William D. III. Title.

HD9000.5 .B313 2002
630'.68—dc21 2001021969

V.P., Editor-in-Chief: *Stephen Helba*
Executive Editor: *Debbie Yarnell*
Associate Editor: *Kate Linsner*
Associate Editor: *Kimberly Yehle*
Production Editor: *Lori Dalberg, Carlisle Publishers Services*
Production Liaison: *Eileen O'Sullivan*
Director of Manufacturing & Production: *Bruce Johnson*
Managing Editor: *Mary Carnis*
Manufacturing Manager: *Ed O'Dougherty*
Marketing Manager: *Jimmy Stephens*
Senior Design Coordinator: *Miguel Ortiz*
Cover Design: *Steven Frim*
Composition and Interior Design: *Carlisle Communications, Ltd.*

Prentice-Hall International (UK) Limited, *London*
Prentice-Hall of Australia Pty. Limited, *Sydney*
Prentice-Hall of Canada Inc., *Toronto*
Prentice-Hall Hispanoamericana, S.A., *Mexico*
Prentice-Hall of India Private Limited, *New Delhi*
Prentice-Hall of Japan, Inc., *Tokyo*
Prentice-Hall Singapore Pte. Ltd.
Editora Prentice-Hall do Brasil, Ltda., *Rio de Janeiro*

ISBN 0-13-014577-7

Dedication

To Cecil and Mary Kay Baker, for teaching me
the important lessons in life and
Louise, Natasha, Michelle, and Scott,
for their love and laughter.

Gregory A. Baker

To my parents, James and Isabelle,
who instilled a love for learning and
my wife, Katharine,
who makes my life worthwhile.

Orlen Grunewald

To my late wife, Gail, whose love and
encouragement for 43 years allowed me to
pursue my career in teaching.

William D. Gorman

CONTENTS

CHAPTER 11 HUMAN RESOURCES MANAGEMENT 311

PREFACE

OVERVIEW

Our goal in writing this book is to provide students and instructors with a book that approaches the management of food and agribusiness firms from a managerial perspective, in a highly understandable style. This book covers many of the areas of expertise that today's managers must master—finance, marketing, operations, forms of business ownership, organizational management, and human resources. Today's food and agribusiness industry managers must also confront a host of issues unique to the industry. These include the vagaries of nature and weather, the politics of agricultural policy and international trade, food safety risks, environmental risks, and issues associated with emerging technologies. In organizing and presenting the material we have paid special attention to the distinct challenges faced by managers in the food and agribusiness system.

Collectively, we have over 75 years experience teaching and practicing the management of food and agribusiness firms. We have taught students at all stages of their careers—undergraduate students, graduate students, middle managers, and executives. And we have taught in a variety of settings—public and private universities, and in Colleges of Agriculture and Schools of Business. The depth and diversity of our experiences have provided us with the perspective from which we have written this book.

TARGET AUDIENCE

This book is written primarily for students in an introductory course in agribusiness management or food business management. Such a course is often required to lay the foundation for students majoring in agribusiness management or a related major. Many students of other agricultural disciplines, such as Animal Science, Agronomy, Soil Science, Horticulture, or Agricultural Engineering also take an introductory course in agribusiness or food business management to broaden their education or because they have management aspirations. Because of the comprehensive nature of the book, it will also serve as a good resource for anyone who wants a broad overview of business management as it applies to food and agribusiness firms.

INSTRUCTOR'S MANUAL

The instructor's manual, which is available to educators using this book, provides instructors with a variety of supporting materials for teaching with the book. Instructors, both new and experienced, teaching in the field of food and agribusiness firm management will find the manual a highly useful accompaniment to the textbook. The instructors' manual includes:

- Teaching outlines for each chapter
- PowerPoint slides for use in presentations (these may be modified by the instructor to suit specific needs)
- Problem assignments designed to reinforce key concepts
- Case teaching notes
- Answers to review questions
- Sample quizzes and answers
- Sample examination questions and answers

KEY FEATURES

Managerial Perspective

Successful food and agribusiness firm managers must first and foremost be good managers. We believe that managers in the food and agribusiness industry must be trained as well as their counterparts in other industries. No amount of technical skills will compensate for the lack of sound management skills. For this reason we have organized the textbook along the same lines as a business management text. We cover specific management tools including brand management, human resources management, operations management, and financial management. The unique characteristics of managing in the food and agribusiness industry are addressed through the application of the concepts and in the examples.

Focus on Application

Another distinguishing feature of this textbook is the emphasis on the application of the concepts. The business management tools that we have included in the book are everyday tools used by successful business managers. Their value lies in the application of the techniques we cover to problems faced by real-world businesses. For this reason, we clearly explain the purpose of each tool and illustrate its use through examples. Students are given the opportunity to reinforce their learning of the most important concepts through the case studies.

Case Studies

The case studies that lead off every chapter, except the introductory chapter, are a distinguishing feature of this textbook. They serve to indicate the importance and relevance of the subject matter contained in each chapter and as a mechanism for integrating and applying each chapter's content. The use of case study analysis and discussion is a time-tested strategy for teaching business management. Students and professors will find the use of our case studies a refreshing change from the traditional course lecture.

Readability

We anticipate that most students reading this book will have little or no background in business management or agribusiness management. We also understand that our readers' knowledge of agriculture will vary from very little to extensive. Because this book is intended for use in an introductory course, we introduce each business management concept as if the reader is seeing it for the first time. New terms are identified in italics and defined for the reader. Examples of complex concepts are provided.

Stand-Alone Chapters

We have designed each chapter to be as independent from other chapters as possible. We have done this for several reasons. Most importantly, instructors often want to present the material in an order different from the way it is presented in the book. Second, instructors often choose not to, or don't have sufficient time to, cover every chapter in the book. Lastly, because this book can also be a resource for food and agribusiness firm managers, it is important that the chapters be largely self-contained. For these reasons, we have written the book so that each chapter can be easily understood without having read prior chapters. The lone exception to this rule is Chapter 3, Financial Statements. Because an understanding of financial terms is critical to understanding many other business management concepts, we recommend covering this chapter early.

Organization of the Book

The book is organized along the lines of the major functional areas of business management. Chapter 1 provides an introduction to the subject of managing food and agribusiness firms. Chapter 2 discusses the forms of business ownership. This is followed by four chapters addressing the financial aspects of managing a business. Chapter 3 covers financial statements, Chapter 4 covers financial analysis and budgeting, Chapter 5 discusses sources of financing, and Chapter 6 addresses capital budgeting and investment analysis. The next two chapters are devoted to marketing, with Chapter 7 addressing strategic marketing and Chapter 8 covering the marketing decisions known as the marketing mix. Chapter 9 addresses the operational aspects of management. In Chapter 10 we discuss organizational management. Lastly, human resources management is covered in Chapter 11.

ACKNOWLEDGMENTS

We would like to thank the numerous people who contributed to the publication of this book. In particular, we would like to thank the students who have used earlier versions of this manuscript and who provided feedback (whether we asked for it or not). We wish to acknowledge and thank our colleagues who reviewed the text and offered their suggestions for improvement: John W. Siebert, Texas A&M University, College Station; and Robert H. Usry, North Carolina State University. We are grateful, too, for the assistance of Lydia Duran, who typed several drafts of this manuscript.

COMMENTS

We tried to make this textbook as comprehensive as possible without being unduly long. The authors of any introductory textbook must balance the concern for comprehensiveness versus conciseness. This involves determining what to include and what to omit, what to address in detail and what to simplify. We hope we have chosen well. We have attempted to ensure that the book is free of errors—factual, grammatical, and otherwise. If you have comments, we would like to hear from you. Please send them to Gregory A. Baker by fax: 408-554-5167, e-mail: gbaker@scu.edu, or mail: Food and Agribusiness Institute, Leavey School of Business, Santa Clara University, 500 El Camino Real, Santa Clara, California, 95053-0396, U.S.A.

Gregory A. Baker

Orlen Grunewald

William D. Gorman

CHAPTER 1

INTRODUCTION

LEARNING OBJECTIVES

In this chapter we will cover the following topics:

- Importance of the food and fiber system in the U.S. economy
- Overview of the food and agribusiness industry
- Description of the input supply, production, and processing and distribution sectors
- Emerging trends and challenges in the food and agribusiness industry

INTRODUCTION

The importance of the food and agribusiness industry can be measured in many ways. Food is a basic necessity for all people. Most countries are concerned about food security; they want to ensure an adequate quantity of food for their citizens as well as guarantee the safety of the food supply. In developed countries, such as the United States, the food and agribusiness system does much more than fulfill one of life's basic necessities. It is a key component of our lifestyle, providing basic nutrition, pleasure, convenience, and many other attributes that today's consumers have come to expect and demand.

In the following sections, the importance of the U.S. food and agribusiness system is explored and the major components of the system are described. We close the chapter with a discussion of some of the major trends and challenges facing food and agribusiness firms at the start of the twenty-first century.

GROWTH OF THE FOOD AND AGRIBUSINESS SYSTEM

In the formative days of the United States, the late 1700s, 90 percent of Americans lived on farms. Most families were largely self-sufficient in terms of meeting their basic food needs. They produced most of what they needed right on the farm. Then, as now, the production of food and fiber involved the same basic steps, although the specific tools and technologies employed were probably different. Farmers produced a variety of crops, including grains, fruits, and vegetables, and would typically raise their own meat. Processing occurred on the farm. Fruits and vegetables were canned for storage and use throughout the year, grain was ground, and meats were dried, salted, or smoked in order to preserve them. Farmers also provided most of their own inputs. Seed from one year's crop was saved to be planted the following year and manure from the animals was used to fertilize the land. Part of the land had to be used to grow feed for draft animals to pull implements and provide transportation. This self-sufficiency, however, came at a cost. After the family's basic food needs were met, there was often very little left over to take to the market to trade for other items that were not produced on the farm.

Our modern economic system has evolved largely because of increases in the efficiency with which we perform basic activities such as food production. At the farm level, economic gains are due to specialization and the substitution of capital for labor. Most commercial farms are highly mechanized, efficient enterprises specializing in the production of one or a few commodities. A midwestern farmer, for example, may produce corn or soybeans as a cash crop.

FIGURE 1–1

Percent of Consumption Expenditures on Food, by Country, 1994

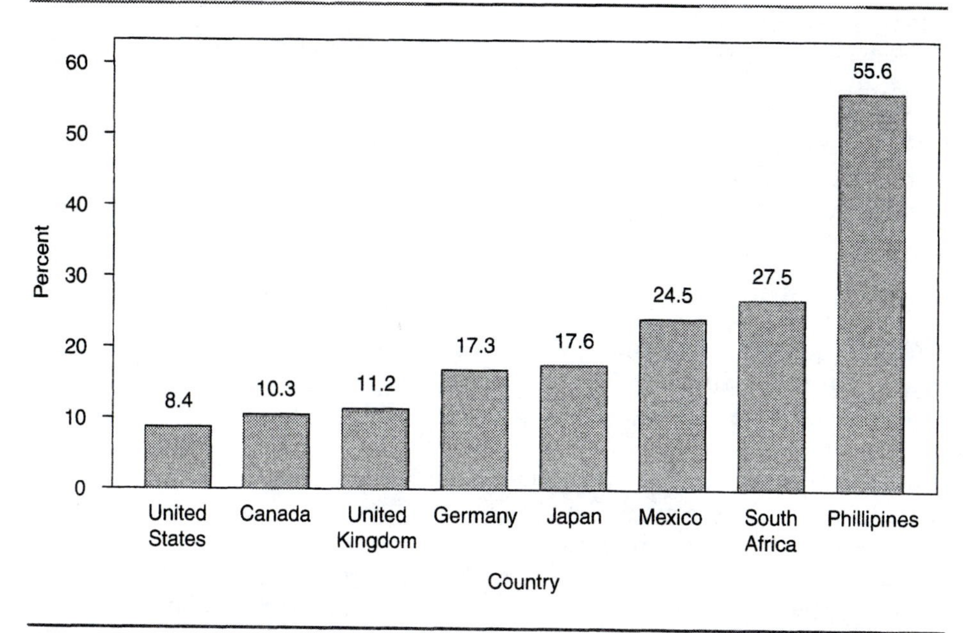

Source: Food Consumption, Prices, and Expenditures, 1970–97, United States Department of Agriculture, Table 101.

Very little, if any, of the crop remains on the farm. All of the seed is likely bought from a seed dealer and most of the crop is probably sold for cash.

By producing necessities, such as food, more efficiently, more workers have been freed to produce the luxuries that make life easier and more enjoyable. Most Americans no longer live and work on farms but live in cities and rural communities and work in industries such as manufacturing, telecommunications, or service. Consequently, Americans spend relatively little of their income on food, compared to people in other countries (Figure 1–1).

The U.S. food and agribusiness industry is the world's largest. It accounts for approximately $1 trillion, or about 13 percent, of the U.S. Gross Domestic Product. It employs approximately 23 million people, 17 percent of the nation's work force. The food and agribusiness industry also contributes to the U.S. and world economy in other ways. With exports of approximately $50 billion, the United States is the world's largest exporter of agricultural products. Moreover, U.S. exports comprise approximately 70 percent of the world trade in corn, 60 percent of the world trade in soybeans, and 30 percent of the world trade in wheat.

THE MODERN FOOD AND AGRIBUSINESS SYSTEM

Broadly speaking, the modern food and agribusiness system is comprised of three sectors. These sectors, also known as the agricultural supply chain, include the input supply sector, the farm sector, and the processing and distribution sector. Figure 1–2 illustrates some of the major industries and firms in each of the three sectors. Although it is necessarily incomplete because of the complexity of the system, the figure is representative of the industries and firms making up the food and fiber system.

Input Supply Sector

The input supply sector is made up of those firms that supply inputs to be used in agricultural production. The input supply sector contributes approximately $300 billion to the U.S. economy, or roughly 30 percent of the value added by the food and fiber system.

The major inputs used by producers include land, buildings, water, feed, seed, livestock and poultry, fertilizer, pesticides, machinery, energy, labor, credit, and management information. Figure 1–3 summarizes the major production expenses incurred by farmers. Of course, the major inputs used by a producer depend on the type of crop and production system. A Washington apple grower will have very different input needs than the operator of a California nursery. Likewise, a Texas cattle rancher will purchase different inputs than a North Carolina poultry producer.

The input supply sector is responsible for much of the productivity gains in the food and fiber system. Increased yields due to improved genetics in the seed and livestock industries, more effective chemicals, more efficient machines, and better management information are some of the major factors contributing to the productivity improvements. In recent years there has been a great deal of consolidation in the input sector, particularly among seed, agricultural chemical, and farm machinery firms. Today, a small number of firms control most of the production in these industries. However, the distribution of the products to producers is still largely handled by many small, locally based dealers. Below, we describe some of the most important and distinctive members of the input supply industry.

Seed. The seed industry is highly concentrated with a few firms producing most of the seed for any given crop. However, different companies compete in the market for different types of seed. For example, Pioneer Hi-Bred and Monsanto are major competitors in seeds for row crops, while Seminis is the world's largest producer of fruit and vegetable seed. Most seed is sold through independent dealers, or distributors, that cover a local area.

FIGURE 1–2

A Simplified Diagram of the U.S. Food and Agribusiness System

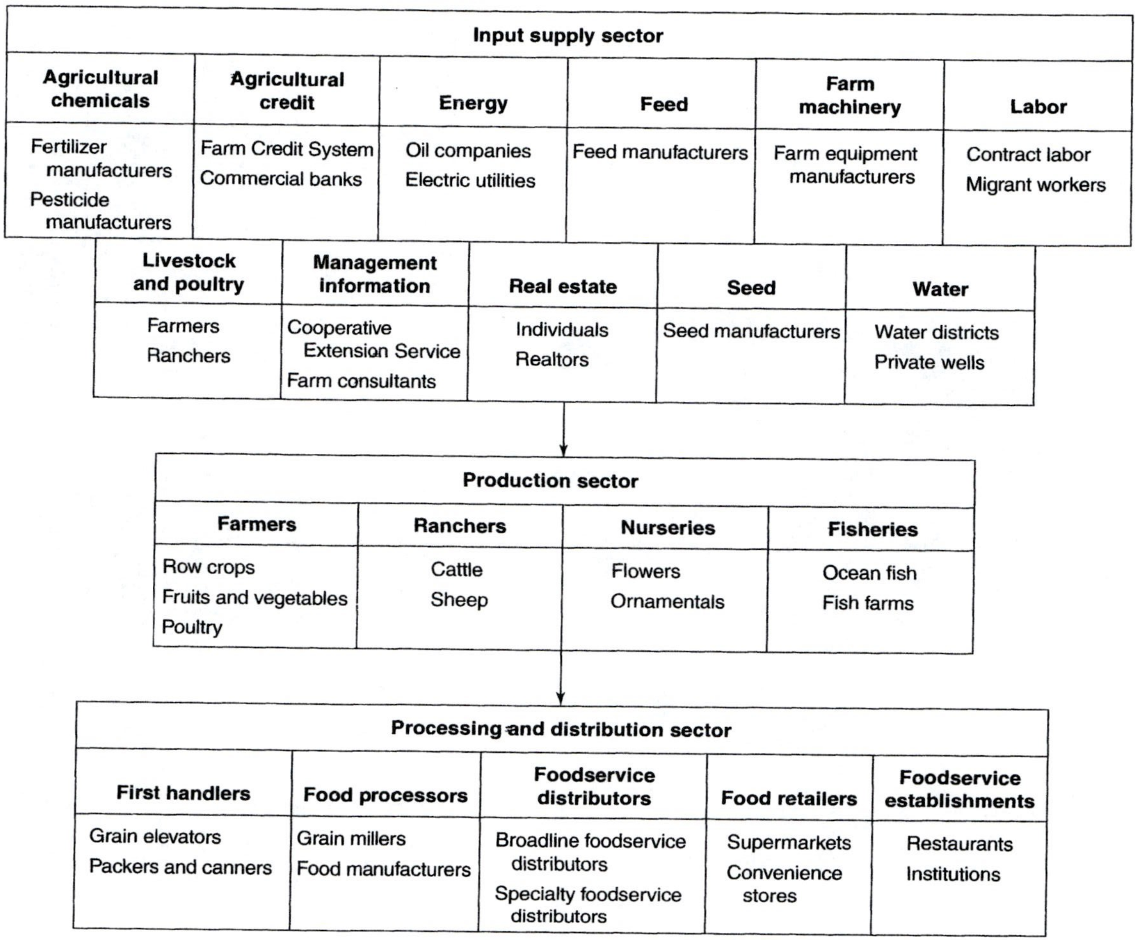

Input supply sector

Agricultural chemicals	Agricultural credit	Energy	Feed	Farm machinery	Labor
Fertilizer manufacturers Pesticide manufacturers	Farm Credit System Commercial banks	Oil companies Electric utilities	Feed manufacturers	Farm equipment manufacturers	Contract labor Migrant workers

Livestock and poultry	Management information	Real estate	Seed	Water
Farmers Ranchers	Cooperative Extension Service Farm consultants	Individuals Realtors	Seed manufacturers	Water districts Private wells

↓

Production sector

Farmers	Ranchers	Nurseries	Fisheries
Row crops Fruits and vegetables Poultry	Cattle Sheep	Flowers Ornamentals	Ocean fish Fish farms

↓

Processing and distribution sector

First handlers	Food processors	Foodservice distributors	Food retailers	Foodservice establishments
Grain elevators Packers and canners	Grain millers Food manufacturers	Broadline foodservice distributors Specialty foodservice distributors	Supermarkets Convenience stores	Restaurants Institutions

FIGURE 1–3

Purchased Farm Production Expenses, 1998

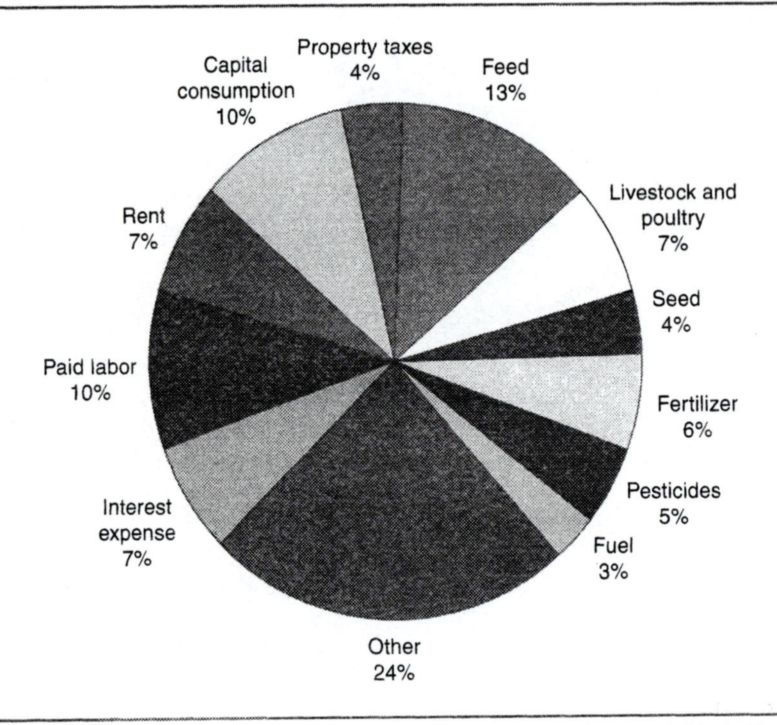

Source: Agricultural Statistics 2000, United States Department of Agriculture, Table 9–40.

Fertilizer. The fertilizer industry is a complex, international network of firms including raw material suppliers, intermediate manufacturers, and manufacturers of finished products. In general, the production of primary products is concentrated near the sources of the raw materials. Fertilizers are distributed through a large number of local and regional distributors.

Pesticides. Pesticides include chemicals to combat weeds, insects, disease, and other plant pests. The major agricultural chemical companies that produce agricultural pesticides are large, diversified corporations that produce a wide range of chemicals. For example, Monsanto, the maker of RoundUp, in addition to producing a wide range of pesticides, has recently diversified into the seed industry. Pesticides, like fertilizers, are distributed through a large number of local and regional distributors.

Machinery. The farm machinery industry, like the seed and agricultural chemical industry, is dominated by a few very large firms, including John Deere and Case IH. The equipment manufacturers have extensive dealer networks comprised primarily of independently operated dealerships to distribute their equipment. Many smaller companies produce equipment for specialized needs.

Feed. The feed industry has evolved relatively recently because most feeds used to be produced on the farm where they were consumed. Today's feed industry supplies producers with both ingredients and premixed rations. Most large feed users purchase ingredients and mix the rations to meet their own needs. Small feed users tend to rely more heavily on premixed rations sold through local or regional distributors.

Credit. Credit is supplied through several sources including the Farm Credit System, commercial lenders, institutional lenders, and individuals. The Farm Credit System, an agency of the U.S. government, and commercial lenders provide both long-term loans for real estate and durable assets, and short-term loans for operating expenses. Institutional lenders, such as large insurers, provide primarily long-term loans for real estate. Individuals are also a source of credit for real estate, typically as a means of financing the sale of their property.

Management Information. This may be the fastest growing area in the input supply sector. Managers have a large array of resources at their disposal. Knowing how to best employ these resources is critical to the success of the producer. Historically, much of the information on technological changes and management alternatives has been developed and disseminated through the land grant universities and the Cooperative Extension Service. Today, agricultural chemical and seed companies often provide information in conjunction with product sales. Furthermore, firms specializing in pest management or farm management provide the farmer with management services for a fee. Much of the farm management information is distributed through the Internet or by dedicated satellite systems.

Production Sector

The production sector, often referred to as the farm sector, includes all firms directly involved in agricultural production, or the growing of food and fiber. It includes farms growing row crops, fruit and vegetable farms, orchards, vineyards, cattle ranches, poultry farms, egg producers, feedlots, greenhouses and nurseries, and fish and seafood operations.

The production sector is comprised of approximately 2.2 million farms with almost one billion acres in production. This sector adds roughly $70 billion in value to the U.S. economy each year, or about 7 percent of the total value added by the food and fiber system. The number of farms has held steady and even increased the last few years, after many years of consolidation and declining farm numbers. The average farm size is approximately 430 acres.

Averages, however, tell only part of the story. The production sector is increasingly characterized by two types of operations: the small producer, who earns most of his or her income off of the farm, and the large commercial

FIGURE 1–4

Farm Size Characteristics, 1998–99

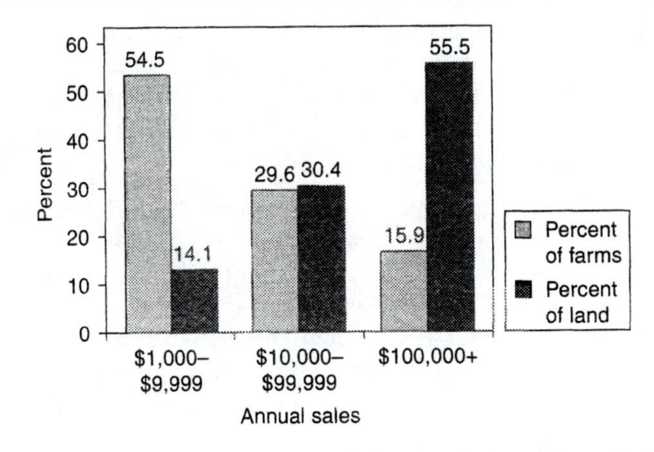

Source: Agricultural Statistics 2000, United States Department of Agriculture, Table 9–3.

producer. Figure 1–4 illustrates this bimodal distribution. The smallest farms (with sales of $1,000 to $9,999) represent 54.5 percent of the farms but only 14.1 percent of the land in production. On the other hand, the largest farms (with sales of $100,000 or more) account for only 15.9 percent of the farms but 55.5 percent of the land in production.

The production sector continues to exhibit increases in productivity. Over the last decade productivity has increased by approximately 2.5 percent per year. One of the principal means of achieving these productivity gains has been through the use of improved inputs. Figure 1–5 illustrates how the mix of inputs at the farm level has changed over the years. While it is still true that farmers continue to substitute capital for labor, it is apparent that most of this substitution comes in the form of the increased usage of agricultural chemicals.

Several trends are likely to continue to have an impact on the production sector. The extremely competitive nature of the sector will force producers to continue to search for ways to become more productive, both through improved inputs and by efficiency gains. This will also mean that the average farm size will increase and farms will continue to be very specialized as they focus their efforts on one or two related activities. Furthermore, the worldwide movement toward lower government involvement in agriculture and decreased government support for farm prices and incomes will make farmers more vulnerable to the volatile nature of farm output and prices.

FIGURE 1–5

Major Farm Inputs Usage Indices, 1966–1996

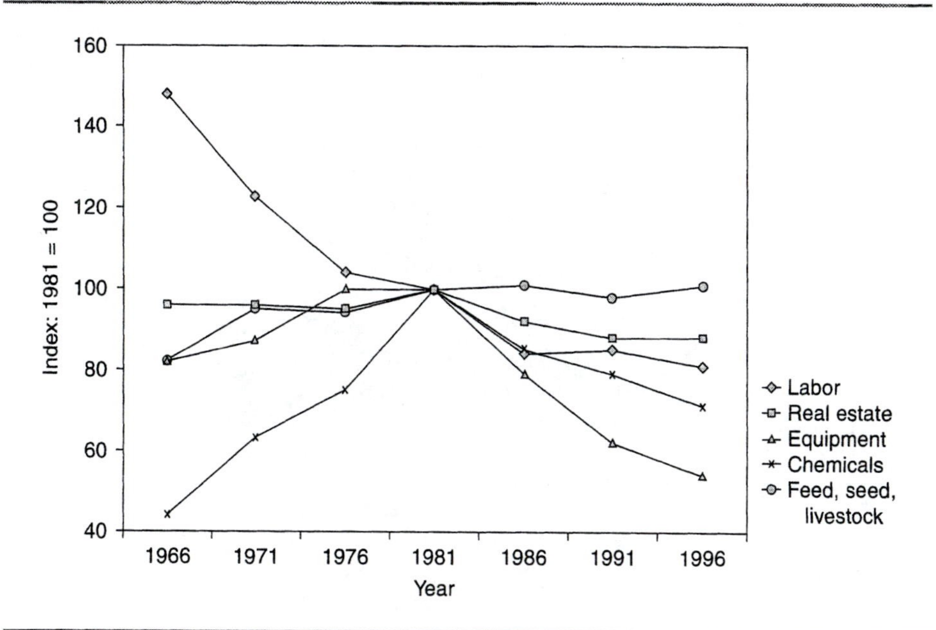

Source: Agricultural Statistics 1980, 1991, 1995–96, and 2000, United States Department of Agriculture, Tables 9–24, 545, 551, and 630, respectively.

Processing and Distribution Sector

The processing and distribution sector is by far the largest component of the food and fiber system. It adds approximately $630 billion in value to the U.S. economy, or 63 percent of the value added by the food and fiber system.

Another way to look at the contribution of the processing and distribution sector is to examine the marketing bill. The marketing bill includes the cost of marketing food products after they leave the farm. In 1998, marketing costs accounted for 80 percent of every dollar U.S. consumers spent on food, meaning that farmers received only 20 cents of the consumer's food dollar (Figure 1–6). This is lower than at any other time in history. The increasing proportion of marketing costs is driven primarily by consumers' demand for convenience, high food quality, and food safety in the foods they buy.

The processing and distribution sector is a complex network of handlers, processors, shippers, warehousers, wholesalers, retailers, and foodservice establishments that are responsible for delivering food to the consumer in the right form, at the right place, at the right time. The supply chain differs for each

FIGURE 1–6

The Consumer's Food Dollar, 1998

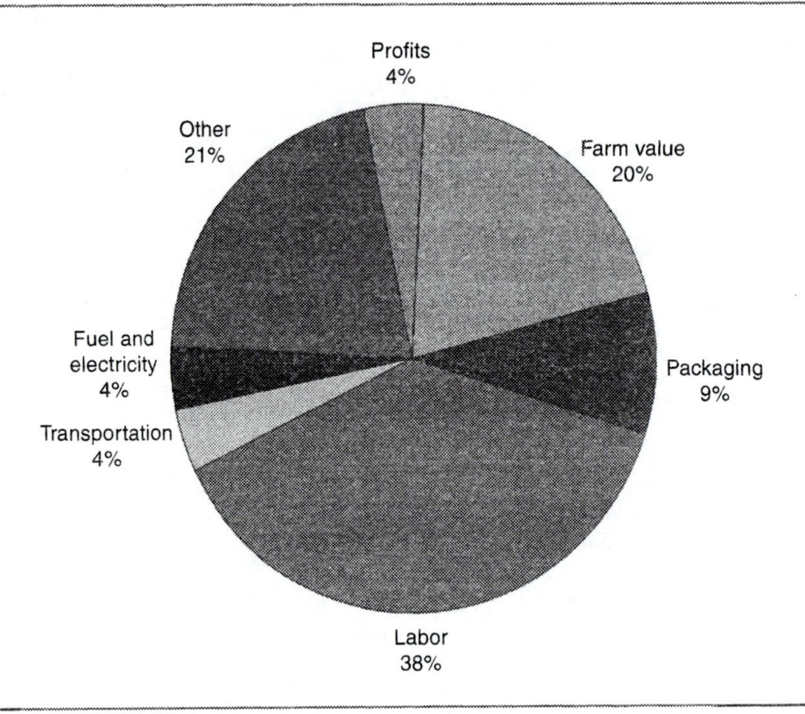

Source: Agricultural Statistics 2000, United States Department of Agriculture, Tables 9–27 and 9–28.

product, and a product may take multiple paths depending on its ultimate use. For example, fluid milk requires relatively little processing. After it is produced at the dairy farm it is shipped to the dairy processor where it is pasteurized, homogenized, and packaged. It may then be shipped directly to the retail store for purchase by consumers. The value added by the processing and distribution sector is relatively little for fluid milk. On the other hand, processed corn used as a sweetener for soft drinks follows a much more elaborate path on its way to the end consumer. The farmer initially sells corn to a local grain elevator. In turn the grain elevator sells the corn to a corn processor where it may be wet milled into a product called high fructose corn syrup. From there it would go to a soft drink manufacturer where it would be used as a sweetener. The soft drink manufacturer may sell the product through a distributor who would supply the retail store that sells the product to consumers. In the sections that follow, we describe several of the key segments of the processing and distribution sector.

Food Processors. Food processors are primarily responsible for transforming raw agricultural products into the products we find in the supermarket. They

BOX 1–1

Top 20 U.S. Food Processors, 1999

1. Phillip Morris Cos., Inc.
2. ConAgra Inc.
3. PepsiCo Inc.
4. Cargill Inc.
5. The Coca-Cola Co.
6. Mars
7. Archer Daniels Midland Co.
8. IBP Inc.
9. Anheuser-Busch Inc.
10. Sara Lee Corp.
11. H.J. Heinz Co.
12. Bestfoods Co.
13. Nabisco Holdings Corp.
14. Nestle U.S.A. Inc.
15. Tyson Foods Inc.
16. Dairy Farmers of America
17. Kellogg Co.
18. Campbell Soup Co.
19. General Mills
20. The Pillsbury Co.

Source: The Food Institute Report, June 5, 2000, American Institute of Food Distribution, Inc.

transform tomatoes into ketchup, wheat into bread, and grapes into wine. They also transform many different products into complete meals that consumers need only pop into a microwave. This sector has experienced rapid growth in the last few decades and in recent years has undergone substantial consolidation as the biggest companies have purchased the most popular brands from smaller companies. The top 20 food processors are listed in Box 1–1. By the time you read this, you will probably notice that some of these companies have been bought out by competitors!

Foodservice Distributors. Foodservice distributors supply foodservice establishments with foods and beverages as well as other products. The largest companies are broadline suppliers that supply a wide variety of products to meet the needs of restaurants, hospitals, schools, and other foodservice establishments. A host of smaller companies provide specialty products, such as fresh fish or meats to the foodservice industry. The largest foodservice distributors are listed in Box 1–2. Although most of these companies are not household names, you will probably recognize their names from the delivery trucks you've seen that bear their names.

Food Retailers. Food retailers sell food products to the final consumer. In the last decade this segment has been transformed by new, nontraditional entrants

BOX 1–2

Top 10 U.S. Foodservice Distributors, 1999

1. Sysco Corp.
2. U.S. Foodservice, Inc.
3. Alliant Foodservice, Inc.
4. PYA/Monarch, Inc.
5. Performance Foods Group
6. Gordon Foodservice
7. Food Services of America
8. Shamrock Foods Co.
9. Reinhart Food Service
10. Ben E. Keith Foods

Source: The Food Institute Report, March 6, 2000, American Institute of Food Distribution, Inc.

into the market. Wal-Mart is now the largest food retailer in the world. Wholesale and warehouse clubs have also grown at the expense of the supermarkets and forced them to rethink the way they do business. Just as the emergence of convenience stores in the 1970s compelled supermarkets to expand their hours and offer express lanes for shoppers purchasing only a few items, the wholesale and warehouse clubs have forced supermarkets to offer shoppers the opportunity to buy some products in bulk at a substantial discount. The 10 largest supermarket chains are listed in Box 1–3.

Foodservice Establishments. Foodservice establishments provide consumers with prepared food and drink. They have grown rapidly in the last two decades in response to consumers' desire to spend less time preparing meals. Today, almost half of the consumers' food dollar (47 percent) is spent eating out, commonly referred to as food away from home. Even supermarkets have gotten into

BOX 1–3

Top 10 U.S. Supermarket Chains, 1999

1. Kroger Co.
2. Albertson's
3. Safeway
4. Ahold USA
5. Winn-Dixie
6. Delhaize America
7. A&P
8. Pathmark
9. Hannaford
10. Penn Traffic

Source: The Food Institute Report, May 1, 2000, American Institute of Food Distribution, Inc.

BOX 1–4

Top 10 U.S. Restaurant Chains, 1999

1. McDonald's Corp.
2. Tricon Global Restaurants (Pizza Hut, Taco Bell, KFC)
3. Diageo PLC (Burger King)
4. Wendy's International
5. Darden Restaurants, Inc. (Red Lobster, Olive Garden)
6. Doctor's Associates, Inc. (Subway Sandwiches)
7. CKE Restaurants (Hardee's, Carl's Jr.)
8. International Dairy Queen, Inc.
9. Allied Domecq (Dunkin' Donuts, Baskin-Robbins)
10. Domino's Inc.

Source: The Food Institute Report, June 19, 2000, American Institute of Food Distribution, Inc.

the game by introducing fully or partially prepared meals (Home Meal Replacements or HMRs).

Foodservice establishments include full-service restaurants, fast-food restaurants, coffee bars, and bars, as well as institutional foodservice establishments that provide food in hospitals, hotels, schools, and jails. The 10 largest U.S. restaurant chains are listed in Box 1–4.

EMERGING TRENDS AND CHALLENGES

The food and agribusiness system is a highly coordinated system. In the above sections we have discussed the various components of the system, the input supply sector, production sector, and the processing and distribution sector as if they were separate and independent elements. In reality there is a high degree of coordination and often integration of the system's players. Actions taken by one firm or industry will have an impact on other firms and industries. Because of the highly competitive nature of the system, firms have to coordinate their actions with other firms in order to efficiently deliver the high-quality products and services expected by their customers. Next, we discuss some of the most important emerging trends and challenges that managers of food and agribusiness firms will confront.

Globalization

In the first half of the twentieth century most countries produced most of the food needed to meet the dietary needs of their people. A country's self-sufficiency in food was considered to be an important political goal. In the second half of the twentieth century the food and fiber system became truly global. Increasingly, country borders are no more significant to a food or agribusiness company than are state borders within a country. Most companies do not think of themselves as a U.S. or French company, but rather as a global company that has its home office in the United States or France. Companies do business where they can get the best deal or make the greatest profit. Globalization brings many benefits to world consumers and allows businesses to expand in regions where they add value by introducing new products or efficiencies to the food system. Continued globalization will depend on the ability of governments to agree on issues such as eliminating trade barriers in the form of tariffs and quotas, and agree on uniform international standards for "hot button" issues such as biotechnology and food safety.

Biotechnology

Biotechnology may be defined as the manipulation of a living organism's genetic endowment to produce beneficial products. It includes both traditional breeding and genetic engineering techniques. Biotechnology holds both the promise of great good or great harm depending on your perspective. Proponents argue that benefits include better, healthier, and cheaper foods and reduced impact on the environment. Opponents warn of hidden dangers in foods, such as severe food allergies to which unsuspecting consumers may be exposed, and increased environmental damage. There is little agreement on what constitutes biotechnology and how it and its products should be regulated. Some of the major issues that must be confronted by participants in the food and fiber system include: a broadly accepted definition of biotechnology, how biotechnology and its products should be regulated, and consumer skepticism of the products of biotechnology.

Information Technology

The information age has changed and continues to change the way that all parties in the food chain do business. Internet information exchange systems and Internet business transactions have greatly increased the efficiency of the food system. Some firms have been eliminated, others have been born, and the activities of virtually all firms have been impacted in some way. One example of how information technology has transformed the food industry is supply chain management. Participants in supply chain management systems share information on costs, products, prices, delivery needs, and expected sales, and they sometimes have shared-profit arrangements. Since information may be exchanged quickly and easily, all parties can achieve the cost savings associated with vertically owned food companies without all of the problems associated with managing a large company. The changes brought about to date by the information age are likely just the tip of the iceberg. Information technology will continue to have a major impact on the way food and agribusiness firms do business for many years to come, often in unpredictable ways.

Food Safety

The safety of the food supply is a major concern for consumers both in the United States and worldwide. Consumers want to know that food is what it purports to be and that it is free of harmful ingredients and pathogens. Several high profile incidents in products as diverse as fruit juice and ground beef have raised the level of consumer consciousness on this issue. A major question is, "To whom will consumers look to ensure the safety of the food supply?" Will it be governments or the companies that produce the food? Can the food system be efficient if many different agencies administer regulations that differ by country?

SUMMARY

The U.S. food and fiber system is the largest in the world, producing products worth approximately $1 trillion and employing approximately 23 million people. The food and fiber system is comprised of three sectors. The input supply sector provides inputs that will be used in the production of agricultural products and accounts for about $300 billion of the value added by the food and fiber system. Input supply firms include feed producers, seed producers, agricultural chemical manufacturers, farm machinery manufacturers, agricultural credit suppliers, and farm supply dealers. The production sector includes those businesses directly involved in agricultural production, such as farmers, ranchers, and nurseries. It adds approximately $70 billion in value. The largest component of the food and fiber system is the processing and distribution sector with $630 billion in value added. Firms in this sector include food processors, foodservice distributors, food retailers, and foodservice establishments.

The food and fiber system is an extremely complex network with a high degree of integration and coordination among the participants in the system. Although the industry is mature, it continues to be impacted by many factors both internal and external to the system. Some emerging trends and challenges that food and agribusiness managers will confront include increasing globalization of the food and fiber system, the impact of information technology and biotechnology, and food safety issues.

REVIEW QUESTIONS

1. Describe the evolution of the U.S. food and fiber system.
2. What is the contribution of the U.S. food and fiber system to the U.S. economy?
3. What types of firms comprise the input supply sector?
4. What are the major production expenses of U.S. farmers?

5. Why is the input supply sector responsible for much of the productivity gains in the production sector?
6. In addition to farms, what types of firms are in the production sector?
7. Describe the major changes that have occurred in the farm sector over the last decade.
8. What is the marketing bill and why is it so high?
9. Describe the major categories of firms in the processing and distribution sector.
10. Which emerging trends do you believe will pose the biggest challenges for food and agribusiness managers in the future?

CHAPTER 2

Forms of Business Ownership

Learning Objectives

In this chapter we will cover the following topics:

- Basic features of each form of business ownership
- Major factors that influence the choice of a form of business ownership
- Advantages and disadvantages of each form of business ownership
- Relationships between businesses, including strategic alliances, joint ventures, and franchises
- Basic features of the cooperative form of business ownership
- Fundamental differences between cooperatives and other forms of business ownership

CASE STUDY

Ann Martin is a Michigan apple grower and has managed her family's farming operation for about 15 years. Although she makes a good living and has managed to save up about $150,000, she would like to expand into the processing side of the business. She recently completed a feasibility study for a medium-sized juice processing plant and determined that approximately $500,000 will be needed to carry her through the first two to three years. After evaluating the profit potential and risks involved, she has decided to move forward with her plan.

Her next major decision is to determine the form of business ownership. Ann has received advice from several friends. A neighboring farmer told her that she should organize the business as a sole proprietorship, just like the farm. He said, "If you incorporate, you'll spend a lot more time keeping records, and besides you'll just end up paying more money to Uncle Sam." One of her other neighbors suggested that she form a corporation because that would be the easiest way to get the additional financing she would need. A lawyer friend from town suggested that she form a limited liability company, saying, "It is the only way to go, because if your business were a sole proprietorship and you were sued, you could be forced to sell your farm to pay off your debts."

Ann's dilemma is not unique. Choosing a form of business ownership is an important decision. She must learn about each of the available forms of business ownership, decide what factors are most important to her and her business, and understand how the forms of ownership differ with respect to these factors. She must then weigh the pros and cons of each form of ownership to reach a decision. After reading this chapter, you should be able to give her some advice.

INTRODUCTION

Most food and agribusiness firms are set up as sole proprietorships, partnerships, corporations, or cooperatives. Some owners have chosen one of the relatively new, hybrid forms of business ownership, the limited liability partnership or limited liability company. These forms of business ownership have many factors in common; however, it is the differences that cause owners to choose one form over another. The most important factors influencing the choice of business ownership are:

- Tax liability
- Personal liability that the owners must assume for business debts or losses
- Access to capital to finance the business
- Ability to control the management of the business
- Ease with which the business can grow
- Stability of the business
- Ease with which partial or total transfer of ownership can be made
- Ability to attract high-quality employees
- Issues such as the cost of organizing the business, required paperwork, state and federal regulations, and any special objectives for which the business may be organized

The following sections describe the principal forms of business ownership and their distinguishing characteristics. The laws governing businesses vary depending on the state and country. The material presented in this book is based on the laws and regulations of the United States and the most common state laws and regulations.

SOLE PROPRIETORSHIP

A *sole proprietorship* is a form of business ownership in which the business is completely owned and controlled by one person. This is the most common form of business, particularly among food and agribusiness firms (see Box 2–1), and it is the simplest to establish. Starting the business establishes a sole proprietorship. The only laws or regulations that must be met are those affecting firms in a particular industry. For example, the owner of a restaurant must obtain the proper license, pay the appropriate fees, and follow the local and state laws governing restaurants.The owner of the business is known as the proprietor.

BOX 2–1

Number of Businesses and Sales in Agriculture, Forestry, and Fishing by Type of Business, 1995

	Number of Businesses		Sales Receipts	
	Thousands	Percent	Billion Dollars	Percent
Nonfarm proprietorships	510	65	19	14
Partnerships	129	16	13	10
Corporations	148	19	101	76

Source: Statistical Abstract of the United States 1998, U.S. Department of Commerce, page 540.

Advantages

A major advantage of this form of business is that there is no income tax charged to the business. The profits of the firm are taxed only once, as personal income to the proprietor. On the other hand, corporations are taxed twice, once as a business entity and a second time as personal income when profits are paid to the owners as dividends. However, this tax structure may be a disadvantage, as we shall see later.

Another important advantage of the sole proprietorship is its simplicity. Little or no paperwork is required to form the business. By the same token, the sole proprietorship is dissolved when you stop doing business. This is very appealing in a world where government regulations abound and it can take months to complete the proper paperwork, not to mention the cost of accounting and legal fees, to form or dissolve other forms of business. The sole proprietor has no boss; no one else shares the responsibilities of decision making or the liabilities that may result from poor judgement. A related advantage is that there are fewer laws and regulations to comply with in a sole proprietorship than there are in a partnership, corporation, or limited liability company. This may result in lower operating costs for the business in terms of personnel requirements and compliance costs.

The flexibility of the sole proprietorship is also attractive to many people, and this can provide an important competitive advantage because it allows a firm to respond quickly to changes in the business environment, which is especially important in a rapidly changing world. Another advantage of the sole proprietorship is the ease with which it can be changed to a more complex business structure, such as a partnership, corporation, or limited liability company, as the business grows or when other considerations make such a move desirable.

Disadvantages

A major disadvantage of the sole proprietorship is that the proprietor is personally responsible for all losses incurred by the business unless specific legal procedures are taken for important transactions. If the losses exceed the value of the firm's assets, the proprietor must then cover the liabilities from his or her personal assets or declare personal bankruptcy. In this sense, Ann's lawyer friend was right. If the business fails, Ann may be forced to sell her farm or other personal assets to meet the liabilities incurred by the business.

Another consideration is the amount of capital needed by the firm. The financial resources of the proprietor are generally limiting to a sole proprietorship. Sole proprietorships are typically smaller than other business forms as can be seen by the share of sales receipts among sole proprietorships compared to partnerships and corporations (see Box 2–1). The proprietor typically acquires some of the assets of the business with his or her own assets and then borrows the remainder. The lender, however, will require some form of collateral or security for the loan. Ultimately, whether the money is the proprietor's or is borrowed, the proprietor's financial resources will limit the total amount of available capital. This can be a disadvantage when large sums of money are needed to finance a business venture.

Gaining access to more capital often means involving more people in the ownership of the business and, therefore, changing the form of business. Other forms of ownership, such as the partnership or corporation, allow several individuals to pool their financial resources and expand a business beyond the financial limits of most individuals. Many businesses start out as a sole proprietorship and add additional owners only when additional capital is needed. On the other hand, successful sole proprietorships often find that their access to capital expands with the success of the business. A sole proprietorship with a proven track record generally experiences less difficulty in attracting capital than a startup organized as a partnership or corporation.

Another disadvantage of a sole proprietorship is that the stability and continuity of the business depend on one person, the proprietor. Therefore, if the proprietor dies, the business no longer exists. This may make it difficult to keep the business going if there are no provisions for keeping the business operating when something happens to the proprietor. This affects the operation of the business in several ways. It may make it more difficult to attract qualified personnel to work for the company because highly qualified people may be unwilling to work for a firm where their continued employment depends solely on one person. Moreover, customers who depend on a business for a product or a service are often reluctant to do all of their business with one company, even more so when the continued operation of the company depends on one person.

The lack of stability and continuity may be at least partially offset by careful planning on the part of the proprietor. For example, if the business has at least one employee who is qualified to operate the business, the proprietor can carry life insurance and disability insurance and designate the qualified employee as the operator of the business in the event of the proprietor's death or disability. The principal advantages and disadvantages of a sole proprietorship are summarized in Box 2–2.

BOX 2–2

Advantages and Disadvantages of a Sole Proprietorship

Advantages	Disadvantages
No tax as a business	Unlimited personal liability
Simple and inexpensive to form and dissolve	Growth limited by the owner's assets
Very flexible	Lack of stability and continuity
Few laws and regulations	Difficult to attract qualified personnel
Complete control by the proprietor	Customers may be unwilling to rely on the firm as a source of supply
Easy to change to more complex structure	

Evaluating the Sole Proprietorship

The sole proprietorship is the most common form of business ownership because it is the easiest to form. For many individuals or families who want to start a business that often supplements a full-time job, the sole proprietorship is the logical choice because it requires the minimum amount of paperwork. Furthermore, because no one else is involved there may be no apparent reason to choose another form of ownership.

Since each form of business differs with respect to the taxes that apply and the rate of taxation, it is important in choosing a form of business ownership to consider the overall tax structure in effect for each form of business. A sole proprietorship may be especially appropriate if losses are anticipated during the start-up phase of the business. In such instances, the losses may be passed on to the proprietor, thereby offsetting other income from active sources. According to the Internal Revenue Service, an owner who actively participates in the management of the firm is considered to earn *active income* (or incur an active loss). Furthermore, since the level of income or loss often varies substantially over time, particularly in the early years of a business, a long-term view should be taken in selecting the form of business ownership. The tax implications of a sharp change in the firm's profit level may make it desirable to change the form of business as the business matures. It is always a good idea to discuss tax considerations with a tax accountant before a decision is made regarding the form of business.

Sole proprietorships should be avoided when the business is very risky and the proprietor has substantial personal assets that he or she does not want to risk losing, especially if the risk cannot be covered by insurance in a cost-effective manner or otherwise mitigated.

Common Mistakes

Since a sole proprietorship is easy to form and is subject to few government regulations, individuals frequently regard accurate record keeping as unimportant. There is a tendency to be sloppy in documenting business or personal transactions, partly because it may not be mandated by regulations and partly because personal income and business income are pooled for income tax purposes. However, it is very important to keep accurate records. The income earned from the business is subject to self-employed social security tax.

Another key reason for maintaining good records is that many small businesses grow into larger businesses. Detailed records that may initially seem like unnecessary paperwork may prove to be highly advantageous in selling a business, acquiring future debt capital, or in changing the form of ownership to a partnership or publicly held corporation. Extracting historical records from an old shoebox does not inspire confidence on the part of potential investors.

In some cases, the sole proprietorship is overlooked as a form of business ownership. There is a tendency for people unfamiliar with the forms of business ownership to equate business ownership with a corporation. Thus, the sole proprietorship may not be considered as an option even though it may be best suited to their situation.

PARTNERSHIP

An alternative form of business ownership similar to the sole proprietorship is the partnership. The major difference between a sole proprietorship and a partnership is that the partnership involves more than one person in the operation and management of the business. Two basic forms of partnerships exist: the general partnership and the limited partnership. Additionally, a relatively new, hybrid form of business, called the limited liability partnership, is available to business owners in many states. The general partnership will be discussed first because characteristics associated with it are necessary for understanding the other forms of partnerships.

General Partnership

Legally, there is no limit to the number of partners (owners) who may participate in a *general partnership.* However, it is necessary to check the state laws governing partnerships and the sale of securities if a large number of partners is involved or if the partners live in more than one state. Like the sole proprietorship, all of the earnings of the business are passed directly to the partners. There is no income tax on the earnings of the partnership as a business; however, the partnership must file partnership income tax forms. Each partner can act on behalf of the partnership on business matters, and each partner's actions are fully binding on all partners.

The laws relating to partnerships in the state where the partnership resides govern the rights, obligations, and duties of a partner in a business. Unless specified by a written agreement, the rights, liabilities, and earnings of the partnership are typically distributed equally regardless of the capital or time each partner contributes. Many times, the rights of each partner are spelled out by a *partnership agreement,* which is a contract that specifies operating arrangements and legalities. A partner who has contributed more capital or time will likely want a greater voice in management and a greater share of the profits. Although many states do not require a written partnership agreement, it is strongly recommended that the partners have an attorney draw one up. This ensures that the relationship between the partners is thought out in advance and there will be no misunderstanding among the partners concerning each partner's rights and responsibilities. Likewise, each partner usually bears an equal responsibility for the debts of the firm unless otherwise spelled out in a partnership agreement. A partnership agreement should include, but not be limited to, the items listed in Box 2–3.

BOX 2–3

Items to Include in a Partnership Agreement

Name and address of the partnership
Date of agreement
Names and addresses of partners
Objective of the partnership
Duration of the partnership
Contribution of each partner
Preparation of financial statements
Salary arrangements
Division of profits and losses
Rights and responsibilities of partners
Provisions for the addition and withdrawal of partners
Dissolution of the partnership

An important, often overlooked, provision is one that provides for the addition or withdrawal of partners. This is especially important because the partnership would cease to exist upon the withdrawal or death of a partner unless the partnership agreement specifically addresses this issue. Planning also guarantees that the deceased partner's heirs receive a fair share of the value of the partnership. A properly drawn out and funded *buy-sell agreement* is frequently used by partnerships to meet this contingency. Such agreements deal with the situation where the partnership liquidates or a partner dies, becomes disabled, retires, divorces, or wishes to sell his or her interest in the partnership. Typically, the buy-sell agreement establishes a formula to determine the price of each partner's share of the business and a mechanism for providing the funds needed to make the purchase. A common mechanism that provides this funding is a life and disability insurance policy.

As with a sole proprietorship, the partners in a general partnership are liable for all of the debts of the firm, even if the debts exceed the partners' investment in the partnership. If any partner does not have the financial resources to meet his or her share of the obligation, then the responsibility falls to the remaining partners. In addition, all partners are personally liable for financial judgements stemming from a lawsuit against a partner. Thus, it is advisable that partnerships purchase liability insurance or form a limited liability partnership (discussed later in this chapter) to reduce the impact of lawsuits.

Advantages

A general partnership has two principal advantages over a sole proprietorship: more people are involved and therefore more resources are available. A major limitation of a sole proprietorship is the limited financial resources of the proprietor. By forming a partnership, the financial resources of two or more people may be pooled, thus increasing the resources available to the firm. Another reason to form a partnership is to pool the talents of several individuals. Partners are often sought out because of what they can contribute to the business, not necessarily because of their similarities to the other partners. There is a related advantage in that the addition of another partner allows specialization to occur. This principle has been heavily used in business. Rather than have every partner wear every hat, why not let one wear the hat of the production manager, another of the marketing manager, and another of the financial manager? Specialization allows each person to become an expert at what he or she does. Partnerships are also effective for involving several family members in a business, although this situation is not devoid of potential problems, specifically, interpersonal problems. Partnerships are a common means for allowing family members to pool their talents and financial resources for business purposes.

As with the sole proprietorship, there is no tax on the profits of the partnership as a business, although a partnership can elect to be taxed as a corporation if the partners so choose. All of the profits, unless specified otherwise, are passed directly through to the partners to be taxed on each partner's personal income tax return. The potential profits from the business, the timing of profits being generated, and the tax status of the individual owners should be examined carefully before making a determination as to which form of business ownership is most advantageous from a tax standpoint.

As with a sole proprietorship, few laws and regulations govern partnerships. Little paperwork is involved, and partnerships are easy to form and dissolve. However, partnerships formed for acquiring ownership of real estate are often treated differently than other partnerships. It is advisable to get legal, tax, accounting, and insurance advice when forming the partnership and when writing partnership and buy-sell agreements.

Disadvantages

Like the sole proprietorship, the partnership is characterized by the owners' unlimited personal liability for the debts of the firm. In some states partners can form limited liability partnerships to shield themselves from any legal wrongdoing by their partners, but not their own wrongdoing. In case of financial troubles, it is important that the division of liability be specified in a partnership agreement because in the absence of such an agreement the liability is equally divided.

BOX 2–4

Advantages and Disadvantages of a General Partnership

Advantages	Disadvantages
No tax as a business	Unlimited personal liability
Simple and inexpensive to form and dissolve	Possible lack of stability and continuity
Few laws and regulations apply	Confusion as to managerial decision making
Easier access to financial resources than a sole proprietorship	More difficult than a sole proprietorship to form and operate
Easier to attract qualified personnel than a sole proprietorship	

Another disadvantage of a partnership is that all partners are liable to the extent of their financial resources, even though one partner's financial resources have been exhausted. Thus, those partners with many personal assets have greater financial resources at risk than those with few personal assets. Another problem is that any partner can act for the partnership. It is possible that one partner may find himself or herself responsible for the undesirable action of another partner, although this potential disadvantage can be partially offset with a well designed partnership agreement.

Because the partnership is a collection of individuals, the withdrawal of a partner for whatever reason (death, sickness, or simply a decision to leave) means that the partnership no longer exists. This can place a great hardship on the partnership because it may mean that the assets of the partnership will have to be liquidated to pay off the partner who is leaving. This also serves to underscore the importance of having a buy-sell agreement to specify each partner's responsibilities and share of the debts, earnings, and assets, as well as the rules that will be followed during liquidation of the partnership or the withdrawal of a partner.

A further disadvantage of the partnership relative to the sole proprietorship is that with the addition of more partners, the responsibility for making decisions must be shared. No one person has complete control, which may cause the firm to respond more slowly to changing business conditions. Employees may also feel confused as to who their boss is. Both of these problems can be at least partially offset if there is a clear delineation of responsibility and authority in the partnership agreement. The principal advantages and disadvantages of a partnership are summarized in Box 2–4.

Limited Partnership

The limited partnership was established to address some of the problems of the general partnership (the limited partnership should not be confused with the

BOX 2–5

Advantages and Disadvantages of a Limited Partnership

Advantages	Disadvantages
No tax as a business	Unlimited personal liability for general partners
Limited personal liability for limited partners	More difficult than a sole proprietorship and general partnership to form and operate
Easier access to financial resources than a sole proprietorship	Possible lack of stability and continuity
Easier to attract qualified personnel than a sole proprietorship	Possible confusion as to managerial decision making
	Limited partners must strictly adhere to rules governing limited partnerships

limited liability partnership or limited liability company, both of which will be discussed later in this chapter). Partners are usually brought into a partnership because of their financial resources or because of a specific talent they possess. However, several owners who have equal authority may make management and decision making more difficult. A *limited partnership* is a partnership with at least one limited partner and at least one general partner. General partners have unlimited liability for the debts of the firm and unlimited responsibility for its management. Limited partners contribute capital but cannot be involved in the management of the business. Thus, a limited partner is sometimes referred to as a "silent" partner. The limited partner is not responsible for the debts of the firm beyond his or her investment. In other words, the most the limited partner can lose is the amount he or she has invested.

The limited partnership must file a certificate and a partnership agreement with the appropriate county government office and sometimes at the state level as well. In many states this filing requirement is unique to limited partnerships and is not required of general partnerships. The partnership must abide by all state and federal laws regulating limited partnerships. These laws are complex and must be carefully considered. Limited partners must take care not to participate in the management of the business. If any of these rules are violated, the limited partner risks losing his or her status as a limited partner. The limited partnership has been increasingly popular as a way to raise additional capital for a partnership. It provides limited liability for the new partner without directly involving him or her in the management of the business. The principal advantages and disadvantages of a limited partnership are summarized in Box 2–5.

Limited Liability Partnership

The laws of most states provide for a relatively new form of business known as the limited liability partnership. In a *limited liability partnership,* partners are not liable for the negligent acts of other owners or employees not directly under their control. This form of business is usually used as an alternative to the general partnership because of the liability protection it affords the general partners. In many states the limited liability partnership is restricted to certain professionals, such as accountants, engineers, health care professionals, lawyers, and veterinarians. This form of ownership is fairly similar to the *professional corporation,* which provides liability protection from the malpractice of other owners. Professional corporations, like limited liability partnerships, exist in many states, but only for certain professionals.

Evaluating the Partnership

Partnerships are widely used when it is desirable to combine either the financial or managerial resources of two or more people. As with sole proprietorships, partnerships are often used when losses are expected, because the losses may be passed through to the owners, thereby offsetting other active income for tax purposes. Limited partnerships have been used to attract additional capital to risky business ventures because the limited partners have the potential for a large profit while their potential loss is limited to their investment. Most limited partnerships are established to take advantage of a particular feature of the laws and as such can be highly complex structures. The assistance of legal and tax experts is therefore essential.

Common Mistakes

In forming a partnership, many individuals do not sufficiently consider the goals and personalities of potential partners. Partners with substantially different goals or personality traits often find it difficult to work together effectively, leading to conflict between partners and in some cases the demise of the partnership. Personal relationships are extremely important in a partnership because partners typically work closely together. Incompatible interpersonal relationships are a major disadvantage to a new company. Furthermore, a good social relationship between family or friends is no guarantee of a successful business relationship.

Because most businesses have to borrow funds to finance existing operations or expand the business, the risk-bearing threshold of potential partners should be considered. Some individuals are able to tolerate a substantial debt burden whereas others experience great anxiety over a large debt, even when their financial resources are similar. The feelings of potential partners regarding debt and the management of the business in general are important considerations in choosing partners.

When forming a partnership, most individuals focus their attention on the prospects of the business at the time it is formed. Almost all parties forming a partnership assume all will go well; otherwise, they would not become involved. However, problems do arise in a large percentage of new businesses, and these problems often lead to the dissolution of the business. Therefore, it is essential to evaluate all potential problems and include a buy-sell agreement that covers the procedures and conditions for dissolving the business.

A common misconception about partnerships is that all partners are responsible for the personal debts of their other partners. This is not correct. Another misconception is that the remaining partners are required to accept a new partner as a general partner should one of the general partners sell his or her interest in the partnership. If a partner sells his or her interest in a general partnership, the individual buying that interest does not have a say in the management of the business unless it is ratified by the remaining partners.

CORPORATION

A *corporation* (also known as a C corporation) is a legal entity that is separate from the person or persons who own it. The corporation can legally enter into contracts, sue and be sued, own property, and pay taxes. Each state has its own laws governing corporations; thus, the process of forming a corporation and the rules governing it depend on the state in which the firm is incorporated. Consequently, a corporation can conduct business only within the borders of the state in which it is incorporated unless it applies for permission and registers with other states.

A major feature of the corporation is the limited liability of the stockholders. *Stockholders* are the owners of the corporation, although the corporation is the owner of its assets. Each stockholder's liability is therefore limited to the amount of his or her investment. Another characteristic of the corporation is that it has permanent life. The corporation does not cease to exist with the death or withdrawal of one of the stockholders. *Stock* (a certificate of ownership) is simply transferred from one individual to another. A corporation must pay taxes as a business. Profits are taxed once as a business and a second time when they are paid to the stockholders in the form of dividends. The stockholder must pay personal income tax on these dividends.

The process of incorporation varies from state to state but usually includes the following standard procedures. Individuals interested in forming the corporation must file with the secretary of state a contract called articles of incorporation. Thus, incorporation requires the assistance of an attorney. A fee is usually due at this time. The articles of incorporation generally contain the following information:

- Name(s) of incorporator(s)
- Name and address of the corporation
- Objective of the corporation

- Number of directors
- Amount, kind, and number of shares of stock issued

After the corporation is established, stock is sold and a meeting of the stockholders is called. At this time, a *board of directors* is elected and bylaws are adopted. These bylaws govern the operation of the corporation and other actions necessary for initiating business and internal governing rules for stockholders, directors, and officers.

The stockholders own the corporation and approve amendments to the articles of incorporation, approve major changes in corporate direction, and elect the directors. Stockholders have one vote for each share of stock owned. The board of directors is responsible for protecting the interests of the stockholders and exercises this responsibility by determining policy for the corporation, appointing officers, and voting on important decisions facing the corporation. The board of directors operates on the basis of one vote per director. An individual director does not have to be a stockholder. The officers report to the directors and are responsible for implementing policy as determined by the board. The officers oversee the day-to-day management of the firm, including hiring and directing corporate employees. The officers may be stockholders or they may be hired management with no ownership interest. Both are very common in the food and agribusiness sector, depending on the degree of control desired by the stockholders.

Advantages

A major advantage of the corporation is that the liability of the stockholders is limited to the amount of their investment. This results from the corporation having a legal identity separate from that of its stockholders. Thus, when the corporation is sued or suffers heavy losses, the corporation must bear the liability, not the stockholders. When the corporation's financial resources are exhausted, there is no further recourse to those who are owed money by the corporation.

The corporation has the advantage of perpetual life and the stability that goes with it. The corporation will exist until the stockholders dissolve it. If a stockholder dies, the ownership or stock is simply transferred to another individual. Unlike a proprietorship or partnership, the corporation is not simply an individual or group of individuals. The corporation is an entity in and of itself and, like a family heirloom, it may be passed from one individual to another without changing its identity. Many families have used corporations as an effective means of passing ownership of a family business having sizable assets to children and grandchildren, thus avoiding some of the burden of inheritance taxes. Because of the stability that the perpetual life of the corporation creates, a secure atmosphere can be established for those who do business with the corporation as well as those who are employed by it. It is often easier to hire and retain skilled personnel in a business that is expected to provide long-term stable employment. Corporations also tend to have personnel that are very specialized, although this is probably more a function of the size of the business than the form of ownership.

A feature often associated with corporations is the increased financial resources available to the firm. It should be noted that simply changing from a sole proprietorship or partnership to a corporation does not generate additional resources. Financial resources are only committed to ventures that are viewed by investors as safe while providing a good return on investment. Simply switching from one form of business ownership to another will not change either of these conditions. The corporation does, however, provide a convenient mechanism for generating more capital. Corporations may sell stock if investors can be found who are willing to buy it. In this way, the current or new stockholders may contribute additional capital to the corporation. However, the ownership of existing stockholders is diluted when new investors buy stock, and depending upon the amount of stock sold, the current stockholders may lose control of the management of the corporation. The corporate form of business allows many people to participate in the ownership of the firm without complicating the decision-making process or management of the firm.

Corporations may also issue bonds, which are debt capital rather than equity (ownership) capital. Bonds will be discussed in more detail in Chapter 5.

Although the corporate form of business ownership offers more alternatives for raising capital, forming a corporation can be a major undertaking. Corporations are relatively few in number when compared to sole proprietorships and partnerships. However, they are proportionally much larger than sole proprietorships by almost any economic measure. The fact that the largest businesses in the United States are corporations is mostly a function of the mechanisms available to corporations for facilitating growth, such as the sale of stock and bonds, liability limits, and historical tax laws.

The creation of a corporation may result in advantageous tax treatment compared to other forms of business ownership, depending on the amount and timing of income the corporation generates. This is because corporate and individual marginal tax rates differ, and at some income levels corporate marginal tax rates are lower than individual marginal tax rates. It is a common strategy for the corporation to pay the corporate tax on business profits and retain most of the earnings for growth, resulting in the appreciation of the value of the firm's stock. Stockholders will realize this capital gain when they sell the stock. Personal income taxes on the capital gain are not paid until the stock is sold. Upon sale of the stock, the personal income tax liability is reduced because capital gains on stock held for more than 12 months is taxed at a lower marginal tax rate than other forms of personal income. Corporations are also allowed to carry losses forward and deduct these losses from earnings in future years.

Another area in which corporations receive advantageous tax treatment is the payment of benefits. The cost of term life insurance, health insurance, dental insurance, retirement benefits, generous travel and entertainment allowances, and other benefits for employees is treated as a corporate business expense and is deducted before earnings are determined. Hence, these benefits are not taxed. Because of the complexity of the tax system and the frequent changes in the tax code, an accountant or attorney who specializes in taxes should be consulted to determine if the corporate form of business will result in the most favorable tax treatment.

Disadvantages

The tax structure may also be a disadvantage for corporations. At some income levels corporate marginal tax rates are higher than individual marginal tax rates. When corporate earnings are not retained for growth but are paid out in the form of dividends to stockholders, the total tax bill will be higher than it would be for alternative forms of business such as a sole proprietorship or partnership. The earnings of the corporation are taxed first to the corporation at its marginal income tax rate and second to the stockholder at his or her marginal income tax rate when passed to stockholders as dividends. In this case the earnings are, in effect, taxed twice. Another disadvantage is that any financial losses sustained by the corporation cannot be passed on to stockholders to be deducted against income on their personal income tax forms.

The advantage of a corporation having perpetual life can also turn into a disadvantage if the stockholders decide that they want to dissolve the corporation or turn it into another form of business organization. Dissolving a corporation is a complicated process that requires legal and accounting advice. When a corporation dissolves, there is also double taxation of the firm's assets. First, the corporation pays corporate capital gains tax on profits accruing from the liquidation of assets. The stockholders are then required to pay personal income tax on all of the retained earnings that have been built up in the corporation.

Corporations must comply with a host of laws and regulations that make conducting business more complicated and expensive. A fee must be paid to incorporate, and it is necessary to hire lawyers and accountants in order to comply with the additional rules and regulations.

In recent years, another disadvantage has surfaced in some corporations. In many large corporations, the stockholders do not participate in the management of the business; this problem is commonly referred to as the "*separation of ownership and management.*" Corporate boards of directors and officers sometimes pursue goals that are not perceived as being in the best interest of the stockholders. Directors and officers, for example, may try to increase their income or their power in the corporation rather than increase stockholder income. Ensuring that directors' and officers' compensation is tied to improved performance can minimize this danger. To achieve this goal, corporations frequently give directors, officers, and employees options to buy corporate stock at or below market value so that they benefit when the corporation does well. The principal advantages and disadvantages of a corporation are summarized in Box 2–6.

BOX 2–6

Advantages and Disadvantages of a Corporation

Advantages	Disadvantages
Limited liability of owners	Profits are taxed twice
Stability and continuity	Many laws and regulations apply
Facilitates raising capital	Costly and difficult to form and dissolve
Easy to attract qualified personnel	Separation of ownership and management
Possible favorable tax treatment	

S Corporation

Subchapter S is a provision of the Internal Revenue Service tax code that provides for small business corporations meeting certain guidelines to pass income or losses through to the stockholders in a manner similar to a partnership. There is no income tax on the profits to the S corporation; however, the S corporation must file corporate income tax forms. All profits and losses are passed directly through to the stockholders, who claim business income on personal income tax returns. An S corporation is a corporation (like the C corporation) in every other respect except in paying income tax.

To qualify to receive tax treatment as an S corporation, the corporation must meet many complex provisions of the Internal Revenue Service code. The provisions have been greatly expanded in recent years. Some of the principal provisions include the following:

- *Only U.S. citizens can own stock in an S corporation.*
- *The S corporation must be a "small business corporation," defined as having 75 or fewer stockholders.* A husband and wife are counted as one stockholder.
- *The S corporation can have only a single class of stock.* The S corporation can have common but not preferred stock, although the common stock can be issued with voting and nonvoting rights (see Chapter 5 for details).
- *Ownership in an S corporation is restricted to individuals, estates, trusts, and tax-exempt organizations.* An *estate* is comprised of the assets of a deceased individual. A *trust* places the assets of an individual in the hands of a trustee(s) to be managed for a beneficiary(ies) who is(are) the recipient(s) of the trust's income and assets. Historically, the trustee has been a bank or other corporate trustee, but the trustee can also be an individual. The trust document specifies who is

to receive the income from the trust assets and who is to receive the assets of the trust upon dissolution. Attorneys draft trusts based on state laws.

- *All stockholders must agree to be taxed under subchapter S of the Internal Revenue Service code.* A corporation may elect to change from C status to S status or vice versa if all of the stockholders consent.

There are several situations when it may be advisable to elect to be taxed as an S corporation. One situation is when losses are anticipated. The losses can usually be passed directly to the stockholders to offset other income. Another situation is when the corporation expects to earn substantial income that would be paid out as dividends in a C corporation. By using an S corporation, the corporate tax on profits can be avoided and only the personal income tax is paid on the income, thereby avoiding double taxation on dividends.

Another reason to form an S corporation is to facilitate the transfer of the firm to succeeding generations and to continue the firm beyond the present generation. There are several aspects of the corporate organizational form that can facilitate this objective. The corporate entity allows a stockholder to transfer shares of stock while retaining management control. In a corporation with one class of voting stock and simple majority rule, up to 49 percent of the stock may be given away without loss of management control. Even more stock can be given away if the corporation has nonvoting stock. Restrictions can be placed on the transfer of stock to an "outsider" or to "unfriendly" heirs. In this way a corporation cannot be partitioned for sale without majority consent. Finally, it is much simpler to settle an estate if the only stockholder asset is corporate stock rather than real estate, equipment, and, in the case of farms and ranches, livestock and crops.

Because of the complexity of the S corporation regulations, it is always advisable to consult a tax accountant or attorney before choosing this form of business ownership. Because it is an elective provision, failing to meet any one of the many guidelines may result in revocation of the favorable S corporation tax status.

Another disadvantage of the S corporation is that it is often difficult to retain earnings for growth. Since all profits are attributable to the stockholders in proportion to their ownership in the firm, the stockholders are responsible for paying personal income taxes on their share of the profits. This results in pressure for the S corporation to distribute a large percentage of its earnings to its stockholders to provide them with the cash to pay the taxes. The difficulty in retaining earnings often makes it hard for S corporations to obtain large amounts of debt capital because they typically have a small equity base (paid-in capital plus retained earnings). The major advantages and disadvantages of the S corporation are presented in Box 2–7.

Nonprofit Corporations

All states permit individuals to form nonprofit corporations in order to receive tax-exempt status by the Internal Revenue Service. The tax-exempt status ex-

BOX 2–7

Advantages and Disadvantages of a Subchapter S Corporation

Advantages	Disadvantages
Limited liability of owners	Many laws and regulations apply
Stability and continuity	Costly and difficult to form and dissolve
Facilitates raising capital	Cumbersome rules pertaining to shareholders
Easy to attract qualified personnel	Separation of ownership and management
Taxed like a partnership	Tax treatment does not favor retaining earnings

empts the corporation from paying taxes on its income and allows individuals and organizations contributing to the nonprofit corporation to take a tax deduction on their contributions. Individuals contributing assets to a nonprofit corporation give up ownership interest in those assets because the assets become permanently dedicated to the specific nonprofit purposes. A nonprofit corporation must be formed for religious, charitable, literary, scientific, or educational purposes. Nonprofit corporations in the food and agribusiness sector provide general benefits to producers, processors, and other organizations, including education on production techniques, market development, legislative activities, and product promotion. Some examples of these nonprofit corporations are the Future Farmers of America (FFA), the Farm Foundation, and the International Food and Agribusiness Management Association (IAMA).

Evaluating the Corporation

Almost all of the largest businesses in the United States are corporations. No other form of business ownership provides as much access to financing, growth potential, and smooth transfer of ownership as the corporation. These are necessary attributes for large food and agribusiness firms that often have countless owners. Owners who do not want to risk their personal assets should the firm fail or who don't want to incur huge financial liability should consider forming a corporation. Whereas the tax status of corporations may be a disadvantage in particular situations, this is often outweighed by the personal liability protection afforded by the corporate form of business.

The S corporation provides an alternative to a limited partnership. S corporations offer limited liability to all owners, but they are taxed much like a partnership. Furthermore, all owners of an S corporation may participate in management, whereas the limited partners may not.

Common Mistakes

The most common mistake in forming a corporation is creating a situation in which you have to complete a lot of administrative paperwork and follow regulations designed primarily for very large businesses when the needs of the business do not justify forming a corporation. Business owners need to be cautious when establishing this form of business organization. The corporate form of business organization cannot be easily changed to other forms, nor can the corporation be dissolved in a financially viable way because the increased value of corporate assets will be taxed twice. Many individuals may form a corporation thinking that this will make it easier to attract funds. However, this is almost never the case. This issue is discussed in greater detail in Chapter 5.

LIMITED LIABILITY COMPANY

Limited liability companies (LLCs) were started in Wyoming in 1988 by legislators who wanted to let small business owners set up flexible businesses that would be taxed like partnerships, have the limited personal liability of corporations, and none of the burdensome restrictions of the S corporation. Consequently an LLC is often referred to as a hybrid corporation/partnership. Currently, all 50 states have adopted legislation for establishing LLCs; however, because it is a relatively new form of business organization, there are many differences among the states in the rules regulating LLCs. The following discussion focuses on the manner in which the majority of the states have adopted LLC legislation.

Forming a Limited Liability Company

There are few drawbacks to forming an LLC. Paperwork requirements include filing articles of organization with the secretary of state and paying a fee. Many states require at least two owners to form an LLC; however, pending legislation in some states is moving in the direction of permitting one owner in all states. If the LLC has more than one owner, an operating agreement and a buy-sell agreement should be drafted. Because most states have default provisions that specify how the LLC will be operated unless specific issues are addressed in an LLC agreement, it is always a good idea to enter into such an agreement to maintain control over the rules that will govern the operation of the LLC.

LLC ownership is extremely flexible. The owner, typically called a member, can be an individual, group, or separate legal entity, such as a partnership or corporation that invests in the LLC. A member holds membership interest in the business rather than shares of stock. Unless otherwise provided in the operating agreement, members vote their membership interest in proportion to their invested capital.

The LLC agreement should specify the terms for continuing the LLC and transferring membership interest. State law specifies that the LLC be dissolved at the end of a fixed period of time unless the LLC agreement specifies some other duration. For this reason most LLC organization agreements specify the duration of the LLC. The LLC will also be dissolved upon the withdrawal or death of a member unless the LLC specifically addresses how the membership interest may be transferred and the conditions for doing so. Unless otherwise specified in the LLC's articles of organization, membership interest may be transferred only with the unanimous consent of the other members.

Advantages

All members of an LLC have limited liability similar to the liability associated with holding stock in a corporation. This means that as a member of an LLC, you are not personally liable for the business debts and court judgements against your business. In other words, you risk only the capital you put into the business. However, you would still be responsible for debts that you personally guaranteed or for any *torts* (civil wrongs) you personally committed.

A principal advantage of the LLC over a C corporation is that the members may elect to have all profits and losses passed through the business to be taxed on the members' personal income tax return, similar to the way in which partnerships are taxed. In this way, all profits and losses are passed through to the members to be filed on their personal income tax returns and partnership tax forms are filed. Current IRS regulations allow both single- and multiple-member LLCs to elect to be taxed as a corporation. In this case, the LLC would file corporate tax forms for business income and members would file personal income tax forms for any income or profits distributed to them. Unlike corporations, which have restrictions on how profits and losses are distributed to shareholders, LLCs can typically distribute profits and losses to members any way they want.

Another characteristic of LLCs is their flexible management structure. Management can be composed of a single member, an outside manager, or a management group consisting of members or nonmembers, or a combination of members and nonmembers. Consequently, it is important to spell out in an operating agreement how the LLC will be managed. An additional benefit of the LLC is that, unlike a corporation, the LLC is allowed to operate across state lines without registering with the foreign state.

Disadvantages

For many small, group-owned agribusinesses, the flexibility and limited liability of the LLC make it an ideal choice for business ownership. However, the decision to form an LLC instead of another form of business ownership may rest on other factors. For example, some states may require professional agribusiness owners, such as veterinarians, to form special professional partnerships or corporations. Business owners might also benefit from the discipline imposed by legal rules regulating corporations that dictate the responsibilities and decision-making power

BOX 2–8

Advantages and Disadvantages of a Limited Liability Company

Advantages	Disadvantages
Limited personal liability	Typically more complex than a sole proprietorship or partnership
No tax as a business	Flexible management structure may require great discipline from owners
Flexible management structure	Does not allow for selling stock to raise capital
Easier to form than a corporation	Formation fees may be substantial

of the directors and shareholders. Although it is easy to transform an LLC into a corporation, if the intent of the owners is to sell shares to the public or to reward productive employees with shares of stock, the corporation would be the preferred form of ownership. Likewise, if you want to provide extensive benefits to employees and owners, the corporate form of ownership has more flexibility and tax-favored fringe benefits.

LLCs also have some disadvantages relative to the sole proprietorship and partnership. One of the major disadvantages is that more paperwork is typically required to form an LLC than for either a sole proprietorship or a partnership. When liability is not a concern, the LLC provides little advantage.

Finally, there are a few states that impose annual fees or taxes on limited liability companies that limit their appeal. As we mentioned in discussions of the other forms of business ownership, legal, tax, and accounting advice is necessary to fully understand the complexities of each form of ownership and to wisely choose the type that best suits the individuals involved as well as their personal and business situation. Box 2–8 summarizes the principal advantages and disadvantages of LLCs.

Relationships Between Businesses

Up to this point only forms of business ownership have been discussed. However, businesses often form relationships with other businesses. These relationships may be formal or informal. In this section three of the most common forms of business relationships, strategic alliances, franchises, and joint ventures, are presented.

Strategic Alliance

The term *strategic alliance* is used to cover a great many types of arrangements between businesses, both formal and informal. Although there is no commonly agreed upon definition, most strategic alliances have several things in common. They involve at least two organizations pursuing some common long-term goals and close coordination between the alliance partners. In the food and agribusiness industry, it is becoming increasingly difficult for individual operators to operate independently. The pressure to decrease costs, provide higher and more consistent quality, reduce inventories, and shorten the lead time in functions as diverse as ordering and product development has encouraged many independent agribusinesses to form strategic alliances with other firms. At the simplest level, strategic alliances are formed because the individuals or organizations involved jointly undertake activities that they could not undertake themselves. Strategic alliances have been formed in the food and agribusiness sector as value-added partnerships consisting of independent agribusinesses that together manage the flow of products and services along the entire value-added chain. An example of a strategic alliance in the foodservice industry is the partnership between Orion Food Systems, owner of many foodservice outlets including Cinnamon Street Bakery, and Cendant Hotel Division, operator of hotel chains including Howard Johnson and Days Inn. In this strategic alliance, Orion is the preferred food service provider to Cendant's chain of hotels.

Joint Venture

A *joint venture* is a formal strategic alliance that is formed when two or more companies formally join together in a single endeavor to make a profit. For example, ConAgra and Kellogg's have created a joint venture to manufacture and market Healthy Choice cereals. Joint ventures are typically created between noncompeting companies using the expertise of each company and combining the financial strength of both companies to enter into business arrangements that benefit both partners. In the case of Healthy Choice cereals, ConAgra provides the Healthy Choice brand while Kellogg's does the manufacturing and distribution. Both companies share in financing and marketing the product.

Joint ventures can be organized as partnerships, corporations, LLCs, or cooperatives. Many small businesses are using their unique research and development capabilities and are establishing joint ventures with larger companies that can provide the financing and marketing clout to get a product to market. Joint ventures are also popular in global marketing. Many U.S. companies establish joint ventures with companies in foreign countries to market their products abroad. Robert Mondavi Winery, discussed in Industry Profile 2–1, is an example of an agribusiness that has used joint ventures to further the goals of its business.

Franchise

When you think of hamburgers, pizza, frozen yogurt, tacos, and submarine sandwiches, you normally think of fast-food restaurants. However, what you

INDUSTRY PROFILE 2–1

ROBERT MONDAVI WINERY

Robert Mondavi Winery, located in Oakville, California, was formed by Robert Mondavi in 1966. After years of working in his family's winery he set out on his own with a vision that California could produce world class wines. Mondavi is widely acknowledged as a leader in the industry. Its efforts to promote Fumé Blanc are largely credited with the growth of the grape variety Sauvignon Blanc and the recognition of California wines as some of the best in the world. Mondavi's brands include Robert Mondavi Winery, Woodbridge, and Robert Mondavi Coastal.

The Robert Mondavi Winery was owned and controlled entirely by the Mondavi family until the 1990s. However, in 1993 the company was faced with a growing need for capital, necessitated by the need to replant many of its vineyards that had been affected by phyloxera, as well as the need for additional capital to finance growth. The company went public with an initial public offering (IPO), in which stock was offered for sale to the public. In addition to raising additional capital, the company was able to reduce its reliance on debt. The Mondavi family also maintained control of the company. Today it controls 92 percent of the voting stock. The company has benefited from the discipline imposed by being a public company. For example, the company has a board of directors that includes outsiders.

In addition to being the first major winery to hold a public stock offering, Robert Mondavi Winery has been a leader in the formation of strategic alliances. Its most famous partnership is with Baron Philippe de Rothschild of Chateau Mouton Rothschild in France. This joint venture began production of the super premium wine Opus One in 1979. More recently, Mondavi partnered with Marchese de' Frescobaldi of Italy in 1995. Its most recent joint venture is with the Eduardo Chadwick family of Viña Errazuriz in Chile. The two wineries plan to jointly produce the Caliterra brand and market it worldwide.

may not realize is that many of these businesses operate under a legal arrangement called a franchise. A *franchise* is a licensing agreement by which the owner (the *franchiser*) distributes a product or service through affiliated dealers (the *franchisee*). A franchise can be operated under any form of business ownership, but most are formed as corporations or LLCs. Franchisers allow franchisees to use their business systems, brand name, and corporate identity for a specified period of time, usually 5 to 20 years, exclusively in a defined geographical area.

Franchising is based on an interdependent relationship between the franchiser and the franchisee. The franchise is established through a franchise agreement governed by state law. In the franchise agreement, the franchiser establishes uniform business practices that must be strictly followed by the franchisee. These business practices involve trademarks, equipment, storefronts, operating procedures, and standardized services or products. The franchiser also provides the franchisee assistance in running the business. This usually includes the following:

- Locating the business and negotiating the lease
- Store design and equipment purchasing
- Initial management training and continuous management support
- Advertising and merchandising
- Standardized operating procedures
- Procuring inputs
- Obtaining financing

In return, the franchisee agrees to comply with the procedures established by the franchiser and pays the franchiser an initial franchising fee plus a percentage of the business's sales.

COOPERATIVE

A *cooperative* is a user-owned, user-controlled business that distributes benefits to its members according to their use of the business. "User-owned" means that the majority of the *patrons* (customers) are *members* (owners) who finance the cooperative. "User-controlled" means that the majority of the patrons must be members who democratically select the board of the directors from the membership. The fact that there are "benefits" to being a member of a cooperative indicates that the primary purpose of a cooperative is to provide and distribute those benefits to its members. There are more than 47,000 cooperatives in the United States, representing a broad range of industries and providing goods and services to more than 100 million people. However, agriculture dominates the cooperative form of business, with grocery wholesaling, hardware and lumber, finance, utilities, and health services comprising the remainder of large cooperatives (Figure 2–1).

Starting a Cooperative

To start a cooperative you must first determine if there is sufficient interest in, and economic need for, the cooperative. This is usually accomplished by inviting leading potential members to gather to discuss the need for the cooperative. Open meetings in which all potential members present ideas for a new cooperative should be held, followed by a survey to determine interest and measure the potential for the success of the cooperative. Next, a committee should be formed to oversee studies that determine the costs and feasibility of the proposed cooperative. The committee should also draft a business plan. The results of the studies and the business plan are then presented

FIGURE 2–1

Industries of the Top 100 Cooperatives, 1996

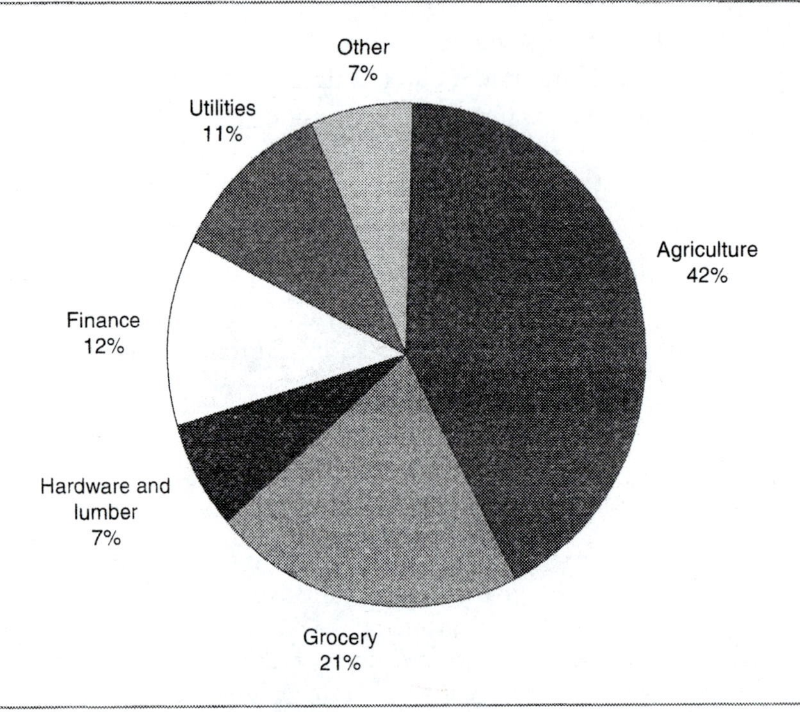

Source: National Cooperative Bank, Washington, D.C., 1997.

to potential members, who vote on the proposal. If the voting majority favors the proposal, a cooperative can be formed as a Subchapter T corporation by submitting articles of incorporation to the secretary of state in the state in which the cooperative will be formed. By-laws are prepared and a meeting is called at which members vote to approve the by-laws and elect a board of directors from the membership.

Objectives of the Cooperative

Agricultural cooperatives, which can be categorized as marketing, farm supply, or service cooperatives (Box 2–9), are started to meet one or more of six primary objectives. *Marketing cooperatives,* which comprise the largest number of cooperatives, are started primarily to increase members' incomes, obtain market access, and broaden market opportunities. *Farm supply cooperatives,* which have the largest cooperative membership, are started primarily to decrease members' costs and improve bargaining power when dealing with farm input suppliers. *Service cooperatives* are founded primarily to obtain products or services for members that would otherwise be unavailable. All three types of

BOX 2–9

Number of Farm Cooperatives and Membership by Business Activity, 1998

Business Activity	Farm Cooperatives		Members	
	Number	Percent	Number	Percent
Farm Service	441	12	180,562	5
Marketing	1,863	51	1,398,356	42
Farm Supply	1,347	37	1,773,659	53
Total	3,651	100	3,352,577	100

Source: Farm Cooperative Statistics, 1998, RBS Service Report 57, Rural Business-Cooperative Service, United States Department of Agriculture, page 2.

agricultural cooperatives also have as their objective improving product or service quality.

Marketing cooperatives are the largest of the three types of agricultural cooperatives, with sales of nearly $77 billion in 1998. Some of the major products they market include grains and oilseeds, milk and milk products, fruits and vegetables, livestock, and other products (Figure 2–2).

A *bargaining association* is a type of marketing cooperative that performs limited marketing functions. Bargaining associations typically do not handle or take ownership of their members' products but represent members by negotiating price, quantity, and grade, or establishing other terms of trade with distributors, processors, or other buyers for their members' products. They often coordinate supplies among producers and arrange shipments to meet larger buyers' demands for quantities and service. By pooling member farmers' production output, bargaining cooperatives can secure higher and more stable prices than members could achieve individually. Some bargaining associations now perform additional functions, such as the transport of products.

Most marketing cooperatives go a step further than bargaining associations, by processing their members' products into value-added commodities and consumer products. These cooperatives market products under some of the most recognizable consumer brand names, such as Land O'Lakes, Sunkist, Ocean Spray, Blue Diamond, Sun-Maid, and Welch's, to name a few. These powerful brand names increase market access on the supermarket shelf and broaden opportunities by offering an array of differentiated food products.

Farm supply cooperatives derived $25 billion in sales in 1997 by supplying petroleum, feed, fertilizer, crop protectants, seed, and other supplies to members (Figure 2–3). By purchasing goods together in bulk, members of farm supply cooperatives can secure volume discounts and thereby reduce the cost of inputs. Cooperatives can also reduce prices by operating at cost when existing suppliers do not price competitively. Many farm supply cooperatives also supply farm equipment, building materials, heating oil, and lawn and

FIGURE 2–2

Relative Importance of Farm Products Marketed by Cooperatives, 1998

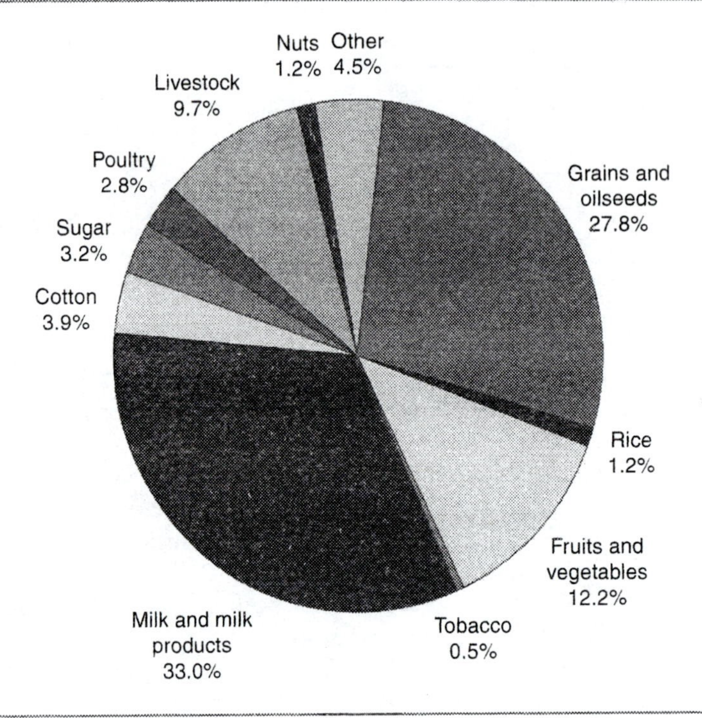

Source: Farmer Cooperative Statistics, 1998, RBS Service Report 57, Rural Business-Cooperative Service, United States Department of Agriculture, page 14.

garden supplies. Additionally, they sometimes provide farm management services such as crop consulting and commodity marketing advice.

Farm service cooperatives provide agricultural services directly related to marketing and purchasing, such as trucking, storage, drying grain, and artificial insemination. Other service cooperatives that provide services to both farmers and rural residents include Farm Credit System banks, rural telephone and electric cooperatives, rural credit unions, and Dairy Herd Improvement Associations. These service organizations meet the needs of members that the existing marketplace does not meet or that can be met cooperatively at a lower price. The advantage of forming a service cooperative is that members can obtain the service "at cost."

Principles of Cooperatives

Agricultural cooperatives are distinguished from other forms of business ownership in that they support and follow a distinct set of principles or guidelines. The

FIGURE 2–3

Relative Importance of Farm Supplies Handled by Cooperatives, 1998

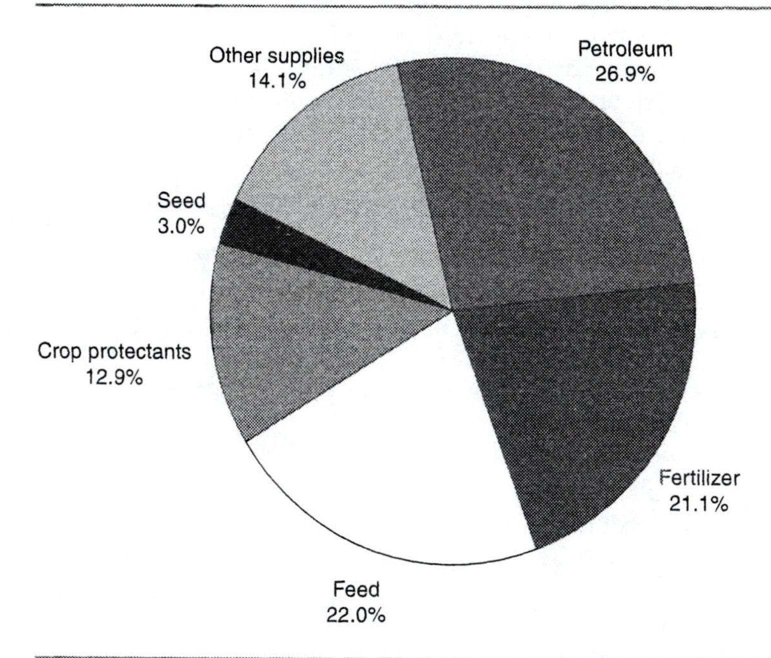

Source: Farmer Cooperative Statistics, 1998, RBS Service Report 57, Rural Business-Cooperative Service, United States Department of Agriculture, page 15.

Rural Business-Cooperative Service, a division of the USDA, recognizes the following guidelines in identifying an organization as an agricultural cooperative:

- Membership is limited to persons producing agricultural and aquacultural products, and to associations of such producers.
- Cooperative members are limited to one vote regardless of the amount of stock or membership capital owned.
- The cooperative does not pay dividends on stock or membership capital in excess of 8 percent a year, or the legal rate in the state in which they are based, whichever is higher.
- Business conducted with nonmembers may not exceed the value of business done with members.
- The cooperative operates for the mutual interest of members by providing member benefits based on patronage (commonly referred to as "operating at cost").

Voting rights have historically been restricted to one-member, one-vote. The original objective was to ensure that all members were equally represented regardless of their volume of business. Most agricultural cooperatives operate under this principle, although it has created problems for some cooperatives. Members who use the cooperative most heavily have in some cases wanted greater control of the cooperative to ensure that the cooperative is responsive to their needs. Although cooperatives that allow voting proportional to patronage or equity investment are in the minority, their numbers are increasing.

Returns on capital are usually limited to 8 percent. State law usually establishes a maximum fixed rate of return. An additional constraint is that marketing cooperatives must not return more than 8 percent on stock to receive the protection of the Capper-Volstead Act (discussed later in this chapter) from antitrust suits, if voting is not done on a one-member, one-vote basis. This is a key feature of cooperatives because it ensures that the users control the cooperative. By keeping the rate of return low, nonusers are discouraged from investing in or attempting to control the cooperative. However, this policy also discourages members from investing equity capital in the cooperative because it limits the return they can receive.

Historically, membership in a cooperative has been open. More recently a new cooperative model, called a "new generation cooperative," has emerged. *New generation cooperatives* differ from traditional cooperatives in that they have a closed membership policy and each member is committed to do a specific quantity of business with the cooperative each year. See Industry Profile 2–2 for a profile on this innovative type of cooperative.

Business conducted with nonmembers must not exceed the business conducted with members. Traditionally, this principle was adopted to make sure that the control of the cooperative remained in the hands of the members and that the cooperative remained responsive to the membership. This principle is not always desirable from the member's perspective because it limits the amount of business the cooperative does, thereby limiting the amount of capital that can be retained by the cooperative for growth. Since membership and patronage change over time it is often difficult to implement this principle in practice.

Finally, benefits should be allocated to members on the basis of use or patronage. What this means is that a cooperative, since it is established for the benefit of its member-users, operates at cost and not to make a profit for its member-investors. Farm supply cooperatives price the supplies or services they sell to members slightly above their estimated costs. Likewise, marketing cooperatives pay members slightly less than their estimated net price from marketing members' products. In this way the cooperative generates income sufficient to cover expenses and allocates the surplus income back to the member-patrons on the basis of their use of the cooperative. In reality this is a narrow view of the benefits accruing to members. Other important benefits provided by cooperatives include gaining access to markets, supplies, or services to which members would otherwise have no access, providing better quality products or services, and increasing competition by providing farmers an alternative to for-profit agribusiness firms in areas where they operate.

INDUSTRY PROFILE 2–2

NEW GENERATION COOPERATIVES AND U.S. PREMIUM BEEF

New generation cooperatives represent the newest type of marketing cooperatives and are a type of strategic alliance. Their major focus is value-added processing and production of food and fiber products for which there is a niche market—a departure from the commodity marketing focus of many of their predecessors. U.S. Premium Beef (USPB) is a new generation cooperative. USPB members formed the cooperative in order to purchase an equity share of Farmland National Beef Packing Company. By purchasing Farmland National, USPB was able to add value to their members' cattle through further processing through Farmland National's facilities and marketing through Farmland's Black Angus Beef and Certified Beef brands. This enabled members of USPB to earn additional profit by processing their cattle and selling branded beef products. As part of a value-added chain, these cattle producers also benefit from additional information because members receive carcass quality data on their cattle and customer purchase data from beef sales. These data are crucial to cattle producers in production planning to meet customers' changing purchasing patterns.

New generation cooperatives (NGC) have three distinguishing characteristics. First, an NGC differs from a traditional processing/marketing cooperative in that members purchase delivery shares that give them exclusive marketing rights for their crops. When an NGC is started, the original price of common stock in the cooperative is tied to a delivery share or marketing right. The purchase of a share of common stock guarantees the member the right and creates a legal obligation to market a fixed quantity of raw product to the cooperative. Once all the shares of common stock are sold to producers, the cooperative has raised sufficient equity to build and operate the processing plant and has contracted a sufficient amount of raw product to run the plant and successfully market its products.

Second, membership in the cooperative is restricted to producers who invest the original equity capital when the cooperative is created. There are a number of reasons for limiting (closing) membership. By limiting membership the cooperative ensures that each member will remain committed and be involved in the cooperative. It also ensures that the rewards to membership are distributed in proportion to delivery rights. When members deliver raw products to the cooperative, they receive a portion of the current market price. At the end of the year, profits from the sale of products sold by cooperative are calculated and are divided among the members in proportion to the amount of product they delivered. Future expansion of the cooperative is financed in the same way as the cooperative was originally financed; that is, members supply new equity through the purchase of delivery shares.

Finally, members are allowed to transfer (sell) equity capital and delivery shares to others at any time. The value of their stock thus appreciates and depreciates like a public corporation depending on the profitability of the cooperative and the demand for equity shares. This has led members to view the NGC as an investment opportunity as well as a value-added marketing opportunity. As such, NGCs are said to play an "offensive" (as opposed to a defensive) role in the marketplace by providing producers a profit-oriented investment that aggressively targets new markets for their products.

Organizational Membership Structures

The Rural Business-Cooperative Service categorizes agricultural cooperatives into centralized, federated, and mixed cooperatives. Of the 3791 agricultural cooperatives in 1997, 3682 were centralized organizations, 67 were federated organizations, and 42 were mixed organizations.

Centralized Cooperatives. *Centralized cooperatives* are owned by their individual members and typically serve an area ranging from a county to a region that might include several states. Centralized cooperatives are heavily involved in both the marketing of commodities and the supply of inputs to farmers. Two well known centralized cooperatives are Sun-Maid Growers of California, a large raisin cooperative, and Sunkist Growers, Inc., a large California-based citrus cooperative. Centralized cooperatives serving a large region typically operate local facilities in the areas they serve.

Federated Cooperatives. When two or more cooperatives join together to form a cooperative, the result is a *federated cooperative.* The members of federated cooperatives are other cooperatives, either centralized or federated. Most federated cooperatives are owned by local centralized cooperatives. Indiana Farm Bureau Cooperative Association, Inc., and Farmland Industries, Inc., are examples of this type of organizational structure. By combining several local associations, it is often possible to increase bargaining or buying power, perform some activities more efficiently, or enter into new production activities. The organization of the federated cooperative is much like that of the centralized cooperative. The member cooperatives elect directors, who in turn hire the management.

Mixed Cooperatives. *Mixed cooperatives* have both individuals and other cooperatives as members. Mixed cooperatives are relatively few in number and have characteristics of both centralized and federated associations. Mixed cooperatives often emerge as a result of a federated cooperative taking control of one of its member cooperatives that is experiencing financial difficulties. Southern States Cooperative, Inc., is a mixed cooperative. It provides wholesale services to its member cooperatives and retail services to individual members.

Antitrust Laws

At the beginning of the 20th century, the United States was rapidly entering the Industrial Age. Farmers and many other owners of family businesses complained about the market power of industrial companies in railroads, meatpacking, and manufacturing. Under the laws at that time, it was not illegal for companies to form agreements to regulate prices or competition. Congress responded to this situation by passing the Sherman Act of 1890, which made it illegal for companies to restrain trade in interstate commerce or to form monopolies. In 1914 Congress passed the Clayton Act, which forbade companies from engaging in exclusive dealing and prohibited mergers that would substantially lessen competition or tend to create a monopoly. Congress intended to exempt agricultural organizations from the Sherman and Clayton Acts; however, the manner in which the Acts were written made the status of agricultural cooperatives under the federal antitrust laws unclear.

The Capper-Volstead Act

Farmers pressured Congress to clarify the antitrust laws to permit farm organizations to collectively sell farm products on behalf of their members. This effort resulted in the Capper-Volstead Act of 1922. The Act permitted persons engaged in the production of agricultural products to act together in associations to collectively process and market their products, provided that the associations were operated for the mutual benefit of their members. To receive the status of a cooperative, the associations were required to accept the following conditions: (1) one-member, one-vote, or (2) dividends on stock cannot exceed 8 percent per year, and (3) business conducted with nonmembers cannot exceed the business conducted by members. While the Capper-Volstead Act gives producers the right to jointly market their products through cooperatives, it does not exempt them from all antitrust action. Essentially, producers may use cooperatives to offset the market power of buyers of their products. However, they may not fix prices or raise prices to "unreasonable" levels, as determined by the Secretary of Agriculture. By allowing farmers to form cooperatives, Congress hoped to strengthen farmers so that they could better deal with unstable prices and powerful distributors and processors. To further assist farmers, in 1926 Congress passed the Cooperative Marketing Act permitting farmers and their associations to acquire and exchange price, production, and marketing information.

Taxation

It is a popular misconception that cooperatives pay no taxes. Cooperatives pay property taxes, sales taxes, employment taxes, fuel taxes, and utility taxes, just like other businesses. With regards to income taxes, cooperatives are allowed to elect exempt status under Section 521 of the Internal Revenue Code and be exempt from paying taxes on income generated by its patrons. The rationale for this tax treatment is that cooperatives, because they are owned and operated for the benefit of their members, pay taxes only once, at the ownership level, similar to sole proprietorships and partnerships, rather than twice as is the case for corporations where earnings are taxed once at the organizational level and again at the ownership level (the stockholder). Cooperatives that choose exempt status must pay patronage dividends (defined below) to both member and nonmember patrons in proportion to their use of the cooperative. These patronage dividends are then deducted from income for tax purposes. Cooperatives that elect nonexempt status must pay taxes on the net margins generated from nonmember business only. Most cooperatives choose to be nonexempt for two reasons. First, it reduces the record-keeping needs for their nonmember business. Second, it increases the returns to their members because only a portion of the net margins generated by nonmembers is paid in taxes.

A *patronage refund* (or patronage dividend) is the amount paid to patrons from net savings generated by the cooperative from business conducted with or for its patrons. Patronage refunds are paid in cash or as written notices of allocation. A *written notice of allocation* is a patronage refund or a per-unit retain kept by the cooperative and issued to the member as capital stock or certificates of ownership. The retained portion of the patronage refund is allocated to the member's equity account and is paid out at a later date when the member's stock is redeemed. Patronage refunds are distributed based on the quantity or value of business done with or for the patrons under a pre-existing legal obligation. Patronage refunds from marketing cooperatives are added to income, and refunds from farm supply cooperatives are deducted from expenses, on patrons' personal income tax returns.

Challenges Facing Cooperatives

Many cooperatives have faced and will continue to face many challenging problems, some of them common to all agribusinesses and some of them unique to cooperatives. In the last decade, some cooperatives have been unable to offer their members lower prices on purchased inputs or higher prices for their commodities than other competing firms. The days are gone when a cooperative member will do business with the cooperative out of loyalty and members are not easily consoled by the fact that the goods and services they purchase are sold at cost when the cost is much higher than the price they would pay at another firm. Many cooperatives have found their membership dwindling as producers have left to do business with firms that offer them better prices.

Many cooperatives have been criticized for being poorly managed and this problem is often blamed on the cooperative's board of directors, which selects the management. In cooperatives, unlike most corporations, the board is made up exclusively of the cooperative's members. In many situations these individuals have neither the ability nor the experience required for charting the future direction of the business or managing complex business issues. It is difficult to get farmer-board members to devote the time and invest in the training necessary to be a productive board member. Furthermore, the opportunity cost for board members of leading and directing a cooperative is high compared to allocating the time to improve their own farm or ranch business.

A major financial issue for members of many cooperatives is that the cash refunds paid by their cooperatives have often been insufficient to cover their tax liability. This can happen because cooperatives typically raise capital by paying out only a portion of the patronage refund in cash and retaining the balance, which the member receives as stock. For example, assume that the cooperative issued a $100 patronage refund to a member, paying $20 in cash and issuing $80 in capital stock. If the member is in the 28 percent personal federal income tax bracket, he or she would have to pay $28 in federal income taxes but would only received $20 in cash from the cooperative. In this case the member would be required to pay the additional $8 in income taxes from other income sources as well as any personal state income tax owed. If this is not offset by additional income from the cooperative as it redeems (pays cash) for stock it issued to the member in previous years, this could cause a cash shortfall for the cooperative's member.

Cooperative capital stock, unlike stock in large corporations, is not traded on the financial markets because stock ownership is limited to members. This has been the source of great concern to members who quit or retire from farming and the heirs of deceased members because this equity capital is not *liquid* (easily converted into cash). The decision as to how quickly capital stock is redeemed is usually left to the board of directors and is subject to the cooperative's financial condition. Members of unsuccessful or poorly managed cooperatives often have no idea when or if they will ever be able to redeem stock or if the stock will have any value when they attempt to redeem it. Consequently, members and state legislators have pressured cooperatives to formulate equity redemption plans. Managers of cooperatives have thereby been encouraged to manage their finances so that members view the annual stock refund as a benefit rather than a liability.

A related problem is that cooperatives find it much more difficult to raise capital than C corporations because they are unable to sell stock on the financial markets. Cooperatives have traditionally relied heavily on debt financing to finance their growth and operations. This has resulted in many cooperatives being highly leveraged with all of the risks associated with firms that are heavily burdened by debt.

Although we close this chapter by addressing many of the problems facing today's cooperatives, we should emphasize that many cooperatives have been very successful and have continually served their members well. The cooperative form of business continues to offer great opportunity for agricultural producers to enhance their incomes when they are well managed and used in the appropriate business environment.

SUMMARY

The choice of the form of business ownership is one of the most important decisions the owner or owners of a food or agribusiness firm will make. It will determine the amount of taxes paid, the financial risk the owners must assume, the firm's access to capital, and the ease with which growth is attained. The choices available include sole proprietorship, various forms of partnerships and corporations, the limited liability company, and cooperative organization.

Most agribusinesses are organized as sole proprietorships. This is mostly a function of their size and the fact that the sole proprietorship is the simplest and cheapest to organize. It is subject to a minimum number of rules and regulations, and income is taxed only once as personal income of the proprietor. A major disadvantage is that the owner is personally liable for all debts of the firm. Sole proprietorships find it more difficult to raise capital and, because the existence of the firm depends on one person, it may be difficult to attract qualified personnel.

Two or more individuals can form a partnership. Many characteristics, such as the method of taxation and the personal liability of the owners, are similar to the characteristics of a sole proprietorship. Most partnerships are governed by a partnership agreement that spells out each partner's rights, responsibilities, distribution of profits and losses, and the rules for dissolving the partnership. The limited partnership allows additional individuals to invest in the partnership while limiting their risk to the amount invested. A relatively new type of partnership available in some states, called the limited liability partnership, may also be used to limit partners' liability.

Corporations enjoy several advantages over the sole proprietorship and partnerships including limited liability for stockholders. Almost all large businesses are corporations because the ease of transferring ownership and the access to capital facilitate growth through the involvement of large numbers of individuals. A major disadvantage is that regular corporations are taxed twice on income distributed as dividends, although they also have available to them several mechanisms to reduce their tax liability. They are subject to many laws and regulations, and compliance may be time consuming and costly. The S corporation, while subject to many restrictions, gives owners all of the advantages of a corporation but allows them to be taxed in a manner similar to a partnership.

The most recent form of business organization is the limited liability company, or LLC. An LLC offers owners of partnerships (and sole proprietorships in some states) limited liability protection as well as management and tax flexibility. The LLC offers an easy transition to other forms of business ownership if changes are required in the future. Because of these strengths the LLC is becoming a popular choice among owners of food and agribusiness firms.

The cooperative is an alternative form of business organization for agricultural producers. Cooperatives are usually formed to increase producers' income by more effectively marketing producers' products, providing supplies to producers at a lower cost, or by offering better services. Cooperatives typically conduct business at cost, are democratically controlled, limit returns on capital invested to 8 percent, and are financed by their members in proportion to their volume of business. In this way, cooperatives are operated for their members' benefit, unlike most firms, which operate to earn a profit for their investors.

CASE QUESTIONS

1. **What factors would you advise Ann to consider in determining which form of business ownership she should choose? Why is each of these factors important?**
2. **Assume that Ann has hired you to determine the form of business ownership that is best for her. What additional information would you need to know about her? How would you use this information to determine what form of business is best for her?**
3. **What form of business ownership would you recommend to Ann? Justify your recommendation.**

REVIEW QUESTIONS

1. Which factors are the most important to consider in choosing the form of business ownership? Why?
2. Compile a chart comparing some of the basic features of the sole proprietorship, partnership (general), corporation, and limited liability company. Include factors such as the number of owners, how profits are taxed, personal liability, how equity capital is raised, who controls management, and transfer of ownership.
3. Describe a limited partnership and its potential advantages and disadvantages over a general partnership.
4. Identify situations in which an S corporation might be preferable to a regular corporation.
5. Discuss how the choice of a form of business ownership may affect a firm's access to capital.
6. Why might farmers prefer to organize as a cooperative rather than as a corporation to market a product?
7. Identify situations in which an LLC might be preferable to an S corporation.
8. Why would companies want to form a joint venture?
9. Why have franchises become popular with people wanting to start a small business?
10. Describe a strategic alliance that has formed among agribusinesses in your area. How does this strategic alliance use information to add value to its product? What niche market(s) is the company targeting?
11. Pick a commodity in your area and describe how you would form a new generation cooperative to assist the local farmers or ranchers in taking an "offensive" direction in the marketplace.

CHAPTER 3

Financial Statements

Learning Objectives

In this chapter we will cover the following topics:

- The difference between financial statements and records
- The characteristics of a good financial records system
- Accrual and cash basis accounting
- The income statement, balance sheet, and statement of cash flows
- Straight-line depreciation
- Calculating the cost of goods sold
- The interrelationships between the income statement, balance sheet, and statement of cash flows
- Management information record systems

CASE STUDY

Paul Welch recently graduated from Texas A&M University with a degree in agricultural engineering. To put himself through school, Paul worked for his father on the family farm in Plainview, Texas. For many years Paul's father had produced alfalfa and cotton, and occasionally some small grains. Recently, he had experimented with growing vegetables that he sold through a local broker. Paul and his father had discussed the possibility of operating their own packing shed so that they could sell products directly to wholesalers and retailers. The family farm wasn't large enough to provide adequate financial support to the both of them and the packing shed would provide an opportunity for Paul to work with his father. Paul and his father decided that Paul would operate the packing shed as a separate business. He was very good with equipment and had no problem designing and constructing the building, and putting in the equipment, but he was at a loss as to what kinds of records he should keep and how they should be organized. After careful study of this chapter, you should be able to help Paul decide what type of records he should keep for his produce packing shed business.

INTRODUCTION

Businesses maintain *records* for multiple users and many purposes. They are kept for many *stakeholders* of the business including owners, managers, employees, creditors, the government, customers, and suppliers, to name a few of the most important users. Owners require records to determine the value of their investment in the business and to evaluate the business's performance. Managers use records to more effectively manage the business. Creditors require detailed records to determine whether to loan money.

Business records may generally be classified as either financial records or management information records. Financial records form the basis for *financial statements* that are used to record what the business owns, what is owed by the business, and the results of the business's operating activity. *Management information records* are used to report information to managers that they use in making decisions.

FINANCIAL STATEMENTS

Developing financial statements is one of the most important functions *management* must perform. Financial statements are the yardstick by which the success of the business will be measured and provide the means for management to make decisions concerning the operations and growth of the business. They are the historical record of the business's activities and in some respects are much like a scorecard, providing information on the amount of cash available, amount of profit, the value of property and inventory, and the debts that are owed, to mention but a few.

The *owners* of a food or agribusiness firm have a major interest in the financial statements of the business. Although the owners and managers may often be the same people, they generally have quite different interests in the financial statements of the firm depending on which hat they happen to be wearing. The owners, or *investors*, in a business are primarily concerned about the earnings the firm generates. They are also concerned about the security of their investment and the growth of the firm that will provide the base for future earnings.

Whereas the managers of the firm should be primarily concerned with achieving the goals of the owners, they do this through managing the day-to-day operation of the business. They are concerned with developing financial statements to assist them in decision making. *Managers* use financial statements to control expenses, ensure that the firm has enough cash to pay the bills, and make sure that the firm has enough inventory to meet its customers' needs. They are typically evaluated based on the performance of the firm and are often rewarded based on the extent to which the financial statements show that they have achieved specified goals.

People outside of the firm are also interested in the firm's financial statements. *Lenders* are concerned about the financial health of the firm. If they are considering making a loan to the firm, they will want to see that the firm will be able to generate the cash necessary to repay the loan. The *government* is also interested in the firm's records. Because most firms pay income taxes, property taxes, and sales taxes, the proper records must be kept in order to calculate the tax owed to each agency. Although most taxes are reported voluntarily, occasionally firms are audited and find it necessary to justify the figures they reported to the agency. Other groups may also be interested in a firm's financial statements and annual report, such as the business's potential investors, employees, customers, and suppliers, as well as the community in which the business operates.

There are many different financial statements. However, the three most important financial statements, and those covered in this chapter, are the income statement, balance sheet, and statement of cash flows.

Guidelines for Financial Statements

Because so many individuals, groups, and institutions have different interests in a firm's financial statements, it is important that they be prepared using accepted standards. For this reason, businesses use a set of guidelines called

Generally Accepted Accounting Principles (GAAP), established by the Financial Accounting Standards Board (FASB). These guidelines ensure that there is consistency and objectivity in the reporting of financial information so that the diverse users may have confidence that financial statements accurately reflect the firm's financial situation.

A firm's *audited* financial statements are included in the firm's *annual report* to the company's owners. Independent accounting firms conduct the audit by examining and reviewing the financial statements for accuracy and compliance with applicable laws, regulations, and guidelines. All corporations and many other businesses prepare an annual report. In addition to the firm's major financial statements, the annual report typically includes a letter from the chairman of the board (for corporations) describing the firm's operating results for the year as well as its future prospects.

Important Characteristics of Financial Records

As we stated, financial records are kept and financial statements are constructed using GAAP and applicable laws and regulations. Nonetheless, businesses have some discretion in the design of a financial record-keeping system. To the extent that a business can exercise discretion in maintaining its financial records, several goals should be kept in mind. In order to be useful, the records should be *understandable.* Records that may be clear to the current manager may not be clear to the person who has to replace that person during an illness or after retirement. The records should also be detailed enough to be useful in making decisions. Records should be *relevant* and provide the necessary information, but not so detailed that it unnecessarily increases the amount of paperwork. For example, it may be useful to categorize the office expenses into telephone expense, copy machine expense, and supplies, but not to the extent of counting paper clips. Other characteristics of financial records include comparability and consistency. *Comparability* refers to the ability to compare the information produced by different firms, particularly firms within the same industry. A high degree of comparability allows for better comparisons across firms and potentially better decision making. *Consistency* refers to the consistent application of accounting methods by a given firm over time. It is important that financial records have *predictive value,* i.e., that the information is useful in predicting future results and therefore can be used by managers to make decisions based on their predictions of future events. Much of the predictive value of financial information depends on trends that appear in the data. If different methods are used to produce the financial records, the predictive value of the information is diminished. Although it is a good idea to organize the financial statements in the same manner year after year for the sake of consistency, the statements should be modified to meet the changing needs of the firm. This may be necessary as the nature of the firm's business changes or as it expands.

Cash Versus Accrual Accounting

There are two different financial record keeping methods: cash accounting and accrual accounting. In a *cash accounting* system sales are recorded when cash is received from the sale of an item and an expense is recorded when cash is actually paid for the purchase of an item. The emphasis is on recording the income or expense when cash changes hands. An *accrual accounting* system recognizes revenues and expenses at the time they are earned or incurred, respectively, regardless of when the cash for the transaction is received or paid out. The emphasis is on recording the income or expense when it occurs regardless of whether cash changes hands. The difference between cash and accrual accounting might best be illustrated with examples.

Example of a Sales Transaction. Assume that on May 15th, a farmer buys a part for a tractor at a farm equipment dealer. The dealer sends out an invoice to the farmer and requests payment for the part within 30 days. The farmer pays for the part on June 5th. If the implement dealer uses a cash accounting system, no record-keeping entry is required on May 15th because there was no cash exchanged between the dealer and the farmer. The sale is not recorded or recognized until the cash is received on June 5th. It will then be recorded and will be considered as a June sale. However, if the implement dealer uses an accrual accounting system, the sale is recorded on May 15th even though no cash has been exchanged. The implement dealer is considered to have earned the income on May 15th (the invoice date), even though the dealer will not receive the cash until June 5th. In other words, it is considered a May sale.

Example of a Purchase Transaction. Assume that on March 23rd, an advertising firm purchases office supplies from an office supply store. The office supply store bills the firm at the end of the month. The advertising firm pays the invoice on April 18th. Under a cash accounting system no record-keeping entry is made by the advertising firm until April 18th when it will be reflected as an April expense. Under an accrual accounting system the advertising agency records the purchase as an expense on March 23rd even though they will not pay until April 18th. The transaction will be reflected as a March expense.

Choosing Cash Versus Accrual Accounting. The Internal Revenue Service requires businesses whose revenues are generated from the sale of inventory (retail stores, wholesalers, manufacturers, etc.), C corporations, and businesses with gross receipts over $5 million to use the accrual accounting method. Most farmers, small individually operated businesses, and some partnerships and S corporations use the cash accounting method.

The rules governing the accrual accounting system represent an attempt to match revenues with the expenses incurred in earning those revenues so that profits may be accurately measured. In most situations, cash accounting systems do not attempt to accurately relate revenues to those expenses associated with earning those revenues. Cash accounting requires making many adjustments to the financial statements to accurately reflect profits or losses.

Given this significant disadvantage of a cash accounting system, one might reasonably ask why anyone would use cash accounting. The answer is that maintaining a cash accounting system is much easier than maintaining an accrual system. Most individuals can easily learn how to keep cash accounting records in a short time. However, accrual accounting systems require keeping double-entry records and trained bookkeeping skills.

Three Important Financial Statements

There are three major financial statements that appear in annual reports. They are the balance sheet, income statement, and statement of cash flows. The balance sheet is often likened to a snapshot. It is a picture of the firm's financial situation at a point in time, showing what the firm owes and owns. If the balance sheet is a snapshot, then the income statement may be likened to a videotape that shows us what has occurred from one balance sheet to the next. The income statement consists of the revenues and expenses of the firm from which we calculate the firm's profit or loss—the bottom line. The third important statement is the statement of cash flows. The statement of cash flows shows all of the cash flowing into and out of the business for a specified period of time. In addition to these statements, a firm will sometimes include in an annual report a statement of changes in owners' equity and a statement of changes in financial position. In this chapter, only the balance sheet, income statement, and statement of cash flows are discussed.

BALANCE SHEET

The *balance sheet,* or statement of financial position, reports the amount and types of assets that the firm controls and the claims on those assets by owners and creditors at a point in time. Two basic kinds of information are contained in the balance sheet: what the firm owns and what the firm owes. *Assets* are the economic resources that the firm owns to operate and generate future profits. *Liabilities* are claims on the firm's assets by someone outside the firm such as creditors, tax authorities, and others. *Owners' equity* represents the claims on the firm's assets by owners.

Because the entire value of the assets must be owed to someone, either to the owners of the firm or to someone outside of the firm, assets must always equal liabilities plus owners' equity. There are few certainties in life, but along with death and taxes it is a sure thing that assets equal liabilities plus owners' equity. It is this principle that gave rise to the term balance sheet; the assets must be balanced against the claims on these assets described as liabilities and owners' equity. See Box 3–1 for the most commonly used balance sheet classi-

BOX 3-1

Balance Sheet Classifications

Assets	= Liabilities	+ Owners' Equity
Current assets	Current liabilities	Contributed capital
Investments	Long-term liabilities	Retained earnings
Property, plant, and equipment	Other long-term liabilities	
Intangible assets		
Other assets		

fications. Firms are not required to use these specific classifications and may use alternatives to make the balance sheet more descriptive.

Assets

Assets are the first category of items listed on the balance sheet followed by the liabilities and finally by the owners' equity. This ordering of transactions is not a law of nature. It is a convention agreed to by accountants, bookkeepers, and others over a long history of keeping financial records.

It is also conventional to list assets on the balance sheet by how quickly they may be converted into cash, if necessary to do so. This is called the *order of liquidity.* The use of the term liquidity when discussing a balance sheet may seem odd, but it is really appropriately descriptive. It is based on the concept of considering assets like water. Liquids, like water, can be quickly poured through a funnel and into a bottle and be available for use as needed. Assets that are described as being liquid can also be quickly "poured" into the companies checking account and be available for use to pay for items as needed. Assets are typically categorized into five groups: current; investments; property, plant, and equipment; intangible assets; and other assets. However, it is proper to have as many categories of assets as deemed necessary to understand the structure of the business. Many businesses use two principal categories for assets, current and noncurrent. *Noncurrent assets* include all assets that do not fall into the current asset category.

Current Assets. *Current assets* are those assets that will or can normally be converted into cash or used during the current operating period, which is typically a month, a quarter (three months), or a year. Assets found within the current asset classification are listed on the balance sheet in the order of their liquidity: cash, short-term investments in marketable securities, accounts receivable, inventory, and prepaid expenses. A sample balance sheet illustrating current assets for a typical food and agribusiness firm is illustrated in Box 3–2. White Elevator is a corporation that operates a grain storage elevator and sells agricultural input supplies, such as seed, agricultural chemicals, and fertilizer. Notice at the top of the statement the words "as of December 31, 2000"

BOX 3–2

White Elevator Balance Sheet as of December 31, 2000 (in thousands of dollars)

Assets		
Current Assets:		
Cash	$ 34,440	
Marketable securities	750	
Accounts receivable, net	22,410	
Inventories	20,100	
Prepaid insurance	450	
Total current assets		$ 78,150
Property, plant, and equipment:		
Land	4,500	
Building	150,450	
Accumulated depreciation, building	(45,900)	
Equipment	51,900	
Accumulated depreciation, equipment	(16,650)	
Total property, plant, and equipment		144,300
Patent, net		11,700
Total Assets		$234,150
Liabilities and owners' equity		
Current liabilities:		
Accounts payable	$ 24,225	
Notes payable	300	
Collections received in advance	600	
Interest payable	225	
Taxes payable	425	
Wages payable	750	
Dividend payable	100	
Total current liabilities		$ 26,625
Long-term liabilities:		
Notes payable	690	
Bonds payable	14,100	
Total long-term liabilities		14,790
Total liabilities		41,415
Owners' equity:		
Common stock	61,000	
Additional paid in capital	48,650	
Retained earnings	95,585	
Treasury stock (cost)	(12,500)	
Total owners' equity		192,735
Total liabilities and owners' equity		$234,150

appear. The numbers in the balance sheet represent ending balances at the end of the current accounting period and the beginning balances of the next accounting period.

Cash. The *cash* account consists of funds that are immediately available to the firm. Examples of cash are money in the company's cash register, safes, and any cash on deposit in a bank account that is immediately available to the firm.

It is the firm's most liquid asset. The example in Box 3–2 shows that White Elevator had $34,440,000 in cash as of December 31, 2000, for use during the current accounting period.

Marketable securities. Most firms do not keep excess amounts of cash on hand, but prefer to invest funds over and above what they need for their everyday use in securities or bank accounts that yield a return. These investments may be stocks, bonds, or certificates of deposit. These securities are temporary investments that companies intend to convert to cash when needed. The securities appear on the balance sheet at their market value. White Elevator has $750,000 in marketable securities listed on the balance sheet. Unless the manager intends to use a large part of the $34,400,000 in cash in the near future, he or she should consider purchasing more marketable securities.

Accounts receivable. *Accounts receivable* represent monies that are owed to the firm but which have not yet been received. Accounts receivable include *notes receivable*, which are short-term loans the company makes to other people, and *trade receivables*, which occur when the company extends credit to a customer who acquires goods and services from the company. Accounts receivable are expected to be repaid and converted into cash within the current operating period. Companies show accounts receivable at their net realizable value, which represents the amount of accounts receivable currently due that the firm estimates is collectable. Many companies classify accounts receivable according to the length of time they have been outstanding (unpaid). Typical classifications are: (1) less than 30 days, (2) 30 to 90 days, (3) over 90 days, and (4) doubtful accounts. Doubtful accounts are accounts receivable that the company does not expect to collect. The company would disclose the amount of the allowance for doubtful accounts in a note accompanying the balance sheet.

Accounts receivable can be a very sizeable asset for many companies. However, they can also be a burden to a company if they become too large, because they cannot be used to pay bills until the cash is received. At this time White Elevator has $22,410,000 in accounts receivable, net of doubtful accounts.

Inventory. Most firms must carry an inventory in order to carry out their day-to-day business activities. For a firm that produces a good, some inputs of production and the finished product must be kept on hand. A firm that sells bagged feed must keep the ingredients on hand to mix the feed and some feed on hand to meet customers' immediate needs. A firm that simply resells items that it purchases must keep a stock of those items on hand. An example is a farm supply store that must keep its shelves stocked with tools, insecticides, fertilizer, etc. The firm has the option of reporting inventory at its cost or its market value, whichever is lower, at the end of the accounting period. How much inventory does White Elevator have on hand as of December 31, 2000?

Prepaid expenses. Occasionally, an expense will be paid before the product or service is delivered or used in the business. The most common example is insurance. The payment may be made quarterly or biannually in advance of receiving the insurance coverage.

It is not easily seen why this is an asset to the firm because it is not a physical asset. However, consider the case where the firm purchases a physical asset such as inventory. When the inventory is received, at the time of payment, cash is reduced and inventory is increased by the corresponding amount. When inventory is paid for but not received, we record the transaction as prepaid inventory, which indicates that the firm has a claim on the asset, but it is not yet in the firm's possession. Once the inventory is received, prepaid inventory is reduced and inventory increased. While insurance is not a physical asset, prepaid insurance does represent a claim on something of value. If the insurance were canceled before it was used, the firm would most likely receive a cash refund.

White Elevator had prepaid insurance of $450,000 as of December 31, 2000. White Elevator paid its insurance premium of $450,000 in December of 2000. The payment is for first six months of the forthcoming year. Another $450,000 insurance premium must be paid in June of 2001, for insurance coverage from July 1, 2001 through December 31, 2001. How much prepaid insurance would be listed as an asset on the balance sheet as of March 31, 2001?

Investments. Following the current assets section on the balance sheet is the investments section. The investment classification on the balance sheet describes the type and extent of company's long-term investments in stocks and bonds of other companies. A firm will invest in stocks and bonds of other companies when management views those investments to be a better use of the firm's capital rather than expanding its own property, plant, and equipment capacity.

Property, Plant, and Equipment. The classifications found most typically under the property, plant, and equipment section are land, buildings, and equipment. Land represents the land acquired to provide a site for the firm's operational activities. *Land* includes all land used to produce revenue for the company. Since land has an unlimited life, it is not depreciated. Although the current value of land is often many times the original purchase price, it is usually listed at cost because its market value to the firm will likely not be known until it is sold. Land is valued at $4,500,000 in Box 3–2.

The *buildings* classification summarizes the cost of various structures such as manufacturing facilities, warehouses, office buildings, retail outlets, and other distinctions useful to statement users. The *equipment* classification summarizes items such as furniture, fixtures, machinery, and equipment. Buildings and equipment are reflected on the balance sheet at their *book value,* which is their initial cost less accumulated *depreciation.*

Because these assets have a limited life, they decline in value over the period of their use. To reflect this decline in the value of assets, and the associated expense of employing them in the business, assets are systematically depreciated. White Elevator has buildings valued at $150,450,000 less accumulated depreciation of $45,900,000 for a net value of $104,550,000. Note that the accumulated depreciation appears in parentheses on the balance sheet. It is an accounting convention to list negative numbers in parentheses. What is the initial and net value of the equipment?

Depreciation. Accountants have developed many methods of calculating depreciation. In practice, the method of depreciation is dictated by the Internal Revenue Service (IRS). The IRS specifies the useful life (it must be greater than one year) over which the asset may be depreciated and the rules for calculating depreciation. A detailed discussion of various depreciation methods will not be covered. However, the concept of depreciation and its application to the financial statements will be explained using the simplest method of depreciation—straight line.

Straight-line depreciation is computed by taking the initial cost of the asset, subtracting its expected value at the end of its useful life (known as the *salvage value*), and dividing the remainder by its useful life. For example, most office equipment and automobiles have a recovery period (useful life) of five years, and machinery has a recovery period of seven years.

To illustrate the concept of depreciation, consider the following example. Mr. Welch bought a hydro-cooler for his vegetables for $40,000. The useful life is seven years and the estimated salvage value is $12,000. The annual depreciation is:

$$\frac{\text{Initial Cost} - \text{Salvage Value}}{\text{Useful Life}} = \frac{\$40,000 - \$12,000}{7 \text{ years}} = \$4,000/\text{year}$$

The value of the asset is therefore depreciated by $4,000 per year, and the $4,000 in annual depreciation is listed as an expense on the income statement. The book value of the asset at the end of the year is the initial cost less the accumulated depreciation or total depreciation to that time.

$$\text{Book Value} = \text{Initial Cost} - \text{Accumulated Depreciation}$$

After three years, the hydro-cooler in this example is valued at the book value less the accumulated depreciation. The accumulated depreciation is $4,000 + $4,000 + $4,000 = $12,000. Its book value at the end of year 3 is $40,000 − $12,000, or $28,000.

The method of calculating depreciation has received considerable discussion because of its importance to businesses and federal and state governments. Ignoring the possible income tax consequences, the depreciation schedule of an asset should reasonably approximate the useful life of that asset to the business. If an asset has a useful life of 10 years, it is misleading to depreciate it over five years. To do so would incorrectly overstate expenses and understate profits during the first five years. During the next five years, expenses would be understated and profits overstated. On the other hand, allocating the net cost of an asset over 10 years when its useful life is more likely to be five years understates expenses and therefore overstates profits during the first five years. It is precisely this impact on expenses and profits that makes the method of calculation so important. Because income taxes are based on the level of income, it is to the firm's advantage to depreciate an asset as quickly as possible. A higher depreciation expense means lower taxable profits and a lower income tax payment.

Because depreciation schedules are established by the IRS for several broad categories of assets, the book value of a depreciable asset, as reflected on the balance sheet, is not generally representative of the asset's current market value. It is possible that a specialized piece of equipment may have a book value on the balance sheet of over $100,000 and a current market value of zero because no one else has a use for it. The opposite may also be true. For example, a building may be fully depreciated and have a very low book value, whereas its market value could be very great because of its superior location.

Capital leases. *Capital leases* appear on the balance sheet as part of property, plant, and equipment. The concept of leasing assets, which is an alternative to buying assets, is covered in Chapter 5. White Elevator has included capital equipment leases with its other machinery and equipment on the balance sheet.

Natural resources. *Natural resources* are nonrenewable resources such as minerals, timber, or oil. Natural resources are included in the property, plant, and equipment section at the original cost less accumulated *depletion.* White Elevator has no investment in natural resources.

Intangible Assets. Intangible assets appear on the balance sheet following property, plant, and equipment. An *intangible asset* is an asset that has no physical substance, such as goodwill, patents, copyrights, trade names, and trademarks. It is often difficult to determine the value of intangible assets. Nonetheless, patents, copyrights, trade names, and trademarks are of value to the firm. Many firms would pay a great deal of money to obtain certain patents or copyrights or to be able to use a well-known trade name or trademark. Consider how much a company might be willing to pay for the use of the Coca Cola brand for a new soft drink, or the Kleenex brand for a new facial tissue. *Goodwill* is the reputation that the firm has established. If the firm were sold, the right to use the firm's name would be of value to the buyer because of the firm's goodwill.

How does the value of these assets show up on the balance sheet? How are the values established? Let's look at a specific case, a trade name, for example. A trade name, if known and respected in the marketplace, can have considerable value. The concept of franchises such as McDonald's or Arby's is built on this principle. If you start a company and build up a good reputation in the marketplace over time, your company name has value, but it is not listed as an asset on the balance sheet. It is not listed because you did not pay anything for the use of the "good name." You developed it over time by providing a quality product or dependable service. However, if you were to sell your business, potential buyers would presumably be willing to pay for the right to use your market-established name. The amount entered for goodwill on the buyer's balance sheet would be the amount the buyer paid for the business above the price paid for the market value of the other assets in the business such as inventory, equipment, and buildings.

It is sometimes difficult to comprehend how the acquiring company would list the value of the trade name on their balance sheet, while the company that built the name's reputation did not list it as an asset on their balance sheet. The rationale is similar to that for listing investments and land at the cost of acquisition. Any possible profit for establishing the trade name is not realized until it is sold; hence, the recording of profit is delayed until that time.

A patent is another example of an intangible asset. The listed value of a patent is the acquired cost if the patent was purchased from another company or individual. If the company did the research or came up with the idea that led to the granting of a patent, the value of the patent asset listed on the balance sheet is the summation of all of the expenses associated with acquiring the patent. These may include research, supplies, travel, attorney fees, and many other expenses. White Elevator has a patent valued at $11,700,000.

Are intangible assets depreciated? The answer is yes, but the process is called *amortization* instead of depreciation. Intangible assets are amortized over their estimated useful life in the business in much the same way as buildings and equipment. Companies report intangible assets on the balance sheet at the original cost less the amortization taken on the assets up to that point in time.

Other Assets. The *other asset* category is used for assets that do not fit in the previous asset classifications. Included in this classification are items such as noncurrent receivables, deferred charges, and special funds. *Noncurrent receivables* are accounts receivable that are not expected to be collected during the current accounting cycle. *Deferred charges* (deferrals) are long-term prepayments that companies amortize over various lengths of time, depending on how long management feels the company will benefit from the expenditure. An example would be the costs associated with starting the company. Special funds would include savings accounts established to pay off major debts that are due over various lengths of time.

Liabilities

Liabilities represent all claims on the business by people or firms other than the owners of the company. The basic classifications of liabilities on the balance sheet are: current liabilities, long-term liabilities, and other long-term liabilities. *Current liabilities* are those liabilities that are due within the current operating period. *Long-term liabilities* consist of obligations to creditors from borrowing that will not come due within the current operating period, and *other long-term liabilities* include obligations such as capital leases or pension plans.

Current Liabilities. Current liabilities are listed first in the liabilities section of the balance sheet. The major categories of current liabilities are:

- Accounts payable
- Notes payable
- Current portion of long-term debt
- Accrued expenses

Accounts payable. Just as customers of the business buy items on credit, the business may purchase items such as inventory and supplies on credit. Most often, this debt, or at least part of it, will come due during the current operating period and, as such, it becomes a current liability. This kind of credit is especially important to agribusinesses, many of which are subject to large seasonal fluctuations in sales or purchases. The need to purchase large amounts of inventory to gear up for their main selling season can create a tremendous drain on the firm's cash supply. Credit from suppliers is often needed until the season is over and the supplies have been sold and converted into cash. White Elevator currently owes its suppliers $24,225,000.

Notes payable. Notes payable are any loans that are due during the current operating period. These are short-term loans from individuals, banks, or other financial institutions. Such loans are often taken out to be used in one operating period and will be repaid in a subsequent operating period.

Current portion of long-term debt. The company may have long-term loans that will come due during the current operating period. A mortgage, for example, is a long-term loan, but the payment is usually made monthly or quarterly. The portion due during the current operating period is categorized under the current portion of long-term debt for that period. The remaining balance of the mortgage is listed under long-term liabilities.

Accrued expenses. Frequently, expenses are not paid as they are incurred. This is true of taxes, interest, dividends, utilities, and wages or salary. These expenses are called *accrued expenses* because they are incurred over a period of time but are not paid until some future period. Interest on a loan may be paid only once at the end of the year. However, the expense accumulates during the entire year and a balance sheet at mid-year should reflect that the firm owes this money but has not yet paid it.

Wages that are owed but not yet paid is another common type of accrued expense. If the accounting period ends in the early part of the week but wages are paid only at the end of the week, the amount of wages due employees for work completed is an accrued expense. Taxes and utility bills are other common examples of accrued expense items. What is the value of White Elevator's accrued expenses at the end of 2000?

Long-Term Liabilities. *Long-term liabilities* are those liabilities that will not become due within one year or until after the current operating period, whichever is longer. Long-term liabilities include: notes payable, mortgages, and bonds.

Notes payable. Notes payable, under the category of long-term liabilities, represent loans from individuals, banks, or other financial institutions that do not come due in the current operating period. For example, it is a common practice for agribusinesses to purchase equipment from a manufacturer and to finance the purchase with the manufacturer over a period of several years. If the company customarily prepares financial statements on an annual basis, then that portion of the equipment loan due in years two through five would be listed as a long-term note payable. White Elevator has $690,000 in long-term notes payable.

Mortgages. *Mortgages* are notes payable that are secured with real estate. If the debt is not paid as scheduled in the loan agreement, the lending agency (or individual) has the right to take possession of the real estate and sell it to get funds to repay the loan. When purchasing land or a building, it is very common to take out a mortgage to finance part of the purchase of the asset. Mortgages may also be taken out on property that has already been paid for by the firm as a means of generating more cash for operations.

Bonds. *Bonds* differ from long-term notes in that the firm typically borrows the needed funds from the public and not from a specific individual or institution. Bonds are issued by corporations. In return for the cash received from the sale of the bond, the corporation promises to pay to the bondholder a specified amount of interest and to pay the principal at specified times. Most small corporations do not issue bonds to raise cash, because very few people are willing to buy bonds issued by small, unknown companies. Bonds will be discussed in more detail in Chapter 5.

Other Long-Term Liabilities. Companies classify long-term obligations other than amounts borrowed as *other long-term liabilities.* Other long-term liabilities include capital lease obligations, pension obligations, employee benefits, and deferred income taxes. White Elevator has no other long-term liabilities.

Owners' Equity

Owners' equity represents the owners' claims against the assets of the business. The owners' equity is classified as *contributed capital* (often called *paid-in capital*), which reflects resources provided by the owners, and *retained earnings* (also called *undistributed earnings*), which are claims generated by retaining the company's profits.

Contributed Capital. Contributed capital is that money that was invested in the business by the owners. Contributed capital in a corporation is called *stock* and *additional paid-in capital,* in a cooperative it is called *stock* and *patronage dividends,* and in a sole proprietorship, partnership, or limited liability company it is called *contributed capital.* Contributed capital includes capital the owners invest in the business when they start the company as well as additional capital contributed at a later date. The owners of White Elevator have common stock valued at $61,000,000 and additional paid-in capital valued at $48,650,000, representing their initial investment in the business. In addition, the company has bought back $12,500 of stock from the owners, which is designated on the balance sheet as *treasury stock.*

Retained Earnings. *Retained earnings* are earnings that have been generated by the company in the course of its operations that have been reinvested in the firm and not been distributed to the owners. When a firm makes a profit, two things can happen. The profit may be paid out as dividends on stock or otherwise withdrawn by the owners, or the money can stay with the firm. When the earned income stays with the firm, retained earnings increases. Retained earnings can be either positive or negative. If the firm has never made a profit but instead has always operated at a loss, the retained earnings would be negative. Furthermore, if the negative value of the retained earnings exceeded the value of the contributed capital, then the owners' equity would also be negative. However, most ongoing businesses make a profit most of the time and retain some of their profits so that retained earnings is generally positive. As retained earnings increases, the total owners' equity of the firm increases, meaning that the owners' initial investment has increased in value. As of December 31, 2000, White Elevator had retained earnings of $95,585,000. Notice that the sum of total liabilities and owners' equity equals the total assets of $234,150,000.

Expenditures Versus Expenses

Before we discuss the income statement, it is important to distinguish between expenditures and expenses. An *expenditure* occurs when an asset is acquired. This may take the form of inventory for resale, land, equipment, etc. *Expenses* are decreases in the owners' equity arising from the operation of the business during a specific accounting period. Although this sounds difficult, in reality it is not. The need to distinguish between expenses and expenditures arises from the desire to report the expenses of a business in the same period in which the revenues were generated, provided the expenses were incurred to generate those revenues. Almost all expenditures (land is an exception) will become expenses over time because they are used up by the business, either immediately or through the process of depreciation. Several examples follow that will help to clarify this distinction between expenditures and expenses and their impacts on the income statement.

When inventory is purchased, it is an expenditure. The assets of the firm have increased. The asset category called inventory on the balance sheet increases and if cash payment was made, the cash account decreases appropriately. The inventory purchased will be reported as an expense when it is sold. At this time, assets would decrease as well as owners' equity. A corresponding revenue would also be reported in the period in which the expense was reported, thereby increasing owners' equity. The net change in owners' equity would be the difference between the sale price and the purchase price of the inventory items. Wages and other expenses such as electricity, water, etc., are reported as expenses in the period in which they occur, regardless of when they are paid. Buildings and equipment are expenditures when they are acquired by the firm. The asset account for equipment is increased by the purchase price of the equipment. The equipment is depreciated as it is used in the generation of revenue and the amount is recorded as an expense on the income statement. Like all expenses, the above expenses result in a decrease in owners' equity.

INCOME STATEMENT

The purpose of the *income statement* is to measure the earnings (income) generated by the company from business operations during the accounting period, typically a month, quarter, or a year. The income statement is the most widely quoted of the financial statements because it indicates the company's profitability. Other names for the income statement are the profit and loss statement, operating statement, and statement of earnings. In some respects, the income statement is the most important financial document. Although the balance sheet shows how much the owners have invested in the firm, the income statement shows how this changes from year to year. The income statement holds some of the major clues as to the successes and failures of a firm. The way in which the income statement is organized and the principles involved in reporting revenues and expenses reflect this. Income statement classifications appear Box 3–3. A sample income statement for White Elevator is illustrated in Box 3–4. At the top of the statement it says, "for the Year Ending December 31, 2000," denoting that the statement covers a time period of one year ending on December 31, 2000.

BOX 3–3

Income Statement Classifications

Gross sales
Sales returns and allowances
Net sales
Cost of goods sold
Gross margin
Selling expenses
General and administrative expenses
Operating income
Nonoperating income
Nonoperating expenses
Income (loss) before taxes
Income taxes
Net income

BOX 3-4

White Elevator Income Statement for the Year Ending December 31, 2000 (in thousands of dollars)

Net sales		$109,500
Cost of goods sold		65,700
Gross margin		43,800
Operating expenses:		
Wages	$4,500	
Insurance	3,000	
Uncollectable accounts	900	
Miscellaneous expense	4,200	
Depreciation—buildings	5,850	
Depreciation—equipment	3,150	
Amortization—patent	1,050	
Total operating expenses		22,650
Operating income		21,150
Nonoperating income:		
Interest	600	
Dividends	900	
Total nonoperating income		1,500
Nonoperating expenses:		
Interest expense	900	
Loss on sale of equipment	2,250	
Total nonoperating expenses		3,150
Income before taxes		16,500
Income taxes		5,850
Net income		$ 10,650

Sales

The first item listed on the income statement is sales. *Sales* represent the dollar value of all goods and services that have been sold by the firm during the 2000 operating year. Sales are normally presented in one of two ways. They may be entered as three separate lines starting with gross sales (total sales) on the first line, less returns and allowances on the second line, to arrive at net sales on the last line. Or, sales may simply be entered as net sales. The first alternative specifically accounts for any items returned by customers, discounts, or other allowances. White Elevator's net sales were $109,500,000 for the operating year.

Cost of Goods Sold

The cost of goods sold is meant to reflect the expenses incurred by the business for acquiring and producing products sold during an operating period. There are two basic kinds of businesses: (1) those that buy and then resell goods (wholesale and retail firms), and (2) those that manufacture the products they sell. Service firms (banks, accountants, and consultants) and most farming companies do not have a cost of goods sold section.

The method of calculating the cost of goods sold depends on the type of business. For businesses involved in buy-sell operations, it is very straightforward. *Cost of goods sold* is the value of the beginning inventory, plus purchases and expenses such as transportation associated with the purchases, minus the ending inventory and any allowances such as quantity discounts. Thus buy-sell operations use the following formula:

  Cost of Goods in Beginning Inventory
\+ Cost of Goods Purchased
− Cost of Goods in Ending Inventory
= Cost of Goods Sold

This method shows just how much inventory was used during an operating period. A buildup in inventory without a corresponding increase in sales would not be reflected as an expense under the cost of goods sold category in the income statement.

Although many firms keep track of the quantity and value of their inventory throughout the year, the actual value of ending inventory at the close of an accounting period is determined through a physical count of all items on hand. Losses due to theft, spoilage, or mysterious disappearance are recorded as a cost of goods sold (shrinkage) for that period because they don't appear in the ending inventory value.

For businesses involved in manufacturing, the cost of goods sold is the beginning inventory, plus the cost of goods manufactured, minus the ending inventory. The *cost of goods manufactured* includes all *direct labor* (that is, labor directly involved in manufacturing the goods), materials used for manufacturing the goods, and *manufacturing overhead* (including indirect labor, maintenance and repairs, property taxes, gas and electricity, supplies, insurance, and depreciation associated with manufacturing during the period). The formula for calculating the cost of goods sold for a manufacturing firm is:

  Cost of Goods in Beginning Inventory
\+ Costs of Goods Manufactured
− Cost of Goods in Ending Inventory
= Cost of Goods Sold

The cost of goods sold for White Elevator was $65,700,000 for the operating year.

Gross Margin

Gross margin (gross profit) is the difference between net sales and the cost of goods sold. The gross margin must be large enough to cover all other expenses of the business, including operating expenses, interest, and taxes, if the business is to make a profit. It is often useful to look at gross margin per unit of product sold. For some products, the per-unit gross margin is very low. Supermarkets, which operate on a low per-unit gross margin, do a large volume of business in order to generate a sufficient total dollar gross margin. Those businesses that do a low volume of business typically have a high gross margin per unit of sales. Agricultural machinery dealers are an example of this type of agribusiness. They may not sell a great number of machines during the year, but they hope to sell each machine for substantially more than they paid for it.

The cost of goods sold amounted to $65,700,000 on $109,500,000 of net sales in the elevator example. This resulted in a gross margin of $43,800,000 for the operating year. If White Elevator is going to make a profit, all additional expenses must be less than $43,800,000.

Operating Expenses

Operating expenses include the remaining expenses associated with operating the business. *Operating expenses* may be further subdivided into selling expenses and general and administrative expenses. Small businesses, with few expense items, frequently list all expenses under the general category of operating expenses.

Selling Expenses. *Selling expenses* include all expenses that can be attributed to the selling of the firm's product or service. Selling expenses include sales commissions, sales salaries, travel expenses, depreciation of sales force vehicles, advertising, marketing expenses, shipping and handling, and other office expenses directly associated with selling the product. Most selling expenses are considered to be *variable expenses* because the amount spent varies directly with the amount of sales.

General and Administrative Expenses. *General and administrative expenses* include management and office salaries, depreciation of office equipment, office supplies, office expenses such as electricity, water, rent, insurance, legal and accounting expenses, and any other expenses associated with the administration or the operation of the business in general. These expenses are considered to be *fixed expenses* because they remain constant regardless of the level of sales over a period of time.

Some categories of operating expenses may be divided into both selling and administrative expenses. Examples are utilities, telephone, and postage. There is no hard and fast rule that is used to determine how to allocate these types of operating expenses. The most important guideline is that expenses should be categorized so that they are meaningful to management.

Operating Income

Operating income is found by subtracting the operating expenses of $22,650,000 from the gross margin of $43,800,000. White Elevator's operating income was $21,150,000 for 2000.

Nonoperating Income and Nonoperating Expenses

Income and expenses not directly associated with the principal activity of the business are included in the categories of nonoperating income and nonoperating expenses, respectively. *Nonoperating income* typically includes items such as interest income and dividend income. *Nonoperating expenses* usually includes categories such as interest expense and business losses from natural disasters.

To understand the rationale for listing nonoperating income and expenses separately, consider the interest expense category. Why not include interest expense as another item in general expenses along with utilities and general management salaries? Although there is nothing wrong with doing this, it is more difficult to compare direct operating expenses from one operating period to another, or from one business to another, if the interest expense is included. The interest expense is generally related to how the company is financed (debt or equity capital) rather than how efficiently management ran operations during a given period. By listing the interest expense separately and then calculating income, management or the owners may compare the operating results of a firm with no debt (no interest expense) to one that has a large amount of debt.

Income Before Taxes

Income (earnings) before taxes is the remainder of operating income after subtracting nonoperating income and expenses. Income before taxes was $16,500,000 in 2000. This measure of income is calculated before income taxes are deducted, because income taxes are based on the firm's income. A loss occurs when income is negative and is signified by placing the dollar amount in parentheses.

Net Income

After interest and all income taxes are paid, the resulting amount is the firm's *net income* (net earnings or profit). White Elevator made $10,650,000 in net income after taxes in 2000. These profits may be taken out as owner withdrawals or dividends, or left in the company and added to retained earnings on the balance sheet.

STATEMENT OF CASH FLOWS

The *statement of cash flows* illustrates the cash flows of the business arising from operating, investing, and financing activities during a specific period of time. This statement measures the increase and decrease in cash to provide the link between the accrual-based income statement and the balance sheet.

Prior to 1988, firms prepared a statement of changes in financial position to show the sources and uses of working capital used by the company during the period of time covered by the income statement. Confusion occurred because some firms used cash, some cash plus liquid assets, and others used current assets minus current liabilities as a definition of *working capital.* Due to the variety of methods used to prepare the statement of changes in financial position and the resulting confusion, the accounting profession developed the more useful statement of cash flows. The purpose of the statement of cash flows is to determine:

- If the company is able to generate future net cash flows
- If the company can meet its obligations
- If the company needs external financing
- Reasons for differences between net income and cash receipts and disbursements
- The cash and noncash impacts of the company's investing and financing transactions

Organization of the Statement of Cash Flows

The statement of cash flows is divided into three sections: operating activities, investing activities, and financing activities. These sections represent the basic functions of any business and the amounts of cash flowing in and out of the business as a result of activities in each of the areas. Box 3–5 illustrates the cash inflows and outflows for each activity.

Operating activities, as you will recall, involve the sale of goods and services by the firm. Cash inflows from operating activities result primarily from sales to customers, but cash may also come from other sources such as interest and dividends. Cash outflows from operating expenses result from payments made for operating expenses. The operating activities section of the statement of cash flows allows users to assess the firm's ability to generate positive cash flows and to assess the reasons for differences between a firm's accrual accounting income and cash flows.

Investing activities usually involve the acquisition and disposition of property, plant, and equipment, and other long-term investments. Disposing of a long-term asset results in a cash inflow whereas purchasing a long-term asset results in a cash outflow.

Financing activities involve raising funds from and distributing funds to owners as well as borrowing from and repaying creditors. The activities in this section involve the issuing of stocks or bonds resulting in cash inflows or the repayment of equity or debt resulting in cash outflows. Together, the investment and finance sections of the statement of cash flows allow users to assess the firm's ability to meet its financial obligations and to determine if the firm requires additional financing.

Statement of Cash Flows—Direct Method

The cash inflows and outflows represented in the operating activities section of the statement of cash flows may be presented using either the direct or in-

BOX 3–5

Cash Inflows and Outflows

Business activities	Cash inflows	Cash outflows
Operating activities	Cash received from customers	Cash payments to suppliers
	Sale of marketable securities considered cash equivalents	Cash payments to employees
	Interest income	Cash payments for operating expenses
	Dividend income	Purchase of marketable securities considered cash equivalents
		Interest expense
		Income and property taxes
Investing activities	Sale of property, plant, or equipment	Purchase of property, plant, or equipment
	Sale of long-term investments	Purchase of long-term investments
	Sale of short-term investments not considered cash equivalents	Purchase of short-term investments not considered cash equivalents
	Collections of loans from other entities	Loans made to other entities
Financing activities	Proceeds from the sale of company stock	Purchase of treasury stock
		Payment of long-term debt
	Proceeds from issuance of long-term debt	Cash payments on short-term notes payable
	Cash received from short-term notes payable	Payment of cash dividends to owners
	Contributions of capital by owners	Withdrawals of capital by owners

direct method. The *direct method* shows the actual cash inflows and outflows of operating activities, and consequently is the preferred method of accountants. The *indirect method* shows the differences between accrual-based net income and cash flows from operations, and is used by most businesses. The net cash flows from operating activities will be the same for each method.

Box 3–6 illustrates the statement of cash flows using the direct method for White Elevator. Notice that the top of the statement indicates, "for the Year Ending December 31, 2000." How does the statement of cash flows differ from the income statement and the balance sheet? A major difference is that it shows the cash flows resulting from operating the business as compared to the accrual-based income and expenses on the income statement. It shows the

BOX 3–6

White Elevator Statement of Cash Flows for the Year Ending December 31, 2000 (in thousands of dollars)

Net cash flows from operating activities:		
Cash received from customers	$101,250	
Cash paid to suppliers for inventory	(58,125)	
Cash paid to employees	(4,800)	
Cash paid for operating expenses	(6,900)	
Interest paid	(960)	
Taxes paid	(6,075)	
Total net cash flows from operating activities		$24,390
Net cash flows from investing activities:		
Cash received from sale of marketable securities	1,950	
Cash received from sale of equipment	800	
Cash paid for building	(31,700)	
Total net cash flows from investing activities		(28,950)
Net cash flows from financing activities:		
Cash received from stock sold	14,250	
Cash paid for treasury stock	(12,500)	
Cash paid for dividends	(5,000)	
Total net cash flows from financing activities		(3,250)
Increase (decrease) in cash		(7,810)
Add beginning cash balance		42,250
Ending cash balance		$34,440

change in the cash account on the balance sheets from the beginning of the year to the end due to operating, investing, and financing operations.

The inflows and outflows for each of the three typical activities are presented for White Elevator. The cash inflows from operating activities include cash received from customers of $101,250,000. Notice that this figure is less than net sales because of credit sales that will not be collected until the current operating period. The cash outflows include cash paid for inventory, employee wages and salaries, operating expenses, interest, and taxes. Subtracting the outflows from the inflows yields the net cash flow provided by operating activities of $24,390,000.

The next section presents the inflows and outflows resulting from investing activities. Cash inflows include cash received from the sale of marketable securities and equipment. The purchase of a building resulted in a cash outflow of $31,700,000. The net cash outflow from investing activities was $28,950,000.

The cash flows from financing activities are provided in the last section. The cash inflow from financing activities was the proceeds from the sale of company stock. Other forms of financing would be proceeds from loans and bonds. Cash outflows include payments for the treasury stock and dividend payments to owners. The net cash outflow from investing activities was $3,250,000.

The overall change in cash flows is calculated by summing the net cash flows from the three activities. White Elevator has a $24,390,000 increase from operating activities, a $28,950,000 decrease from investing activities, and a $3,250,000 decrease from financing activities, resulting in a net cash decrease of $7,810,000. Because White Elevator had a beginning cash balance of $42,250,000, the resulting ending cash balance was $34,440,000. This ending cash balance should reconcile with the cash balance reported on the balance sheet.

The last step in a statement of cash flows is to reconcile net income to the cash flows from operations. To determine the cash flows for the period the accountant relates income and expense accounts with corresponding accounts in the balance sheet from the start of the period and the end of the period. This analysis is beyond the scope of this textbook. However, in the next section, we will illustrate how the three financial statements relate to each other.

INTEGRATED FINANCIAL STATEMENTS

Many individuals who are not accountants, or at least not well trained in accounting, tend to view the balance sheet, income statement, and statement of cash flows as three independent or at best partially related statements. An explanation of the rationale of the three statements is that the balance sheet is a listing of the firm's assets, liabilities, and owners' equity; the income statement is designed to measure the firm's profit (or loss); and the statement of cash flows traces all cash inflows and outflows. These concepts are correct but, based upon the authors' collective experiences, few students and business managers fully recognize the degree to which the three statements are interrelated. Every transaction that the company makes affects at least one and often all three of the statements.

It is important to think of the three statements as fully integrated parts of a working machine. If one part fails (accounting errors are made), the total machine will not function. It is a common error to assume that given the balance sheet formula, assets = liabilities + owners' equity, the way to make sure the balance sheet balances is to subtract liabilities from assets and what is left over is owners' equity. From a technical viewpoint this approach will lead to a correct answer, but only if no errors are made elsewhere. From an operational standpoint this is not the way it is done. The owners' equity categories of contributed capital and retained earnings are adjusted in the same manner as assets and liabilities are. Contributed capital can only be changed by adding or removing assets from the company. For example, if an owner contributes $10,000 in cash to the business, the cash asset is increased by $10,000 and the contributed capital category of the owners' equity section is increased by $10,000. By the same token, the retained earnings category may be increased by retaining profits (which appear on the income statement), or decreased through cash withdrawals by the owners. The point is that changes in the owners' equity section are interrelated with entries shown elsewhere on the balance sheet and other statements. All changes in the balance sheet from one period of time to another are a direct result of action taken by the company and most are calculated through changes recorded on either or both the income statement and statement of cash flows.

Relationships Between Financial Statements

A set of integrated financial statements for ASCO, a firm that supplies irrigation equipment to farmers, is presented in Box 3–7. The purpose of this example is to show the interrelationships between the three basic financial statements. The time period covered is the 2000 calendar year operating period.

It may be easier to understand this flow of financial information by studying the diagram in Figure 3–1 that represents the relationships between the statements. We begin with the balance sheet at the close of operations on December 31, 1999. This is the beginning balance sheet. It may be called the balance sheet as of the close of operations 12/31/1999 or as of the beginning of operations on 1/1/2000. There is no difference because no business transactions occur when the business is not open.

The next step is to follow the summary of transactions that occurred during the operating year that are recorded on the income statement and statement of cash flows. This leads us to the ending balance sheet as of December 31, 2000. The changes throughout the operating year that were recorded on the income statement and statement of cash flows are reflected in the balance sheet at the end of the operating year. Now we are ready to trace the interrelationships of the three financial statements through selected transactions in the statements listed in Box 3–7.

Cash. The balance sheet at the beginning of the year shows a beginning cash balance of $15,000. This figure is also shown as the beginning cash balance on the statement of cash flows. The statement of cash flows shows an ending year cash balance of $5,000. This figure was arrived at by carefully recording all of the incoming and outgoing cash. The ending cash position of $5,000 as shown on the statement of cash flows becomes the cash asset figure on the balance sheet at the end of the operating year, 12/31/2000. The accountant doesn't just automatically transfer the ending cash position on the statement of cash flows to the ending balance sheet. He or she carefully counts all the cash in the business accounts (including petty cash accounts) and lists it on the ending year balance sheet. The cash total must equal the ending cash balance on the statement of cash flows. If it doesn't, an error has been made. Likewise you may examine all of the cash reported on the statement of cash flows to understand how the final cash balance was derived.

Inventory. The beginning balance sheet shows inventory at $15,000. The cost of goods sold section of the income statement shows the same beginning inventory as on the beginning balance sheet, $15,000. The cost of goods sold section of the income statement shows ending inventory at $20,000. This figure

BOX 3–7

Example of Integrated Financial Statements

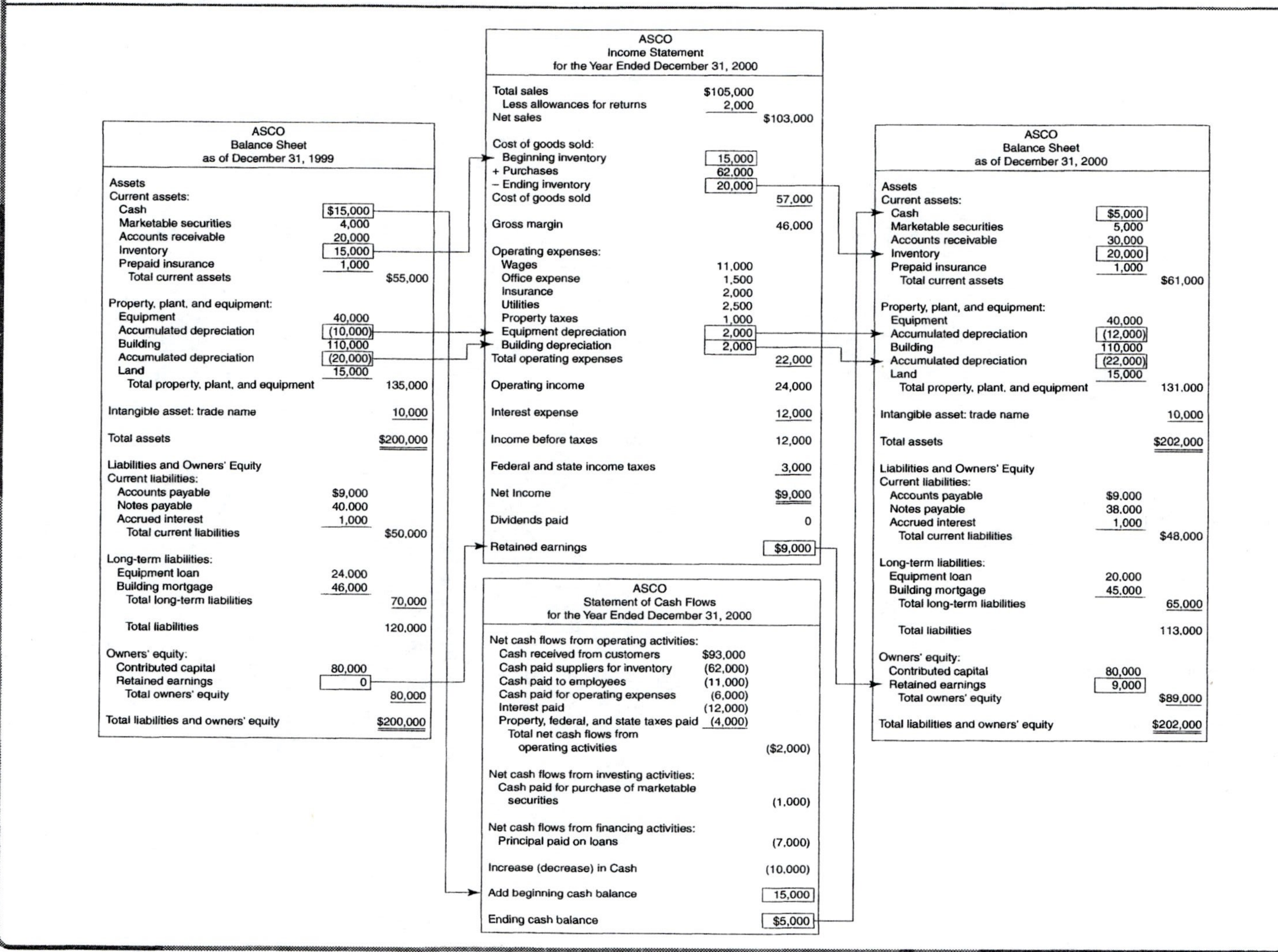

ASCO
Balance Sheet
as of December 31, 1999

Assets		
Current assets:		
Cash	$15,000	
Marketable securities	4,000	
Accounts receivable	20,000	
Inventory	15,000	
Prepaid insurance	1,000	
Total current assets		$55,000
Property, plant, and equipment:		
Equipment	40,000	
Accumulated depreciation	(10,000)	
Building	110,000	
Accumulated depreciation	(20,000)	
Land	15,000	
Total property, plant, and equipment		135,000
Intangible asset: trade name		10,000
Total assets		$200,000
Liabilities and Owners' Equity		
Current liabilities:		
Accounts payable	$9,000	
Notes payable	40,000	
Accrued interest	1,000	
Total current liabilities		$50,000
Long-term liabilities:		
Equipment loan	24,000	
Building mortgage	46,000	
Total long-term liabilities		70,000
Total liabilities		120,000
Owners' equity:		
Contributed capital	80,000	
Retained earnings	0	
Total owners' equity		80,000
Total liabilities and owners' equity		$200,000

ASCO
Income Statement
for the Year Ended December 31, 2000

Total sales	$105,000	
Less allowances for returns	2,000	
Net sales		$103,000
Cost of goods sold:		
Beginning inventory	15,000	
+ Purchases	62,000	
– Ending inventory	20,000	
Cost of goods sold		57,000
Gross margin		46,000
Operating expenses:		
Wages	11,000	
Office expense	1,500	
Insurance	2,000	
Utilities	2,500	
Property taxes	1,000	
Equipment depreciation	2,000	
Building depreciation	2,000	
Total operating expenses		22,000
Operating income		24,000
Interest expense		12,000
Income before taxes		12,000
Federal and state income taxes		3,000
Net Income		$9,000
Dividends paid		0
Retained earnings		$9,000

ASCO
Statement of Cash Flows
for the Year Ended December 31, 2000

Net cash flows from operating activities:		
Cash received from customers	$93,000	
Cash paid suppliers for inventory	(62,000)	
Cash paid to employees	(11,000)	
Cash paid for operating expenses	(6,000)	
Interest paid	(12,000)	
Property, federal, and state taxes paid	(4,000)	
Total net cash flows from operating activities		($2,000)
Net cash flows from investing activities:		
Cash paid for purchase of marketable securities		(1,000)
Net cash flows from financing activities:		
Principal paid on loans		(7,000)
Increase (decrease) in Cash		(10,000)
Add beginning cash balance		15,000
Ending cash balance		$5,000

ASCO
Balance Sheet
as of December 31, 2000

Assets		
Current assets:		
Cash	$5,000	
Marketable securities	5,000	
Accounts receivable	30,000	
Inventory	20,000	
Prepaid insurance	1,000	
Total current assets		$61,000
Property, plant, and equipment:		
Equipment	40,000	
Accumulated depreciation	(12,000)	
Building	110,000	
Accumulated depreciation	(22,000)	
Land	15,000	
Total property, plant, and equipment		131,000
Intangible asset: trade name		10,000
Total assets		$202,000
Liabilities and Owners' Equity		
Current liabilities:		
Accounts payable	$9,000	
Notes payable	38,000	
Accrued interest	1,000	
Total current liabilities		$48,000
Long-term liabilities:		
Equipment loan	20,000	
Building mortgage	45,000	
Total long-term liabilities		65,000
Total liabilities		113,000
Owners' equity:		
Contributed capital	80,000	
Retained earnings	9,000	
Total owners' equity		$89,000
Total liabilities and owners' equity		$202,000

FIGURE 3–1

Relationships between the Balance Sheet, Income Statement, and Statement of Cash Flows

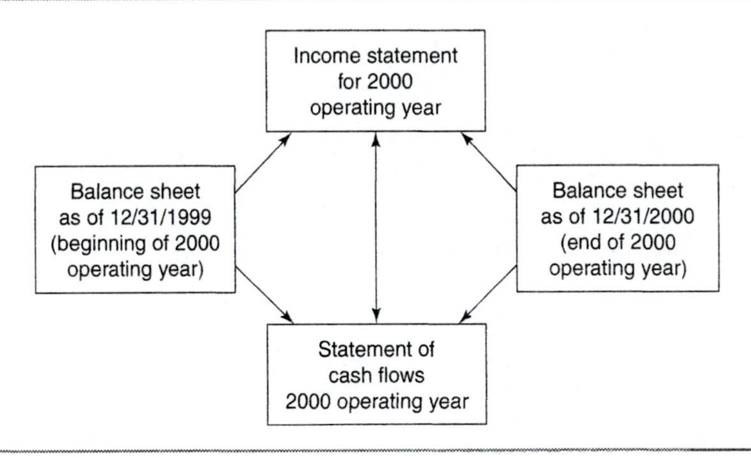

is calculated by taking a physical count of the inventory on hand valued at its acquired cost. This ending inventory figure listed on the income statement is also listed as the value of inventory on the ending balance sheet.

Building. ASCO started out the year with a building worth $90,000 ($110,000 less $20,000 depreciation) as listed on the beginning balance sheet. A total of $2,000 of the cost of the buildings was written off as an expense on the income statement for the 2000 operating year (see depreciation for buildings). The $2,000 in depreciation expense is also reflected on the ending balance sheet. The accumulated depreciation is increased from $20,000 to $22,000, resulting in a book value of $88,000 on the end of the year balance sheet. If depreciation in 2001 is $2,000, what will the book value of the building be at the end of 2001?

Retained Earnings. The balance sheet as of 12/31/1999 shows retained earnings as zero. This means that no profits have been retained in the business. The 2000 income statement shows retained earnings of $9,000. This $9,000 figure was derived from the $9,000 in net income (earnings) for the year. All income earned during the 2000 operating year was retained in the business; none was paid out as dividends. Hence, $9,000 is added to the retained earnings category on the ending balance sheet resulting in a balance of $9,000.

The above items discussed are just a few of the interrelationships among the statements presented in Figure 3–1. For example, the changes in the operating loan account, equipment note, and building mortgage balances from the 12/31/1999 to the 12/31/2000 balance sheets are recorded as principal payments on loans in the statement of cash flows. The operating loan (notes payable) was reduced by $2,000, the equipment loan by $4,000, and the building mortgage by $1,000. These changes total $7,000, the amount listed on the statement of cash flows.

It is very important that managers have a complete understanding of the three basic financial statements and their interrelationships. A manager with a full understanding of the financial accounting statements and how they are derived is in a much better position to discuss financial management decisions with the accounting department, bankers, and owners. A thorough understanding of the interworkings of the financial statements is best obtained by careful study and actually working through several accounting periods using realistic data.

MANAGEMENT INFORMATION RECORDS

Management information records are maintained to assist managers in their decision making. They are utilized in managing the day-to-day operations, and as an aid in long-range planning. It is therefore essential that the record-keeping system be designed with the user in mind. The widespread availability of accounting software has greatly facilitated the maintenance of detailed management records and simplified the performance of analyses designed to answer specific management questions. In the following sections, two of the most important types of management information records are discussed. The examples represent the types of records that might be kept for a produce packing firm like that described in the case at the beginning of the chapter.

Sales Records

Once Paul Welch has his packing shed operating, he will probably find it useful to collect information on the efficiency of his sales program for his vegetable packing shed. Information on the distribution of sales by type of product, similar to the figures listed in Box 3–8, would be useful in analyzing sales to compare performance with his previously established sales goals and in making plans for future years. In terms of volume (crates marketed), broccoli and cauliflower are the two most important products, whereas in terms of dollar sales, asparagus, cauliflower, and broccoli are the most important products. This type of information is readily obtained by keeping an accurate record of the quantity of crates sold and prices received.

Paul might also want to keep track of sales by customer or type of customer. Such sales information may be used to identify his most important customers so that he can make personal calls on them during the off-season, or send Christmas presents to help in maintaining their loyalty. These are just a few examples of the type of sales records that might be kept and potential uses of the information. Although the information relates to a farm likely to be

BOX 3–8

Welch's Vegetable Packing Shed Sales Record for 2000

Product	Average Crates (number)	Price/Unit (dollars)	Sales (dollars)	Percent of Total
Asparagus	27,000	$24.00	$ 648,000	26
Broccoli	120,000	5.50	660,000	27
Cabbage	60,000	5.00	300,000	12
Cauliflower	110,000	7.00	770,000	31
Spinach	12,000	8.00	96,000	4
Total	329,000		$2,474,000	100

found in Texas or California, the same principles apply to a farm supply store in Illinois or a food wholesaler in Florida.

Cost Records and Profitability Analysis

Paul should also maintain cost records so that he may determine if a particular product line is profitable. This is commonly accomplished by calculating what is called the *contribution to profit and overhead.* (The concept of contribution to overhead will be covered in more detail in later chapters.) It is difficult and often highly inaccurate to attempt to allocate fixed costs across product lines that share the same facility. To avoid these potential problems, managers subtract all clearly identifiable expenses associated with a product (variable expenses) from revenues received for that product. The resulting figure is the product's contribution to profit and overhead expenses. An example of this type of analysis for Paul's produce business is shown in Box 3–9. Paul should group his clearly identifiable expenses that can be attributed to a specific product into two categories: raw product costs and variable packing expenses. The raw product costs are what he has to pay the grower for his crop, whereas the variable packing expenses include items such as direct labor, packing crates, labels, and other supplies. As can be seen in Box 3–9, all of the products made a positive contribution to profit and overhead expenses except spinach, which lost $1.00 per crate sold. Such information would allow Paul to evaluate the relative profitability of each product line. To arrive at this type of information, Paul needs to collect information on labor use by product and the cost of each crate or supply item used for each product line. He also needs to record information on the prices that he pays to his growers for each commodity.

Based on the information in Boxes 3–8 and 3–9, Paul might decide that he should discontinue packing spinach because he is apparently losing money on the product line. The records show that he lost $12,000 from handling spinach,

BOX 3–9

Welch's Vegetable Packing, Shed Revenues, Expenses, and Contribution to Profit and Overhead, per crate, 2000

Product	Variable Product Expense	Packing Expenses	Sales Revenue	Contribution to Profit & Overhead
Asparagus	$18.00	$4.50	$24.00	$1.50
Broccoli	2.60	2.55	5.50	.35
Cabbage	2.15	2.35	5.00	.50
Cauliflower	3.95	2.75	7.00	.30
Spinach	5.00	4.00	8.00	(1.00)

based solely on the revenues earned and variable packing expenses (Box 3–10). If he did not handle spinach, he would have had $12,000 additional dollars to cover overhead expenses and hopefully make a profit. However, before Paul makes a decision to drop the spinach line, he would need to evaluate the potential impact on his growers and buyers. Do the spinach growers also supply him with broccoli or asparagus? If he dropped spinach, would they discontinue growing broccoli also? What about his buyers? Would they still continue buying asparagus or cabbage from him if he discontinued selling spinach, or would they transfer their business to a company that offered a more complete line? In other words, before Paul made a decision, he would need to assess the potential impact of other factors that are not included in the figures in Boxes 3–8, 3–9, and 3–10 but that could have a major impact on his sales or product supply.

The examples we have used present some of the management information records that might be kept by a produce packing firm. Other food and agribusiness firms have other needs. Retail food stores collect information on sales per linear

BOX 3–10

Welch's Vegetable Packing Shed Total Contribution to Profit and Overhead, 2000

Product	Contribution to Profit & Overhead (dollars/crate)	Cases Sold (number of crates)	Total Contribution to Profit & Overhead (dollars)
Asparagus	$1.50	27,000	$40,500
Broccoli	.35	120,000	42,000
Cabbage	.50	60,000	30,000
Cauliflower	.30	110,000	33,000
Spinach	(1.00)	12,000	(12,000)

foot of display space by product line. These managers are also interested in sales per person-hour by department. Well-managed farms keep track of all expenses for each crop and many farm managers record yields by field or by soil type.

It is not the intent of this chapter to provide a detailed analysis of management information needs of agribusinesses or how to collect or manage the information. Rather this brief introduction to management information records should serve to illustrate the importance and kinds of records that managers employ in making decisions.

SUMMARY

Financial statements and management information data are used primarily by the owners and management of the firm. However, lenders, the government, and other individuals also have an interest in these records. The financial accounting statements should be organized so that they meet established guidelines, facilitate decision making, and meet the changing needs of the business.

The two types of accounting systems are cash and accrual. Most farms and some small food and agribusiness companies use the cash basis in which expenses and revenues are recorded when cash is paid or received. However, most food and agribusiness firms use the accrual basis of accounting. Under the accrual system, revenues are recorded when they are earned and expenses recorded when they are incurred, regardless of when cash changes hands.

The balance sheet shows the resources and financial obligations of the firm. The resources, which the business owns, are called assets. Financial obligations to people or firms outside of the business are called liabilities, whereas the financial obligation to the owners of the business is called owners' equity. The value of the assets always equals the value of the liabilities plus owners' equity. Assets are listed by order of liquidity, or how quickly they may be converted into cash. Liabilities are listed in the order in which they come due.

Assets that have a useful life of longer than one year are depreciated. The concept of depreciation allows for the systematic accounting of the decline in an asset's value over its useful life and the corresponding expense that is charged on the income statement. The useful life of assets and the methods of calculating depreciation are established by the IRS.

The income statement lists the revenues and expenses of the firm. The first major category of expenses for most firms is the cost of goods sold (or manufactured), which includes the costs associated directly with the purchasing or manufacturing of the product. A second major expense category is operating expenses, which include the costs associated with the operation of the business, marketing of the product, and noncash expenses such as depreciation. The last two expense categories are interest and federal and state income taxes. After all expenses are subtracted from the firm's revenues, the remainder is the profit or loss for the period.

The statement of cash flows supplements the accrual-based accounting information illustrated in the income statement and balance sheet. The statement of cash flows is useful in assessing the firm's ability to generate positive future net cash flows, pay its obligations and debts, pay dividends, and determine its need for external financing. Its purpose is to illustrate the cash flows arising from the operating, investing, and financing activities of the firm. The operating section of the statement of cash flows may be prepared using either the direct or indirect format. The direct format illustrates the actual amounts received or paid for operating activities, whereas the indirect format requires adjustments to accrual-based net income to determine cash flows from operations. The investing and financing sections of the statement of cash flows indicate the cash received from or paid for investing and financing activities, respectively.

Together, the statement of cash flows and income statement along with beginning and ending balance sheets present an integrated, historical record of the business's transactions throughout the accounting period.

Management information records provide the manager with information to improve the management of the firm's resources. They typically include detailed sales and cost data that may be analyzed to determine the impact of each product and input on the firm's profitability.

CASE QUESTIONS

To start the business that Paul and his father will jointly own, as of January 1, 2000, they obtained the following assets:

- 2 acres of land–contributed by Paul and his father and valued at $10,000
- Packing shed building–built by a local contractor at a cost of $55,000
- Washing equipment - $20,000
- Sorting and grading equipment - $30,000
- Packaging equipment - $15,000
- Packing supplies–purchase price $2,000

Paul and his father were able to come up with $40,000 in cash. They had to borrow $80,000 from the local bank. The buildings and equipment were completely paid for prior to January 1, 2000. However, they still owe $2,000 to their supplier for packing supplies.

The Welch's operate their business as a partnership. In the first month, the following transactions occurred:

- Sales - $40,000
- Purchases of raw materials and supplies - $20,000

- Ending inventory (packing supplies) - $1,000
- Hired labor for packing - $2,000
- Utilities (almost entirely for the packing process) - $1,000
- Office expense - $500
- Salaries - $3,000
- Selling expense - $3,000
- Interest - $1,500
- Principal on loan - $1,000
- Property taxes - $500
- Depreciation on building - $100
- Depreciation on equipment - $400
- Income taxes - $2,000

1. Construct an initial balance sheet for the Welch and Son Packing Shed.
2. Explain why you set up the categories for the assets as you did. In other words, if you grouped some items together or reported each separately, justify your reasoning.
3. Construct an income statement for the first month of operation for the Welch and Son Packing Shed.
4. How did you determine what costs were included in the Cost of Goods Sold section?
5. If all earnings are retained in the company, what is the net worth of the company after the end of the first operating month?
6. How much cash does the company have listed on the balance sheet at the end of the first operating period if all sales and purchases are for cash?

REVIEW QUESTIONS

1. List the important users of financial statements and why they are interested in them.
2. What are the most important characteristics of good financial statements and why are each of these important?
3. What are the principal functions of each of the three financial statements?
4. What is the accrual basis of accounting? How does this differ from cash accounting?
5. In what order are assets and liabilities listed on the balance sheet? Explain.

6. Explain the concept of depreciation.
7. A piece of equipment is purchased for $30,000. It has a useful life of 10 years and a salvage value of $5,000. What is the annual depreciation using the straight-line method? What is the book value after eight years?
8. If $10,000 of dividends are paid out in cash to a company's owners, how does this affect the balance sheet? Is the transaction recorded on any other financial statement?
9. Assume a company has net income of $10,000 for the operating year and decides to pay $5,000 in dividends and retain $5,000 in the business. Explain which accounts on the balance sheet are affected.
10. What types of management information records might ASCO keep?

CHAPTER 4

FINANCIAL ANALYSIS AND PLANNING

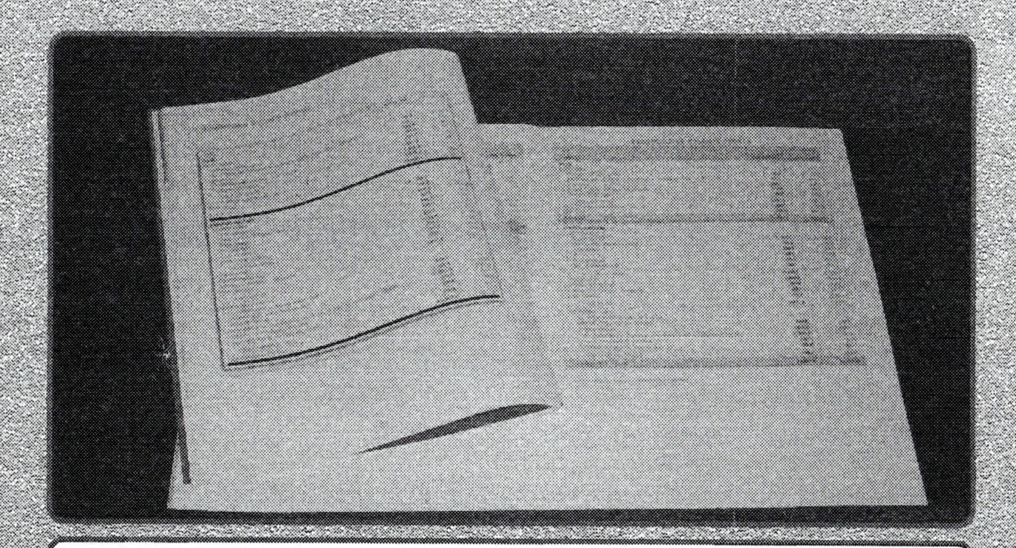

LEARNING OBJECTIVES

In this chapter we will cover the following topics:

- The 3C-statement method of financial analysis
- The use of financial ratios to analyze financial performance
- Determining trends in financial performance
- Break-even Analysis
- Financial planning and control
- The use of budgets as a tool for planning and control
- The budget process

CASE STUDY

The Desert Rose Nursery is a small nursery located in southern California. Jeannie Wilson, who has a B.S. degree in horticulture, is the manager. Jeannie has always loved gardening and yard work and pursuing a college degree in horticulture fit her plan of some day owning a nursery. Upon graduating she was offered the position of manager by some business people who were starting a nursery. They all had high hopes for the venture's success, but they could not devote much time to running the new business. Jeannie thought that this would be the ideal way to get her foot in the door and took the position. She proposed that they let her participate in the ownership of the business and they agreed that if the business was successful, they would work out an arrangement whereby she could be given an ownership stake in the company based on its profitability.

The nursery specializes in outdoor ornamental plants and sells only to wholesalers and retailers. The growth in the nursery business was substantial in the mid- and late-1990s and the Desert Rose Nursery had captured what the owners thought was a reasonable share of the market. It was not quite as profitable as other similar-sized, more established nurseries, but the owners felt that this was not unusual for a new company with high start-up expenses. Jeannie spent most of her time working in the nursery and the results were the production of the high-quality products for which the Desert Rose Nursery was known. Jeannie spent only as much time in the office as was absolutely necessary.

Starting in early 2000, total industry sales flattened out and the Desert Rose Nursery's profits began to decline. Recently, the board of directors asked Jeannie for an explanation of the decline in profitability. They explained that anyone could survive in a growing industry and that the true test of a manager was during an industry recession.

By the end of the chapter, you should be able to help Jeannie assess the firm's present financial situation, analyze its problems, and recommend possible solutions.

INTRODUCTION

Financial managers need good information to be able to predict the effects of possible courses of action in order to make good decisions. In practice, financial managers make predictions about future events using information from the past and present. The preparation of financial statements, discussed in the previous chapter, is just the beginning of successful financial management. These statements provide the framework for measuring the profit or cash flow of the agribusiness. In this chapter, we outline the financial techniques used to evaluate the health of the firm. Is the firm profitable? What are the trends in profitability, and how does the firm's profitability compare with similar firms and industry norms? We follow with a discussion of financial planning where we discuss how to prepare pro forma income statements, balance sheets, and cash budgets. These plans provide the type of information we need to analyze financial decisions.

FINANCIAL STATEMENT ANALYSIS

Financial analysis is used to evaluate the financial condition of a firm by measuring its performance. In order to analyze financial performance, we must dissect the basic financial statements of the firm—the balance sheet, income statement, and statement of cash flows—to understand the financial workings of the business. Many managers begin the process by conducting a 3C-statement analysis. The 3C-statement analysis provides a basic overview of the firm's financial position and includes:

- Comparative analysis in which each item on a financial statement is compared with that same item from financial statements of previous periods
- Change analysis that shows the changes that have taken place in a firm's financial statements over time
- Common-size analysis that shows the size of selected items in one financial statement relative to some base item for one period

Preparation of comparison, change, and common-size statements (known as the *3C-statement analysis*) is a means of gaining insights into the agribusiness's financial statements. Managers use 3C-statement analysis to evaluate their present financial performance relative to the firm's financial statements of previous years, and by comparing the firm's statements to those of competitors or industry norms.

The first step in a 3C-statement analysis is to *compare* two years' (or more) financial statements side-by-side. The purpose of comparing financial statements is to become familiar with the firm's accounting practices and to get a general feel for the firm's financial statements before examining specific aspects of its operation. A comparison of the firm's financial data is useful in evaluating the firm's performance and provides a starting point for financial analysis. Boxes 4–1 and 4–2 illustrate the two-period balance sheets and income statements for Green Valley Farm Supply. Green Valley sells farm supplies including petroleum, fertilizer, feed, herbicides, and pesticides, and provides

BOX 4–1

Green Valley Farm Supply Balance Sheet (in thousands of dollars)

	December 31 2000	December 31 1999
Assets		
Current assets:		
Cash and equivalents	$ 267	$ 280
Accounts receivable (net)	1,108	820
Inventory	1,195	809
All other current assets	99	68
Total current assets	2,669	1,977
Noncurrent assets:		
Property, plant, and equipment (net)	936	700
Intangibles (net)	43	32
All other noncurrent assets	330	219
Total noncurrent assets	1,309	951
Total assets	$3,978	$2,928
Liabilities and Owners' Equity		
Current liabilities:		
Notes payable	$ 657	$ 400
Accrued compensation and employee benefits	115	77
Accounts payable	757	553
Income taxes payable	12	11
All other current liabilities	354	240
Total current liabilities	1,895	1,281
Long-term liabilities:		
Long-term debt	306	313
Deferred tax liabilities	27	11
All other long-term liabilities	61	71
Total long-term liabilities	394	395
Total liabilities	2,289	1,676
Owners' equity	1,689	1,252
Total liabilities and owners' equity	$3,978	$2,928

services to farmers including product delivery, fertilizer application, and chemical application. These financial statements indicate that Green Valley has been profitable the last two years and that it is growing larger. Without further analysis on our part there is not much else we can say about its financial situation at this point.

BOX 4–2

Green Valley Farm Supply Income Statement for the Years Ending December 31 (in thousands of dollars)

	2000	1999
Net sales	$11,410	$7,842
Cost of goods sold	9,116	6,156
Gross margin	2,294	1,686
Selling, general, and administrative expenses	2,043	1,497
Operating income	251	189
Interest expense	86	65
Income before taxes	165	124
Income taxes	41	31
Net income	$ 124	$ 93

The second step in 3C-statement analysis is to determine the *change* in the financial statements. It is relatively easy to conduct a change analysis once you have the statements side by side. Each of the accounts is then netted (the old figures are subtracted from the more recent figures). These differences are changes that take place over the year. Box 4–3 illustrates the changes that occurred in Green Valley's balance sheet between 1999 and 2000. Some notable changes have occurred. Cash and equivalents, long-term debt, and all other long-term liabilities have decreased while all of the other accounts have increased. Green Valley appears to be experiencing a substantial year-to-year growth in business as accounts receivable, inventory, current liabilities, and owners' equity have grown substantially. It is also interesting to note that the increase in owners' equity ($437,000) is greater than the net income for the year ($124,000). Thus, we can infer that the owners have paid in additional capital during the year.

The third step in the 3C-statement analysis is to construct the *common-size statements.* You do this by converting all of the balance sheet and income statement dollar amounts into a percentage of a common base. This puts all of the figures into perspective. To do this you divide each entry in the balance sheet by total assets and each entry in the income statement by net sales. Green Valley's common-size balance sheet is shown in Box 4–4 and the common-size income statement in Box 4–5. To illustrate, we start with the balance sheet. The cash and equivalents portion of total assets is calculated by dividing cash and equivalents of $267,000 by total assets of $3,978,000 or 7 percent. In other words, on

BOX 4–3

Green Valley Farm Supply Statement of Changes in Balance Sheet (in thousands of dollars)

	2000 - 1999	
	Increase	Decrease
Assets		
Current assets:		
Cash and equivalents		$13
Accounts receivable (net)	$ 288	
Inventory	386	
All other current assets	31	
Total current assets	692	
Noncurrent assets:		
Property, plant, and equipment (net)	236	
Intangibles (net)	11	
All other noncurrent assets	111	
Total noncurrent assets	358	
Total assets	1,050	
Liabilities and Owners' Equity		
Current liabilities:		
Notes payable	257	
Accrued compensation and employee benefits	38	
Accounts payable	204	
Income taxes payable	1	
All other current liabilities	114	
Total current liabilities	614	
Long-term liabilities:		
Long-term debt		7
Deferred tax liabilities	16	
All other long-term liabilities		10
Total long-term liabilities		
Total liabilities	613	
Owners' equity	437	
Total liabilities and owners' equity	1,050	

December 31, 2000, cash and equivalents made up 7 percent of Green Valley's assets. Using the same procedure on the income statement, we calculate the cost of goods sold proportion of sales as cost of goods sold of $9,116,000 divided by sales of $11,410,000 or 80 percent. This process is repeated for each entry in the financial statements.

The common-size balance sheet statement indicates that current assets and current liabilities, as a percentage of total assets, have increased from 1999 to

BOX 4–4

Green Valley Farm Supply Common-Size Balance Sheet

	December 31			
	2000	1999	2000	1999
	(thousands of dollars)		(percent)	
Assets				
Current assets				
Cash and equivalents	$ 267	$ 280	7	10
Accounts receivable (net)	1,108	820	28	28
Inventory	1,195	809	30	28
All other current assets	99	68	2	2
Total current assets	$2,669	$1,977	67	68
Noncurrent assets:				
Property, plant, and equipment (net)	936	700	24	24
Intangibles (net)	43	32	1	1
All other noncurrent assets	330	219	8	7
Total noncurrent assets	$1,309	$ 951	33	32
Total assets	$3,978	$2,928	100	100
Liabilities and Owners' Equity				
Current liabilities:				
Notes payable	$ 651	$ 400	17	14
Accrued compensation and employee benefits	115	77	3	3
Accounts payable	757	553	19	19
Income taxes payable	12	11	0	0
All other current liabilities	354	240	9	8
Total current liabilities	1,895	1,281	48	44
Long-term liabilities:				
Long-term debt	306	313	8	11
Deferred tax liabilities	27	11	1	0
All other long-term liabilities	61	71	1	2
Total long-term liabilities	394	395	10	13
Total liabilities	2,289	1,676	58	57
Owners' equity	1,689	1,252	42	43
Total liabilities and owners' equity	$3,978	$2,928	100	100

BOX 4–5

Green Valley Farm Supply Common-Size Income Statements for the Years Ending December 31

	2000	1999	2000	1999
	(thousands of dollars)		(percent)	
Net sales	$11,410	$7,842	100	100
Cost of goods sold	9,116	6,156	80	79
Gross margin	2,294	1,686	20	21
Selling, general, and administrative expenses	2,043	1,497	18	19
Operating Income	251	189	2	2
Interest expense	86	65	1	1
Income before taxes	165	124	1	2
Income taxes	41	31	0	0
Net income	$ 124	$ 93	1	1

2000. An analysis of the component parts of each of these entries indicates that cash and equivalents have declined while inventory, notes payable, and all other current liabilities have increased. The owners' equity, as a percentage of total assets, has declined. Turning to the common-size income statement we also note that the cost of goods sold percentage has increased while the general, selling, and administrative expense percentage has declined. What the common-size statements show is that this farm supply store is attempting to increase sales by carrying higher inventory levels. This has put pressure on the firm's cash and equivalents and increased the firm's use of short-term debt. Has the strategy worked? The firm remains profitable, but we will need to conduct a deeper examination of the financial statements using ratio analysis before we can answer this question.

Common-size statements are used to compare any two financial statements by converting the absolute numbers to percentages. Within a firm, balance sheets and income statements, as well as statements of cash flows and production cost schedules, may be converted to common-size statements to facilitate comparison. Common-size statements can be useful for comparing statements for nonconsecutive years, firms that vary greatly in size, or evaluating a firm relative to industry norms. By expressing dollar figures relative to a base figure, differences in size are eliminated, making comparison much easier. However, caution should be used when comparing statements from different sources, because different standards in the method used to classify items such as expenses or assets can make such analyses misleading.

RATIO ANALYSIS

Ratio analysis is the most widely used type of financial analysis. Ratios measure one variable relative to another. *Ratio analysis* evaluates and analyzes the financial position of the firm by transforming dollar figures into ratios that may be compared to ratios of past periods, other firms, or to industry norms. Ratios may be used in evaluating the financial health of the firm and specific areas of the firm's operation as well as an aid in identifying problem areas. We will discuss four types of ratios:

- *Profitability ratios* examine various aspects of the firm's profitability.
- *Activity ratios* include several measures relating to the use of the firm's resources.
- *Liquidity ratios* measure the firm's ability to meet its short-term credit obligations.
- *Solvency ratios* provide an assessment of the firm's ability to meet its long-term debt obligations.

In the following sections we will illustrate the calculation of financial ratios for Green Valley Farm Supply using the information contained in the balance sheet (Box 4–1) and income statement (Box 4–2) for 2000.

Profitability Ratios

Profitability ratios relate the income of the firm to other important financial figures. These ratios are useful in evaluating the management of the firm's resources. Profitability ratios typically compare the firm's profits either to the resources invested or to sales. The first two ratios presented compare profitability to the resources invested, whereas the last two ratios examine profitability in terms of sales.

Return on Equity. The most commonly used profitability ratio is *return on owners' equity* (ROE), which is also commonly called return on equity or return on investment.

$$\text{Return on Equity} = \frac{\text{Net Income}}{\text{Owners' Equity}}$$

$$\text{Green Valley's ROE} = \frac{\$124{,}000}{\$1{,}689{,}000} = 7.34\%$$

This is the most important ratio to the firm's owners because it measures the return on their investment. The ratio can be interpreted as the return on the owners' initial capital invested in the firm and the earnings retained in the firm over time. Owners and potential investors will look at this ratio to determine whether they should invest additional funds in the firm or possibly sell their investment, because it gives them a percentage return that they can compare with alternative investments.

Return on Assets. A ratio that reflects the return to the firm's total resources is the *return on assets* (ROA).

$$\text{Return on Assets} = \frac{\text{Net Income}}{\text{Total Assets}}$$

$$\text{Green Valley's ROA} = \frac{\$124{,}000}{\$3{,}978{,}000} = 3.12\%$$

This ratio reflects the total return to the firm's owners divided by the total resources invested in the firm. This ratio is often used to examine the efficiency with which a company uses its resources. However, in recent years financial analysts have relied more heavily on a ratio known as the basic earning power.

Basic Earning Power. The *basic earning power* (BEP) ratio is similar to the ROA, except that operating income (earnings before interest and taxes) is used instead of net income.

$$\text{Basic Earning Power} = \frac{\text{Operating Income}}{\text{Total Assets}}$$

$$\text{Green Valley's BEP} = \frac{\$251{,}000}{\$3{,}978{,}000} = 6.31\%$$

It is important to use operating income rather than net income in calculating BEP because we are trying to measure the efficiency of the use of all assets regardless of how they may have been financed or the firm's tax situation. Operating income represents the profitability of the business divided among the creditors (interest), government (taxes), and owners (net income).

The BEP ratio is particularly useful for comparison with other firms, particularly those in the same line of business. If this ratio is less than management expects, further analysis should be conducted to locate the source of the problem.

Gross Margin Percent. Ultimately, it is through the utilization of the resources of the firm that income is generated. For this reason, we examine the ROE and BEP. However, there are several other important steps in the process that may provide insight into why the ROE or BEP were high or low. The *gross margin percent* yields the gross margin generated by each dollar of sales.

$$\text{Gross Margin Percent} = \frac{\text{Gross Margin}}{\text{Net Sales}}$$

$$\text{Green Valley's Gross Margin Percent} = \frac{\$2{,}294{,}000}{\$11{,}410{,}000} = 20.11\%$$

The gross margin percent of 20.11 percent means that nearly 80 cents out of every dollar of sales goes to pay for the cost of goods sold and 20 cents is available for other expenses and profit. The gross margin percent is particularly useful when looked at over a period of time or when compared to similar firms in an industry. If the margin is too low, it is likely that the firm will not have enough earnings to cover selling, general, and administrative expenses and allow for an adequate profit. In general, firms that have a low gross margin percent, such as supermarkets, have a high volume of sales. Firms with a higher gross margin percent usually have relatively lower sales.

Return on Sales. The last profitability ratio we will examine is the *return on sales* (also known as the net income percent or profit margin).

$$\text{Return on Sales} = \frac{\text{Net Income}}{\text{Net Sales}}$$

$$\text{Green Valley's Return on Sales} = \frac{\$124,000}{\$11,410,000} = 1.09\%$$

The return on sales ratio means that Green Valley makes 1.09 cents of profit for every dollar of sales. This ratio is often looked at over time or compared to ratios of similar firms. By itself, it may not have much meaning. But changes in this ratio can indicate problems within the firm. When this happens, further analysis is warranted. A relatively low profit margin indicates that prices are too low, costs are too high, or both. A significantly lower return on sales, when compared to similar firms within an industry, may indicate that there is room for improvement by management. On the other hand, good management of expenses may result in a high profit margin.

Activity Ratios

Activity ratios (also known as turnover ratios) measure how well the firm is utilizing its resources. Specifically, they measure the efficiency with which the firm uses its assets and manages its expenses. When a firm is not very profitable or suffers a decline in profitability, activity ratios may provide an insight as to why it has not been as profitable as was expected. Activity ratios may also be used to monitor the performance of the firm in specific areas and to flag problems before they occur.

Asset Turnover. The *asset turnover* indicates the efficiency with which assets were used to generate sales.

$$\text{Asset Turnover} = \frac{\text{Net Sales}}{\text{Total Assets}}$$

$$\text{Green Valley's Asset Turnover} = \frac{\$11,410,000}{\$3,978,000} = 2.87$$

This ratio indicates that Green Valley is turning over its assets nearly three times within a year. An asset turnover close to or exceeding industry norms may indicate that the firm is operating close to capacity, and that generation of additional sales is possible only with an increase in capital investment. In general, the higher the asset turnover, the more efficiently the assets of the firm are being utilized and the higher the ROE will be. In other words, it is good to generate sales with as few assets as possible. This ratio is useful when making comparisons within the firm over time or with similar size firms in the same industry.

Operating Expense Percent. Another key activity ratio is the *operating expense percent.*

$$\text{Operating Expense Percent} = \frac{\text{Selling, General, and Administrative Expenses}}{\text{Net Sales}}$$

$$\text{Green Valley's Operating Expense Percent} = \frac{\$2,043,000}{\$11,410,000} = 17.91\%$$

It is important that this ratio be monitored from period to period. If this ratio is too high, it could significantly affect the profitability of the firm. An increase in this ratio will only indicate that there is a problem, not what the problem is. Individual expense items, such as labor or marketing expenses, must be examined to determine the cause of the problem.

Inventory Turnover. Inventory management is especially important when the cost of carrying inventory is high, as may be the case when interest rates are high. A commonly used ratio to measure the efficiency with which inventory is managed is *inventory turnover:*

$$\text{Inventory Turnover} = \frac{\text{Cost of Goods Sold}}{\text{Average Inventory}}$$

$$\text{Green Valley's Inventory Turnover} = \frac{\$9,116,000}{(\$1,195,000 + \$809,000)/2} = 9.10$$

This ratio measures the number of times inventory turns over in one year. The higher the inventory turnover ratio, the shorter the period of time inventory is held. High inventory turnover may indicate better liquidity or superior merchandising. It may also indicate a shortage of needed inventory for sales. Low inventory turnover is an indication of poor liquidity, unneeded or obsolete inventory, or a planned inventory buildup in anticipation of a shortage. In contrast, too high an inventory turnover can indicate that the firm may run out of stock. Both the benefits and costs of carrying inventory must be weighed to determine the proper level of stocks.

Note that average inventory is used in the calculation of this measure. Inventories may vary greatly during the year, especially in agribusinesses that are highly seasonal. The average inventory is calculated by averaging the amount of inventory at the beginning and end of the period for which the inventory turnover is calculated. It is important to use the cost of goods sold for the same period for which the average inventory is calculated.

The annual inventory turnover ratio can be converted into the number of days that the inventory is held by dividing the turnover into the number of days in the period.

$$\text{Average Holding Period} = \frac{\text{Days per Period}}{\text{Inventory Turnover}}$$

$$\text{Green Valley's Average Holding Period} = \frac{365 \text{ days}}{9.10} = 40 \text{ days}$$

Green Valley's inventory is held an average of 40 days before it is sold.

Accounts Receivable Turnover. The firm's credit policy may be examined by looking at the *accounts receivable turnover.*

$$\text{Accounts Receivable Turnover} = \frac{\text{Net Sales}}{\text{Average Accounts Receivable (net)}}$$

$$\text{Green Valley's Accounts Receivable Turnover} = \frac{\$11{,}410{,}000}{\$1{,}108{,}000} = 10.30$$

Accounts receivable (net) is all accounts receivable from trade, net of allowance for doubtful accounts. The ratio measures the number of times accounts receivable turn over during the year. Thus, Green Valley turned over receivables approximately 10 times during the year. The higher the receivables turnover ratio, the shorter the time between sales and cash collections. A low receivables turnover ratio indicates either that a large amount of credit has been extended for a long period of time, or that the firm is having difficulty collecting the debt. A good balance must be struck between using credit as a tool to increase sales and the cost of having too liberal a credit policy.

By dividing the accounts receivable turnover ratio into the number of days in the period, we obtain the average number of days that it takes to be paid.

$$\text{Average Collection Period} = \frac{\text{Collection Period}}{\text{Accounts Receivable Turnover}}$$

$$\text{Green Valley's Average Collection Period} = \frac{365 \text{ days}}{10.30} = 35 \text{ days}$$

Green Valley collects payments an average of 35 days after the sale date.

Net Sales to Working Capital. The *net sales to working capital* ratio is calculated by dividing net sales by working capital (current assets less current liabilities equals net working capital).

$$\text{Net Sales/Working Capital} = \frac{\text{Net Sales}}{\text{Net Working Capital}}$$

$$\text{Green Valley's Net Sales/Working Capital} = \frac{\$11{,}410{,}000}{(\$2{,}669{,}000 - \$1{,}895{,}000} = 14.74$$

Working capital reflects the ability of the firm to finance current operations and is used to measure how efficiently working capital is employed. Consequently, it is a popular ratio used by the firm's creditors. A low ratio may indicate an inefficient use of working capital whereas a high ratio may signal that the firm is overextended.

Liquidity Ratios

Liquidity is a measure of the firm's ability to meet current obligations as they come due. Liquidity analysis includes ratios that measure cash as well as assets that can be converted easily to cash. When analyzing the liquidity of the firm, two sections of the balance sheet are used: current assets and current liabilities. The reason is simple; in an emergency, it is the current assets that must be used to meet the short-term or current credit obligations.

Current Ratio. The two most common measures of liquidity are the current ratio and the quick ratio. The *current ratio* is the ratio of current assets to current liabilities.

$$\text{Current Ratio} = \frac{\text{Current Assets}}{\text{Current Liabilities}}$$

$$\text{Green Valley's Current Ratio} = \frac{\$2{,}669{,}000}{\$1{,}895{,}000} = 1.41$$

This ratio should always be greater than 1 since the firm's current assets should always be great enough to meet its current liabilities. The higher the current ratio, the greater the firm's ability to pay current obligations. A strong current ratio means that current assets are numerically greater than current liabilities. However, the composition and quality of current assets, is a critical factor in the analysis of the firm's liquidity that is not measured by the current ratio. For example, cash balances are the most liquid of the current assets followed by marketable securities. The least liquid current assets are accounts receivable, which represent past sales that should be collected soon, and inventory that is waiting to be sold.

Quick Ratio. The quick ratio, also known as the acid test ratio, is a refinement of the current ratio and is a more conservative measure of liquidity. The *quick ratio* is measured by subtracting inventory from current assets and dividing this amount by current liabilities.

$$\text{Quick Ratio} = \frac{\text{Current Assets} - \text{Inventory}}{\text{Current Liabilities}}$$

$$\text{Green Valley's Quick Ratio} = \frac{\$2{,}669{,}000 - \$1{,}195{,}000}{\$1{,}895{,}000} = 0.78$$

This ratio includes only those current assets that can most quickly be converted to cash. Inventory is not included because future sales are uncertain and may frequently be on credit. Therefore, inventory is the least liquid of the current assets. Generally, any value of less than 1 implies that the firm is dependent on inventory to liquidate short-term debt. As with the current ratio, the quick ratio should be closely monitored for changes.

Solvency Ratios

Solvency ratios (also known as leverage ratios) measure the firm's ability to service long-term debt. *Solvency analysis* is used to determine if the firm will be able to pay its debts in the future and thus stay in business. To protect creditors in the event that a firm cannot pay its debt, lenders will look at the amount of assets the firm has available to cover debts. Creditors not only get a claim on earnings ahead of owners, but also have a claim on assets if the firm becomes bankrupt. For this reason, assets are important to creditors.

Debt-to-Equity. One of the most popular measures of the firm's ability to meet its long-term credit obligations is the *debt-to-equity* ratio.

$$\text{Debt-to-Equity} = \frac{\text{Total Liabilities}}{\text{Owners' Equity}}$$

$$\text{Green Valley's Debt-to-Equity} = \frac{\$2{,}289{,}000}{\$1{,}689{,}000} = 1.36$$

The debt-to-equity ratio expresses the relationship between capital contributed by creditors and capital contributed by owners. The higher the ratio, the greater the risk to creditors, and conversely, the lower the ratio, the more protection the owners provide to creditors. Highly leveraged firms (those with heavy debt in relation to owners' equity) are more vulnerable to business downturns than those firms with less debt. Acceptable solvency ratios vary greatly depending on the requirements of a particular industry and should only be used to make comparisons within an industry. Some creditors have been eager to loan money to highly leveraged start-up companies such as Internet and technology companies because they receive higher returns in terms of the interest they charge on loans. Conversely, creditors typically require companies in older industries such as the food and agribusiness industry to have lower debt-to-equity ratios. In general, however, creditors are more likely to lend money to companies with low debt-to-equity ratios than to highly leveraged companies.

Debt Ratio. Sometimes, the amount of debt is measured relative to the total assets. This is referred to as the *debt ratio.*

$$\text{Debt Ratio} = \frac{\text{Total Liabilities}}{\text{Total Assets}}$$

$$\text{Green Valley's Debt Ratio} = \frac{\$2{,}289{,}000}{\$3{,}978{,}000} = 0.58$$

Notice that the debt-to-equity ratio and the debt ratio provide essentially the same information because debt plus owners' equity equals total assets. Green Valley's debt-to-equity ratio tells us that debt is 1.36 times equity and the debt ratio tells us that creditors are supplying 58 percent of assets. If Green Valley becomes bankrupt, the assets would have to be liquidated for 58 cents on the dollar to satisfy creditors.

Coverage Ratio. The coverage ratio (also known as times interest earned) is a measure of the firm's ability to service debt. The *coverage ratio* is calculated by dividing earnings (profit) before interest and taxes by interest expense.

$$\text{Coverage Ratio} = \frac{\text{Operating Income}}{\text{Interest Expense}}$$

$$\text{Green Valley's Coverage Ratio} = \frac{\$251{,}000}{\$86{,}000} = 2.92$$

The ratio is a measure of the firm's ability to meet interest payments. A high ratio indicates that the firm will have little difficulty meeting its interest obligations. Creditors use the ratio to determine the firm's ability to take on additional debt. Green Valley's coverage ratio indicates that the firm's profit available for interest expenses is almost three times as large as its current interest expense. Thus, Green Valley could suffer a substantial decline in profits and still be able to meet its interest expense obligations.

Applications of Ratio Analysis

Now that we have an idea as to how ratio analysis is conducted and a basic understanding of what the relationships mean, we are ready to discuss the application of these calculations. Three types of ratio analyses are normally conducted: industry norm analysis, time analysis, and decomposition analysis.

Industry Norm Analysis. How do we know if a financial ratio is right on target or if it is too low or too high? A firm's financial ratios should be compared to the average ratio of firms in the same industry. Thus, to conduct an *industry norm analysis* of Green Valley, we will compare its financial ratios with the average farm supply store. Industry ratios and financial statement information are available for numerous industries from Dun and Bradstreet's *Industry Norms & Key Business Ratios*, Prentice Hall's *Almanac of Business and Industrial Ratios*, and RMA's *Annual Statement Studies*. Because some financial ratios are calculated using different formulas, it is always important to examine the underlying formula before making comparisons.

Let's assume that Green Valley's manager has approached a bank to secure a temporary increase in working capital of $50,000 to be repaid in four months. The loan officer will base his or her decision on the ratio analysis we completed for Green Valley. Green Valley's financial ratios and norms for the farm supply/marketing industry are illustrated in Box 4–6.

BOX 4–6

Green Valley Farm Supply's Financial Ratios and Farm Supply Industry Norms

Ratio	Green Valley	Industry Norm[1]
Current	1.41	2.0 (Strong) 1.3 (Median) 1.1 (Weak)
Quick	0.78	1.1 (Strong) 0.7 (Median) 0.4 (Weak)
Net Sales/Working Capital	14.74	8.7 (Strong) 19.6 (Median) 53.3 (Weak)
Coverage Ratio	2.92	5.2 (Strong) 2.7 (Median) 1.5 (Weak)

[1] ***Source:*** RMA, The Association of Lending & Credit Risk Professionals, ***Annual Statement Studies 1999-2000***. Page 790, 1999.

The loan officer would be interested in financial ratios that focus on Green Valley's ability to service additional short-term debt. These ratios include the current ratio, quick ratio, net sales to working capital ratio, and the coverage ratio. To determine whether Green Valley's ratios provide sufficient protection for the bank, the loan officer will compare each of Green Valley's ratios to the median (midpoint) of the farm supply/marketing industry. The loan officer would note that Green Valley's ratios are all stronger, but very close to the industry median. From the loan officer's point of view this is a good sign. If Green Valley has been a good and reliable customer, the loan officer would look upon this information favorably and would likely fund the loan request.

Trend and Graphical Analysis. Trend and graphical analyses are used to analyze a firm's financial performance over time. They are simple but powerful instruments that can be combined with the other financial management tools. The analysis of trends can provide some useful insights into past occurrences. Constructing a table of the financial figures of interest may identify trends. For example, Green Valley's operating expenses as a percent of net sales are presented in Box 4–7. Trends may also be used to give an indication of the future. Although predictions based on past occurrences are simplistic, they may serve as an expectation of what might occur if management takes no action. The addition of graphical analysis may help to visualize a trend. The numbers

BOX 4–7

Green Valley Farm Supply's Operating Expense as a Percentage of Net Sales, 1996–2000

Year	Operating Expense as a Percentage of Net Sales
1996	15.9 %
1997	15.8 %
1998	16.0 %
1999	19.1 %
2000	17.9 %

FIGURE 4–1

Green Valley Farm Supply's Operating Expense as a Percentage of Net Sales, 1996–2000

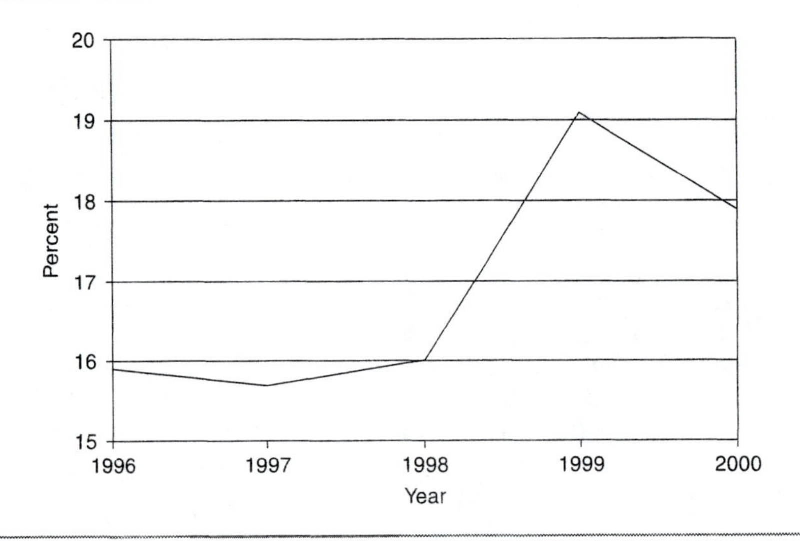

in Box 4–7 are graphically displayed in Figure 4–1. The graph illustrates that operating expenses as a percent of sales have been generally increasing since 1996.

Decomposition Analysis. *Decomposition analysis* involves separating the ratios into their individual components and examining each in detail. The purpose of decomposition analysis is to determine the quality of the ratio. Since each ratio is comprised of many different elements, it is possible for two ratios of equal value to have components that vary greatly in size. For example, the current ratio can be separated into its various components (i.e., cash, accounts receivable, inventory, accounts payable, prepaid expenses, etc.). In decomposition analysis a detailed investigation is made of each component individually with the objective of determining whether each component is on target.

BREAK-EVEN ANALYSIS AND CONTRIBUTION TO OVERHEAD

In the first part of this chapter we demonstrated techniques for conducting a financial analysis of the firm. As we have discussed, one of the key measures of a firm's success is its profitability. To understand the effects of various actions on the firm's profitability, many food and agribusiness firms employ break-even analysis and contribution to overhead analysis to understand the relationship between prices, costs, and volume and to analyze the response of revenues, costs, and profits to changes in sales volume. We begin our discussion of these tools with a definition of some terms and key relationships.

Some costs, known as *variable costs*, vary directly with changes in production volume. Variable costs include materials, manufacturing supplies, direct labor, utilities, maintenance, packaging, freight, and sales commissions. Variable costs per unit are constant, but total variable costs increase as the number of units sold increase. In contrast, *fixed costs* are independent of the number of units sold. Fixed costs include executive salaries, depreciation, rent, property taxes, insurance, and interest. The fixed costs must be paid regardless of the firm's level of production.

Per unit contribution to overhead (per unit CTO) is defined as the per unit price less the per unit variable cost:

$$\text{Per Unit CTO} = \text{Price per Unit} - \text{Variable Costs per Unit}$$

In other words, the per unit CTO is the amount per unit that is left to pay the overhead or fixed costs once variable costs are deducted from the per unit price.

A product's total *contribution to overhead* (CTO) is the per unit CTO times the number of units sold:

$$\text{CTO} = \text{Per Unit CTO} \times \text{Units Sold}$$

CTO should always be positive, that is, price should always be set so that variable costs are met; if not, the firm is losing money with every sale, and it could minimize its losses by simply not selling the product at all. An exception is when a firm is forced to sell at a price below variable costs, in the short run, just to stay in business. Another exception is when selling at a price below variable costs makes sense because of the impact on the CTO of other products. For example, some products may be sold at a loss because customers want to purchase them along with other products that are very profitable for a firm.

The *break-even quantity* is the quantity of sales at which sales revenues equal total operating costs. The break-even quantity is equal to fixed costs divided by the unit price less the variable cost per unit:

$$\text{Break-Even Quantity} = \frac{\text{Fixed Costs}}{\text{Price per Unit} - \text{Variable Costs per Unit}}$$

At this volume of sales, profits, as defined as operating income, are zero. Also:

$$\text{Break-Even Quantity} = \frac{\text{Fixed Costs}}{\text{Per Unit CTO}}$$

Break-even and contribution to overhead analyses are illustrated in the following example of a sod manufacturer. The price of sod is $0.20 per square foot. The company's fixed cost of production is $10,000 per year and variable cost is $0.15 per square foot. Last year the firm sold 500,000 square feet of sod.

The per unit CTO is calculated as:

$$\text{Per Unit CTO} = \$0.20/\text{sq. ft.} - \$0.15/\text{sq. ft.} = \$0.05/\text{sq. ft.}$$

The interpretation is that for each square foot of sod the firm sells, it generates $0.05 over the variable costs of producing the sod.

Total CTO for the firm is:

$$\text{CTO} = \$0.05/\text{sq. ft.} \times 500{,}000 \text{ sq. ft.} = \$25{,}000$$

That is, the firm sold 500,000 sq.ft. of sod and each square foot of sod generated a contribution to overhead of $0.05 for a total contribution to overhead of $25,000.

Lastly, in order to determine whether the firm made a profit, the firm's fixed costs must be subtracted from the CTO:

$$\text{Profit} = \$25{,}000 - \$10{,}000 = \$15{,}000$$

This can be verified by calculating the firm's total revenues and subtracting its costs, both fixed and variable costs. In this example, revenue is equal to $100,000 ($0.20 per square foot multiplied by 500,000 square feet). Fixed costs are $10,000 and variable costs total $75,000 ($0.15 multiplied by 500,000 sq.ft.). After subtracting fixed and variable costs from revenue, profit is $15,000.

We can also use the information in the above example to calculate the break-even quantity:

$$\text{Break-Even Quantity} = \frac{\$10{,}000}{\$0.20 - \$0.15} = \frac{\$10{,}000}{\$0.05} = 200{,}000 \text{ sq.ft.}$$

The break-even quantity, the quantity at which total revenue equals total costs, is 200,000 square feet of sod in this example. If the firm sells fewer than 200,000 units, it will suffer an operating loss and if it sells more than 200,000 units, it will make an operating profit.

The break-even quantity can also be determined graphically. Figure 4–2 illustrates a break-even graph based on the information provided in the algebraic method. The fixed cost of $10,000 per year is plotted as a horizontal line that intersects the vertical axis at $10,000. Total revenue at each level of output is calculated as price multiplied by quantity. The total revenue line is constructed by plotting the total revenue at each quantity. The total cost at each level of output is calculated by adding the variable cost at each level of output,

FIGURE 4–2

Break-Even Graph for Sod Manufacturer

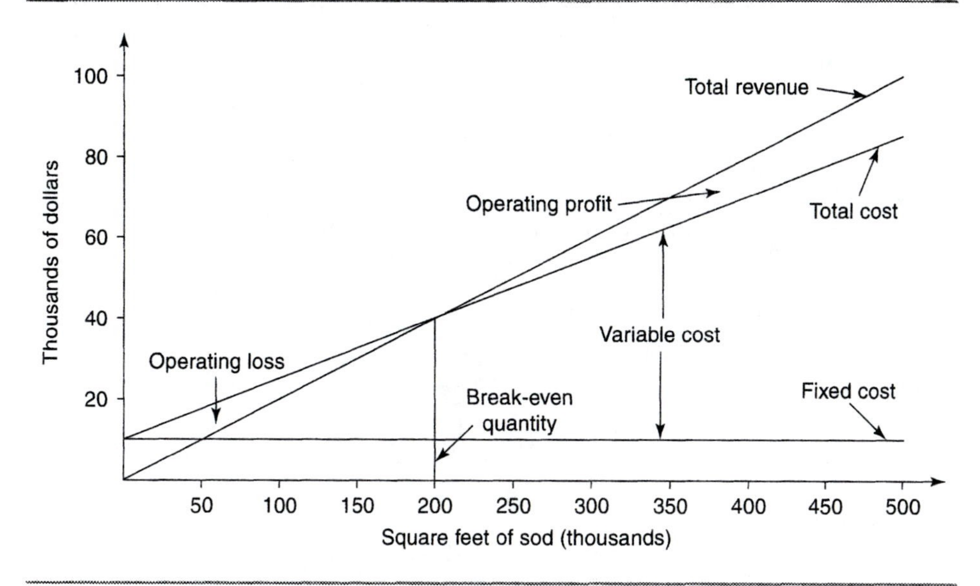

which is calculated as the variable cost per unit multiplied by the quantity, to the fixed cost. The break-even quantity is found at the intersection of the revenue line with the total cost line. In this example, the break-even quantity is 200,000 sq. ft. At this point operating profit is exactly zero because the firm's revenue exactly covers its total costs. It can also be seen that at any quantity above the break-even quantity the firm will make an operating profit. Conversely, at quantities below the break-even quantity the firm will have an operating loss.

Break-even analysis and contribution to overhead analysis have several important applications. They are commonly used in evaluating product profitability, establishing prices, and evaluating new ventures or new products. These three uses are discussed next.

For firms that sell multiple products, it is useful to calculate the contribution to overhead for each product. As we have discussed earlier, the per unit CTO should, as a general rule, be positive. Products with a negative CTO should be eliminated unless they contribute to the firm's profitability in some other way. Products that contribute little to the firm's profit may be targets for improvement or elimination. It may also be useful to conduct a break-even analysis for each product to determine whether a product's sales are sufficient to cover the fixed costs associated with the product. This may be easier said than done because the allocation of fixed costs to specific products can be difficult and is usually arbitrary, particularly when several products share common resources. Many methods for allocating fixed costs to various products have been used, including volume or dollars of sales and utilization of floor space. Ultimately it is a judgement call.

Contribution to overhead analysis also provides a simple mechanism to evaluate price changes. Economists tell us that profits are maximized where marginal cost (the cost of producing and selling one additional unit of product) equals marginal revenue (the revenue generated from selling one additional unit of product). However, this principle is not easily applied. For most firms, marginal cost is relatively constant over the relevant range of production and therefore is approximately equal to variable cost. Marginal revenue is more difficult to estimate because it requires estimating the demand curve for the firm's product. By using contribution to overhead analysis, a manager can formulate a much simpler equation than estimating a product's demand curve. To evaluate a price decrease, you calculate how much sales must increase to justify the lower price. Similarly, to evaluate a price increase, you calculate how much sales could decline before profitability declined. In this way the price decision is narrowed to a choice between two alternatives.

For example, in the sod example we could ask the following question. If the price of sod is raised from \$0.20 to \$0.21 per square to foot, how far can unit sales decline before profit decreases?

At a price of \$0.20 per square foot, CTO was \$25,000.

$$\text{CTO} = (\$0.20/\text{sq. ft.} - \$0.15/\text{sq. ft.}) \times 500{,}000 \text{ sq. ft.} = \$25{,}000$$

The firm's profit is \$15,000 or the CTO of \$25,000 less the fixed costs of \$10,000. Since the fixed costs have not changed, the total CTO must remain unchanged at \$25,000 to maintain the same level of profit.

In order to calculate the level of sales that will maintain a constant CTO, we must rearrange the equation:

$$\text{CTO} = \text{Per Unit CTO} \times \text{Units Sold}$$

to get:

$$\text{Units Sold} = \text{CTO}/\text{Per Unit CTO}$$

We then calculate the per unit CTO at a price of \$0.21:

$$\text{Per Unit CTO} = \$0.21/\text{sq. ft.} - \$0.15/\text{sq. ft} = \$0.06/\text{sq. ft.}$$

Thus to maintain a CTO of \$25,000, unit sales must be:

$$\text{Units Sold} = \$25{,}000/\$0.06/\text{sq.ft.} = 416{,}667 \text{ sq. ft.}$$

At a level of sales of 416,667 sq. ft. profits will remain unchanged at \$25,000. Anything less than this will lead to decreased profits.

Break-even analysis is also used in evaluating new products or ventures. For example, an entrepreneur would estimate that a new venture, such as the sod business described, would have to generate sales of 200,000 sq. ft., the break-even quantity, to be successful.

Limitations of Break-Even Analysis and Contribution to Overhead

Break-even and contribution to overhead analyses are useful tools, but it is important to keep their limitations in mind. First, fixed costs are only fixed over some relevant range of output. In the long run, all costs are variable. The firm can go out of business and sell all its assets or it can expand by purchasing more land and buildings. Thus, break-even analysis is valid only up to the firm's production capacity. Second, we assumed in both break-even and contribution to overhead analyses that the selling price is constant. In practice, price and quantity are often related, and at some point the firm must lower prices to achieve a higher level of output. Third, variable costs may not be constant either. For example, as a firm expands its production and approaches its capacity, labor costs may increase because overtime must be paid in order to generate the additional output. Fourth, any classification of costs as either fixed or variable often is an oversimplification. For example, a firm's electric power bill may be largely fixed, but may also vary with the level of output. Moreover, fixed costs and even some variable costs are sometimes difficult to estimate by product when products share common resources. For this reason caution must be used in interpreting the results of any analysis because the results are only as good as the assumptions used to obtain them.

FINANCIAL PLANNING AND BUDGETING

Financial analysis, which we covered both in Chapter 3 and at the beginning of this chapter, focuses on analyzing the performance of the firm from a historical perspective. From this material we learned that good record keeping provides the basis for good decision making and problem diagnostics. Another task of financial managers is to look ahead or plan. *Financial planning* involves goal setting, forecasting, budget preparation, and the establishment of a control system to monitor the financial performance of the agribusiness. The remainder of this chapter covers introductory aspects of financial planning, which represents an overall financial plan for the firm.

Budgeting

A *budget* is defined as a forecast of future events. Budgets perform three basic functions for a firm:

- Budgets are used to indicate the amount and timing of the firm's financial needs.
- Budgets are used to evaluate performance.
- Budgets are used to determine if the firm is meeting its goals by comparing budgeted figures with actual outcomes.

Perhaps the most important aspect of budgeting is that it reflects the allocation of resources within the organization. Since it is through the direction of resources that the business's goals are achieved, it is essential that managers use the budget process to direct resources to their most productive uses. Unfortunately, this is a missed opportunity for many firms. Budgets are often prepared with little planning. Last year's budgets are often modified with across-the-board increases or decreases. To be effective, the budgeting process requires positive, careful scrutiny by managers.

The Budget Process

The steps in the budget process are outlined in Figure 4–3. A person or group should be designated to be responsible for budget preparation. Committees are often established with representatives being selected from each of the firm's organizational areas.

Since the budgets must be compatible with the long-term plan, it is a good idea to review the organization's long-term goals and objectives. An assessment of the firm's current situation should include the firm's resource base, the firm's standing in the market, information relating to competitors and the industry in general, and the firm's strengths and weaknesses. Forecasts of likely future conditions that will affect the food or agribusiness firm should then be obtained or developed. Such forecasts are likely to include information on the general economic environment, industry conditions, competitors' actions, and prices of inputs and outputs.

Based on the information obtained in the first four steps in the budget process (see Figure 4–3), the budget committee can develop strategies and specific actions. The firm's activities are conducted according to the plans detailed in the budgets. This should include the establishment of annual goals and objectives, as well as strategies and specific actions to be taken to achieve these goals. Controls should be set up to monitor the progress toward the established goals and objectives. This may consist of periodically preparing status reports of the firm's activities, which can be compared with the budgets. Monthly or quarterly budgets are particularly useful for this purpose. Deviations from the budget may indicate the need for corrective action and possibly the need for reformulating strategy.

Budgets are primarily used in the preparation of financial plans. They provide managers benchmarks to evaluate the performance of employees responsible for carrying out those plans, and when appropriate, to control their actions. In general, a firm will utilize four types of budgets: physical budgets, operating budgets, capital budgets, and cash budgets. *Physical budgets* are constructed for physical items such as unit sales, personnel, unit production, inventories, and physical facilities. These budgets are also used to generate operating, capital, and cash budgets. The *operating budget* is comprised of the projected revenues and expenses of the firm and includes revenue projections,

FIGURE 4-3
Steps in the Budget Process

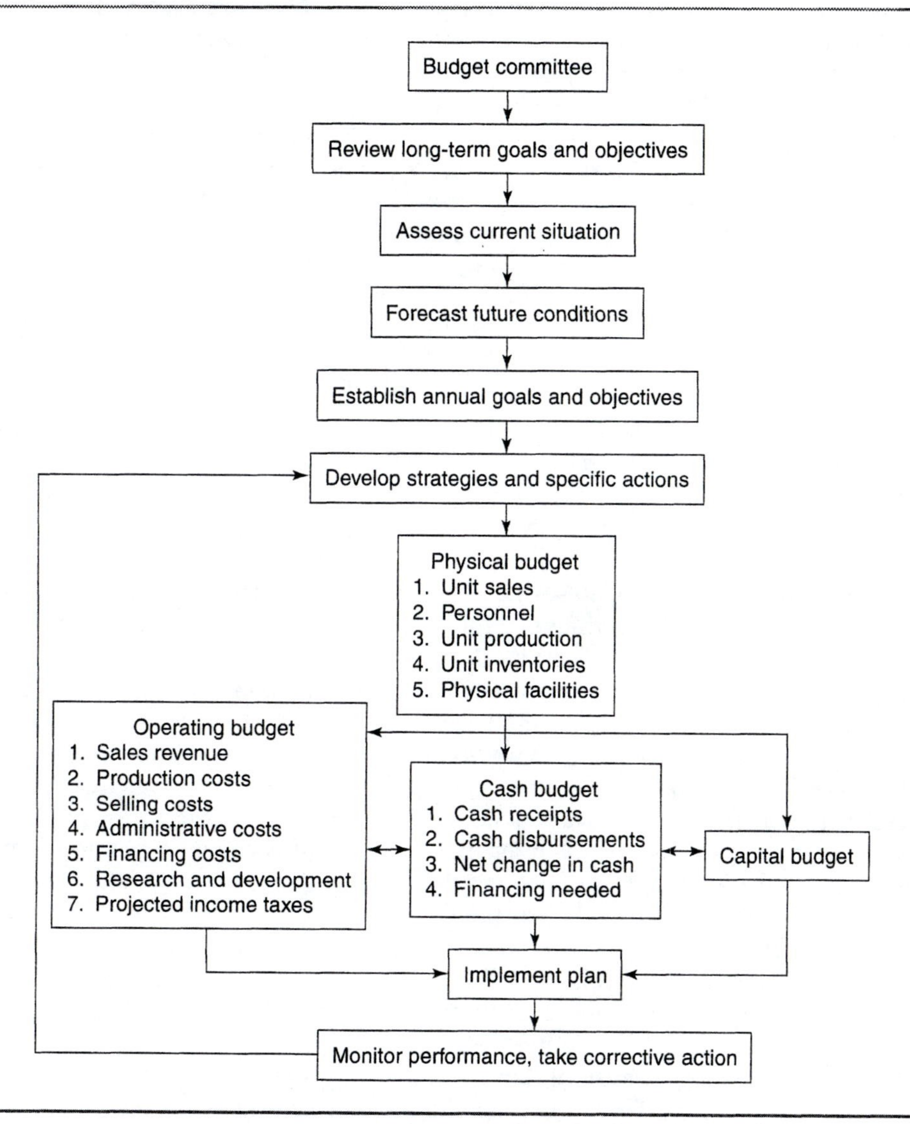

estimates of manufacturing or production costs, selling costs, administrative costs, financing costs, research and development costs, and income taxes. The operating budgets provide the basis for estimating profits or losses. Many firms develop component budgets for each of the major revenue sources and expense categories. The *capital budget* includes planned expenditures for capital items such as buildings, land, machinery, and equipment. The capital budget will be discussed in detail in Chapter 6. All the information from the physical, operating, and capital budgets is then converted to a cash basis to prepare the cash budget. The *cash budget* represents a detailed plan of future cash inflows and outflows. A *pro forma balance sheet,* which shows the expected financial position at the end of the year, is sometimes also prepared.

We have assumed that the budgets will be prepared on an annual basis. This is typical for most agribusinesses that have a planning cycle of one year. However, weekly, monthly, or quarterly budgets are often prepared in conjunction with the annual budget. This is especially important for agribusinesses whose business is highly seasonal. Lending institutions may also require monthly cash budgets.

Sales Forecast

A major challenge in financial forecasting is predicting future sales, which in turn provides the basis for predicting the level of investment in inventories, accounts receivable, plant, and equipment. Thus, a common starting place for most firms in the planning process is the *sales forecast.* A sales forecast may be developed from a number of different sources, including sales trends, expert opinions, marketing and sales staff estimates, statistical forecasts, and customer surveys. Traditional financial forecasting then takes the sales forecast as a given and makes projections of its impact on the firm's various expenses, assets, and liabilities. Two common methods used for making these projections are the percent of sales method and the component budget method.

The *percent of sales method* involves estimating the level of expense, asset, or liability for a future period as a percent of the sales forecast. The percent used can come from a recent financial statement where each item is expressed as a percent of current sales. Sometimes the percent used is an average computed over several years or an analyst's judgment. Figures for use in the percent of sales method are obtained from the common-size income statement found in Box 4–5. In this example, each item in the firm's income statement is converted to a percentage of 2000 sales. The forecast of next year's operating budget for each item would be calculated by multiplying this percentage by the net sales forecast for the 2001 planning period.

This method offers a relatively low-cost and easy-to-use first approximation for estimating the firm's financial needs for a future period. However, the danger in using this approach is that it is easy to make modifications by increasing or decreasing all figures by a fixed percentage. This defeats the planning process because past patterns of expenses and revenues are used for forecasting future needs rather than the future objectives of the firm. Consequently, the component budget method is generally regarded as a more appropriate financial tool from a strategic perspective.

The *component budget method* uses the sales forecast as a base from which to develop separate budgets for each expense item. For example, the production schedule must be developed in conjunction with the sales forecast so that sufficient product is available to meet the projected needs. The production schedule in turn determines the materials and labor budgets. Other examples of component budgets which may be developed are the selling and advertising budget, office expense budget, and the interest expense budget. Component budgets can be expressed in physical units as well as dollars. For example, the labor budget may express the labor requirements by type of labor and the wage rate with a dollar summary that can be used in other budgets. Most importantly, the component budget should reflect the strategic objectives of the firm. For example, increased operational efficiencies would be reflected in the materials and labor budgets, whereas new marketing programs would show up in the selling and advertising budget as well as the sales budget.

The Cash Budget

The *cash budget* represents a detailed plan of future cash flows and is composed of four elements: cash receipts, cash disbursements, net cash for the period, and new financing needed. It is one of the manager's most important responsibilities, because cash management ensures that the business will have enough cash on hand to meet its financial obligations. Conversely, good cash management is necessary to keep surplus cash invested. This is particularly important when interest rates are high and the opportunity cost of idle cash is great.

We demonstrate the construction and use of the cash budget using Green Valley Farm Supply as an example (Box 4–8). Green Valley's sales are highly seasonal, peaking in April through June. Roughly 10 percent of Green Valley's sales are for cash, 75 percent are collected one month after the sale, 10 percent are collected two months after the sale, 4 percent are collected three months after the sale, and the remaining 1 percent represents bad debts.

Green Valley attempts to coordinate its purchases with its forecast of future sales. Purchases generally equal 65 percent of sales and are paid for the month following delivery. For example, April sales are estimated to be $1,200,000, thus April purchases are 0.65 × $1,200,000, or $780,000. Correspondingly, payments for purchases delivered in April will be made one month later in May. Direct labor and operating expenses are estimated to be 15 percent of sales. Consequently, April direct labor and operating expenses are 0.15 × $1,200,000, or $180,000. Selling, general, and administrative expenses are fixed at $160,000 each month. Additional planned cash expenditures are quarterly estimated income tax payments in February and May, and $56,000 to purchase equipment in January. Green Valley will pay $7,000 in interest each month on its long-term debt for the period of January–June 2000. Together these projected revenues and expenditures yield Green Valley's cash budget for the six-month period ending in June 2001.

Green Valley started the period in January with a cash balance of $267,000 that was obtained from the balance sheet ending December 31, 2000, found in

BOX 4–8

Green Valley Farm Supply Monthly Cash Budgets for the period January 1, 2001, through December 31, 2001 (in thousands of dollars)

	January	February	March	April	May	June	Total (Jan.-June)
Planned sales	$ 820	$ 770	$1,160	$1,200	$1,200	$1,250	$6,400
Cash receipts:							
Cash sales (10%)	82	77	116	120	120	125	640
One month ago (75%)	750	615	578	870	900	900	4,613
Two months ago (10%)	90	100	82	77	116	120	585
Three months ago (4%)	36	36	40	33	31	46	222
Total cash receipts	958	828	816	1,100	1,167	1,191	6,060
Beginning cash balance	267	229	217	192	191	207	267
Total available cash	1,225	1,057	1,033	1,292	1,358	1,398	6,327
Cash disbursements:							
Purchases	650	533	500	754	780	780	3,996
Direct labor and operating expenses	123	116	174	180	180	188	961
Selling, general, and administrative expenses	160	160	160	160	160	160	960
Income taxes		24			24		48
Capital equipment	56						56
Interest expense	7	7	7	7	7	7	42
Total cash disbursements	996	840	841	1,101	1,151	1,135	6,064
Net cash balance	229	217	192	191	207	263	263
Financing (repayments)	0	0	0	0	0	0	0

Box 4–1. The net cash balance of $229,000 for January is calculated by subtracting total cash disbursements from total cash available for the month. Additional financing may be required if the net cash balance falls below an amount necessary to maintain a minimum cash balance established by the firm. Green Valley's management has established a minimum cash balance of $100,000, which has been maintained throughout the period. Consequently, no additional financing was necessary. If the ending cash balance falls below $100,000, Green Valley will borrow additional funds to restore the ending cash balance to $100,000. The ending cash balance of $229,000 for January becomes the beginning cash balance for February.

Cash Budgeting Frequency

The frequency with which the cash budget is prepared depends primarily on the seasonality of the food or agribusiness firm. To be useful, the cash budget must reflect the timing of the expected cash inflows and outflows so that cash surpluses or deficits can be predicted and managed. For example, food and agribusiness firms that are highly seasonal may find it necessary to prepare their cash budgets monthly, whereas a business that has fairly constant sales may find it sufficient to prepare quarterly cash budgets. Firms with severe liquidity problems are often forced to prepare frequent cash budgets, sometimes on a weekly or even a daily basis, depending on the severity of the problem.

The Operating Budget

The *operating budget*, or *pro forma income statement*, represents a statement of planned profit or loss for a future period. The operating budget comprises the firm's principal activities, including sales, production, and the other expenses associated with producing revenue for the firm. It is often broken down into its components in order to provide sufficient detail for management purposes. The operating budget is similar in appearance to the income statement. The key difference is that an income statement represents revenues and expenses that have already occurred, whereas the operating budget is a projection of revenues and expenses in the future.

An operating budget for Green Valley Farm Supply is illustrated in Box 4–9. We will use information from the cash budget found in Box 4–8 in constructing the operating budget; however, you could construct the operating budget first and use information from it to construct the cash budget. The manner in which you construct the budgets is a matter of personal preference.

Net sales, found by summing the six monthly sales projections (January through June) from Box 4–8, total $6,400,000. Cost of goods sold is computed as 80 percent of net sales (as calculated from the common-size income statement found in Box 4–5), or $5,120,000. Selling, general, and administrative expenses, found by summing the six monthly projections in the cash flow budget,

BOX 4–9

Green Valley Farm Supply Operating Budget for the Six-Month Period Ending June 30, 2001 (in thousands of dollars)

Net sales	$6,400
Cost of goods sold	5,120
Gross margin	1,280
Operating expenses:	
Selling, general, and administrative expenses	960
Depreciation expense	135
Total operating expenses	1,095
Operating income	185
Interest expense	42
Income before taxes	143
Income taxes	48
Net income	$ 95

amount to $960,000. Depreciation expense cannot be obtained from the cash budget, because it does not constitute a cash flow. Thus, this expense must be determined from the depreciation schedules of Green Valley's plant and equipment. We estimate those expenses to be $135,000 for the six-month period.

Subtracting the above operating expenses from gross margin (gross profit) leaves earnings before interest and taxes (net operating income) of $185,000. The interest expense of $42,000 is then deducted from net operating income to obtain earnings before taxes of $143,000. Income taxes are estimated to be $48,000 for the period. Finally, subtracting the estimated taxes from earnings before taxes results in net income for the period of $95,000.

In this example, the construction of the operating budget was illustrated using primarily historical data and relationships. Although this is acceptable as a first approximation, it should be carefully analyzed and adjusted to reflect the goals and objectives of the firm.

FINANCIAL CONTROLS

The processes of planning and control cannot be separated. It is difficult, if not impossible, to control the financial management of the firm if no planning is done, because there is no basis on which to judge the firm's performance. Likewise, planning without control is a wasted opportunity because we never go back and see whether our plans were successful and, more importantly, why they were or were not successful.

Financial control involves the monitoring of the various financial elements of the firm. Controls may be used in three ways, the first being to measure success. At the end of the year, we may measure our success toward reaching the profit goal of $100,000. If we had a profit of $80,000, we could say that we achieved 80 percent of our goal.

Controls may also be used to monitor the progress towards a goal. If our sales goal was $500,000, at the end of six months we may monitor our progress by noting that our sales for the first six months were $300,000 and that this was more than satisfactory progress towards the goal. Budgets are often used for this purpose. They allow managers to make the mid-course corrections necessary to attain the goal.

The third way in which controls are used is as warning signals, much like the oil warning light in a car. It only comes on when the engine is in danger because the oil pressure is low. These "red flags" are used to warn managers when something is wrong, such as when expenses are too high, or when the firm's debt level is unacceptable. This type of control requires the development of norms or standards. Under normal conditions, we may not be aware of the control; it is only when something out of the ordinary occurs that it is brought to our attention.

SUMMARY

The 3C method of analysis, utilizing comparison, change, and common-size statements, provides an overview of the financial position of the firm. Two financial statements are placed side-by-side, the change from one period to another is calculated, and the statements are converted to a common size. The 3C analysis is useful in analyzing changes from one period to another and in comparing performance from one year to another. Common-size statements are particularly useful for comparing the financial statements of two firms that differ greatly in size.

Ratio analysis is used to analyze and evaluate performance by converting dollar figures into ratios, thereby facilitating comparisons between periods or firms. Financial ratios are used to measure the firm's profitability, its use of resources, and its ability to meet its short-term and long-term financial obligations.

Successful financial management entails goal setting, budget preparation, and a system of controls to monitor the progress toward goals and evaluate performance. Budgets play an important role in the planning process and establishment of goals. Most food and agribusiness firms develop physical, cash, operating, and capital budgets. Budgets also provide the basis for financial control. They are used to monitor the firm's performance during an operating period, indicate whether corrective action is needed, and serve as a basis for evaluating financial performance.

CASE QUESTIONS

The following questions are based on the information presented at the beginning of the chapter and the income and balance sheets for the Desert Rose Nursery for 1999 and 2000 presented in Boxes 4–10 and 4–11.

1. What tools would you choose to help Jeannie analyze the financial situation of the Desert Rose Nursery and why?
2. What are the major financial problems of Desert Rose Nursery? What evidence supports your conclusions?
3. What steps would you recommend that Jeannie take to improve the financial management of the firm?

BOX 4–10

Desert Rose Nursery Income Statements for the Year Ending December 31

	2000	1999
Net sales	$1,100,000	$1,200,000
Cost of goods sold	370,000	400,000
Gross margin	730,000	800,000
Operating expenses:		
Selling	160,000	130,000
General and administrative	220,000	200,000
Depreciation	50,000	50,000
Total operating expenses	430,000	380,000
Operating income	300,000	420,000
Interest expense	160,000	104,000
Net income before taxes	140,000	316,000
Federal and state income taxes	50,000	116,000
Net income	$ 90,000	$ 200,000

BOX 4-11

Desert Rose Nursery Balance Sheets

	December 31 2000	December 31 1999
Assets		
Current assets:		
Cash	$ 50,000	$ 100,000
Accounts receivable	50,000	50,000
Inventory	1,070,000	600,000
Total current assets	1,170,000	750,000
Property, plant, and equipment:		
Equipment	150,000	150,000
Less depreciation	(130,000)	(110,000)
Building	970,000	970,000
Less depreciation	(460,000)	(430,000)
Land	40,000	40,000
Total property, plant, and equipment	570,000	620,000
Total assets	$1,740,000	$1,370,000
Liabilities and Owners' Equity		
Current liabilities:		
Accounts payable	$ 450,000	$ 50,000
Total current liabilities	450,000	50,000
Long-term liabilities:		
Building mortgage	220,000	320,000
Total long-term liabilities	220,000	320,000
Total liabilities	670,000	370,000
Owners' equity:		
Contributed capital	500,000	500,000
Retained earnings	570,000	500,000
Total owners' equity	1,070,000	1,000,000
Total liabilities and owners' equity	$1,740,000	$1,370,000

REVIEW QUESTIONS

1. What are the three components of the 3C method and how are they used in conducting financial analysis?
2. List and describe the four types of financial ratios.
3. What are the principal differences between return on equity and basic earning power? How would you determine which ratio would be most appropriate?

4. Explain the difference between a liquidity ratio and solvency ratio.
5. Compare the kinds of information obtained by a 3C analysis relative to ratio analysis.
6. Explain how financial ratios can be used as controls.
7. Explain why the unit price less the variable cost is called per unit contribution to overhead in break-even analysis.
8. Describe the components of financial planning and the importance of each component.
9. What are the key budgets that every firm must prepare and their basic characteristics?
10. Why should the projected ending cash balance in the cash budget be greater that $0 for each month?
11. Prepare a new cash budget for Green Valley so that a minimum cash balance at the end of each month is $250,000.

CHAPTER 5

Financing the Business

Learning Objectives

In this chapter we will cover the following topics:

- The importance of using appropriate means to finance different types of assets
- Differences between equity financing and debt financing
- Opportunities and risks associated with various methods of financing businesses
- Determining whether to lease or buy an asset
- Differences among financing ongoing operations, financing the expansion of existing firms, and financing new ventures
- Financing cooperatives

CASE STUDY

Alberto Alvarez is from a family whose ancestors have lived in southwestern New Mexico for nearly three centuries. The family is involved in production agriculture, growing alfalfa, cotton, lettuce, onions, and pecans. Some of Alberto's fondest memories are of listening to his great-grandfather talk about when the family grew grapes and operated a small winery in their community around the turn of the century, before Prohibition. His great-grandfather said there were many small vineyards and wineries along the Rio Grande in the 1800s.

Alberto is 28 years old. Since graduating from New Mexico State University with majors in agricultural business management and horticulture, he has worked for his father in the family farming operation and farm supply firm. He has gained a lot of experience since graduation and enjoys the farm supply business, but is restless. He wants to start his own business. His interest in grape growing and wine making has increased, much of it as a result of the 10-week internship he completed with a small California winery in the Napa Valley during the summer following his junior year in college.

Recently, there has been a substantial renewed interest in the wine industry in New Mexico. Several years ago, a group of Swiss investors started building a large vineyard and winery in a community about 60 miles to the west of Alberto's home. At about the same time, another group, composed mostly of French investors, started a large vineyard and winery 80 miles to the north. While working in California, Alberto befriended a French student who was also working as an intern at the winery. This past fall he visited his friend in France and together they toured several of the famous grape-growing regions of France and Italy. After returning from Europe, he was convinced that he wanted to start his own grape-growing and winery business in his local community. A feasibility study indicated the venture could require as much as $7.5 million in investment and operating capital. Although the family had sizable assets, they were tied up in the farming and farm supply businesses.

Alberto visited with the president of the bank that handles the family's farming and farm supply firm accounts. The banker said his bank would consider loaning some funds for the wine business, but first Alberto would have to demonstrate that the project was financially feasible and that the business would have a substantial portion of the required assets financed through equity capital.

How should Alberto proceed with his project? After carefully studying the information in this chapter, you should be able to understand how existing and new businesses are financed and advise Alberto how to proceed.

INTRODUCTION

The term *financing* is used to describe many different activities. In this chapter, financing is used to describe how food and agribusiness firms provide the funds needed to acquire the assets necessary to operate the firm. Each company, including Alberto's proposed vineyard and winery, must address the financing questions specific to its own business. Understanding the financial alternatives and their risks helps owners determine the financial structure best suited to their firm. This information is important to creditors and investors because it helps them evaluate whether to lend or invest funds in a business.

The form of business organization an owner chooses affects the amount and kind of capital available to the firm, the degree of risk the owner assumes, and the control the owner will exercise over the business. For example, if Alberto chose to start his business as a sole proprietorship, his personal assets would be the primary capital available to the business. Alberto would either have to contribute his own cash or assets or use his assets as collateral to secure a loan. He would have total control, but he would be putting his personal assets at risk. On the other hand, Alberto might choose to form a limited liability company. This might give him access to capital from additional investors and limit his liability to the assets he invests in the business.

Regardless of whether Alberto chooses to finance his vineyard and winery with equity (ownership) capital through a sole proprietorship, partnership, S corporation, or limited liability company, he must have some idea of the financial feasibility of the business. To do this, Alberto must have historical balance sheets, income statements, and statements of cash flow for the farming and farm supply businesses. In addition, he will have to complete a business plan and pro forma balance sheets, income statements, and statements of cash flow for the proposed vineyard and winery. The historical statements indicate to potential investors the management ability and integrity of the managers. The pro forma statements indicate the firm's potential to generate a profit and, more importantly, whether the business will be able to pay back money to the investors out of the firm's earnings. To give the project credibility, it is also important that a certified public accountant develop the financial statements.

The rate of return on a venture that investors require before contributing capital will depend primarily on the risk involved. All businesses and investments involve some degree of risk. Investors look at risk as the likelihood that a business or investment will fail, not make as much money as anticipated, or lose money, including the capital that they have invested. To attract investors, businesses with higher degrees of risk must offer higher potential returns to compensate investors for the risk involved. In addition, if the firm uses debt to finance its operations, the risk to the owners increases. Owners must satisfy the terms of any debt agreements before receiving any returns from the business. The greater the debt the firm incurs, the greater the risk to the owners.

EQUITY VERSUS DEBT CAPITAL

Agribusinesses acquire funds from three possible sources: (1) owners' contributions, (2) earnings generated by operating the business and retained by the firm, and (3) debt. These three sources can be classified as either equity financing or debt financing. With equity financing, funds are obtained in exchange for ownership in a company. Equity funds are raised when owners place funds into the firm or when the owners reinvest the firm's earnings in the business. With debt financing, funds are obtained in exchange for a liability to repay the borrowed funds.

Equity capital is the term used to describe the funds put into or retained in the firm by the owners. These funds can include cash, land, equipment, or other assets contributed by the owners of the business. Equity capital may also be accumulated by retaining earnings from previous operating periods. Sources of equity capital are listed in the owners' equity section of the balance sheet.

The owners, anticipating that profits will be forthcoming, put up equity capital. Hence, payment for the use of equity capital used in the firm is not mandatory. If the business is profitable, the owners will be rewarded. If the business is not profitable, the owners will not receive benefits. However, in the long run, the owners expect the return on equity capital to be higher than the return on debt capital to compensate them for putting their capital at risk.

Debt capital is accompanied by the obligation that it be repaid, is of a fixed term, and is repaid with interest in return for the use of the money. *Debtors* (those to whom the debt is owed) typically have first claim on the assets of the firm. Debt capital is fundamentally different from equity capital because equity capital earns a return only if the business is profitable. The fixed term of the debt also contrasts with equity capital, for which there is no repayment schedule. The use of debt capital requires that cash be available to make the scheduled principal and interest payments, regardless of the profitability of the firm.

The use of debt capital has two additional advantages. The first advantage is that interest paid on the use of debt capital is tax deductible. This contrasts with the dividend payments of corporations that are not tax deductible. Secondly, raising capital through the use of debt capital does not dilute the owners' ownership interest in the firm. Despite these advantages, debt financing can be risky because the failure to repay debt can result in insolvency and bankruptcy, and ultimately the loss of the firm's assets.

The combination of debt capital and equity capital is referred to as a firm's *capitalization structure*. By examining the liabilities and the owners' or stockholders' equity on a firm's balance sheet, the capitalization structure of the firm may be ascertained. An examination of the pro forma balance sheet (Box 5–1) for Alvarez Vineyards indicates that stockholders will provide 54 percent of the firm's capital (calculated as total stockholders' equity divided by total liabilities and stockholders' equity), and debt holders will provide 46 percent of the firm's capital. Of the debt capital, 89 percent is expected to be long-term debt and 11 percent will be current, or short-term, debt.

BOX 5–1

Alvarez Vineyards and Winery, Inc., Pro Forma Balance Sheet for the Year Ending December 31, 2001

		Percent of Assets/Liabilities and Owners' Equity
Assets		
Current assets:		
Cash	$ 12,000	
Accounts receivable	30,000	
Inventory	180,000	
Supplies	9,000	
Total current assets	231,000	11%
Property, plant, and equipment:		
Vineyard equipment	13,500	
Winery equipment	646,500	
Buildings	750,000	
Land and improvements	300,000	
Vineyard	252,000	
Total property, plant, and equipment	1,962,000	89%
Total assets	$2,193,000	100%
Liabilities and Owners' Equity		
Current liabilities:		
Accounts payable	$ 12,000	
Notes payable	69,000	
Accrued interest	27,000	
Taxes payable	3,000	
Total current liabilities	111,000	5%
Long-term liabilities:		
8% bonds due 2010	180,000	
Mortgages	570,000	
Equipment loan	47,000	
Equipment lease obligation	100,000	
Total long-term liabilities	897,000	41%
Total liabilities	1,008,000	46%
Owners' equity:		
Stockholders' equity:		
Common stock, 200,000 shares authorized; 75,000 shares issued	750,000	
Capital in excess of par value	375,000	
Retained earnings	60,000	
Total owners' equity	1,185,000	54%
Total liabilities and owners' equity	$2,193,000	100%

EQUITY CAPITAL

The kind of equity capital raised depends principally on the form of business ownership. Sources of equity capital for financing expansion or refinancing an existing successful agribusiness are limited to a great extent by the size of the business and form of business ownership. Most food and agribusiness firms are relatively small and are not incorporated. They are typically formed as sole proprietorships, partnerships, or limited liability companies. In sole proprietorships and partnerships the equity capital invested by the owners is called contributed capital, whereas it is called membership interest in limited liability companies. The ownership shares are held by a relatively small number of people and are not publicly traded. Similarly, many small- and medium-sized firms are organized as corporations and have stock that is closely held by a few individuals and is not traded on public exchanges. On the other hand, large corporations, which require a substantial amount of equity capital from a large number of owners, issue stock that can be bought and sold on public exchanges without the consent of the other owners.

Stock

Stock that is owned by shareholders—whether individuals, groups, or businesses—constitutes ownership in a corporation. In its articles of incorporation, the corporation requests permission from the state to sell stock, called *authorized stock*. At the time the stock is issued, the corporation assigns an arbitrary value (called *par value* or *stated value*) to each share. The par value is typically low relative to the price of the stock because most states require corporations to retain equity equal to the amount of par value times the number of shares issued to protect the corporation's creditors. The corporation sells a portion of the authorized shares (called *issued shares*) to individuals, groups, trusts, or businesses (called *investors*) through an *initial public offering*. An investment banking firm handles the transaction by putting together a group of other investment banking firms (called a *syndication*), which sells the new offering. A successful large corporation can raise hundreds of millions of dollars of equity for expansion or for retiring debt capital, if the company so chooses.

An example of accounting for the sale of stock is illustrated in the hypothetical balance sheet (Box 5–1) for Alvarez Vineyard. As shown under the Stockholders' Equity heading, the corporation issued 75,000 shares of common stock at $10 par value worth $750,000. However, each share sold for $15. The balance sheet shows the amount of capital in excess of par value from stock issuance as $375,000. Proceeds from the sale of stock are used to fund property, plant, and equipment and reduce the need for long-term debt.

After the stock is issued, shareholders are free to buy and sell the shares on secondary markets, such as the New York Stock Exchange, American Stock Exchange, and NASDAQ. The price of the shares of the corporation's stock on the exchange will fluctuate depending on the stock's valuation by buyers and sellers in the secondary market. Changing market values reflect investors' perceptions of the corporation's financial condition, expected growth, and its future profit potential. The exchanges exist primarily for the benefit of shareholders, although they do facilitate the issuance of new stock.

The number of shares issued by the corporation is not affected by this market activity because investors are trading existing shares of stock. It should also be noted that the balance sheet of the corporation is not affected. The number of shares issued and the value of those shares as reflected on the balance sheet changes only when the corporation sells more of its authorized shares or buys back some of its issued shares. The corporation is allowed to buy back its own stock because it is a legal entity. When a corporation buys back its stock on the secondary market, it reduces the number of outstanding shares (shares issued and held outside the corporation) and increases the amount of *treasury stock* (repurchased stock) it holds. When the corporation buys treasury stock, it gives shareholders cash for shares of stock. Thus, this transaction will show up on the balance sheet as a reduction in owners' equity and as an equivalent reduction in cash, not as an asset or owners' equity. Treasury stock is retired and does not yield voting or dividend rights.

Stock may be issued with a variety of voting rights, dividend privileges, and claims to assets upon liquidation. The two broad classes of stock are *common stock* and *preferred stock*. Generally, common stock carries with it voting rights, whereas preferred stock does not. However, common stock can be issued without voting rights (Class A shares) or with voting rights (Class B shares). *Voting rights* allow shareholders to participate in management by participating in the election of members of the board of directors, voting to permit the issuance and the purchase of company stock, and voting on major changes in corporate direction. The principal characteristics of common and preferred stock are highlighted in Box 5–2.

Both common and preferred shareholders may share in corporate profits through *dividend* payments. Dividends are almost always paid in cash, although occasionally they may be paid as additional shares of stock. Preferred stock has a predetermined dividend rate that is *cumulative,* which means that the preferred shareholder's dividends that are not paid accumulate over time. Missed dividends and the current dividends must be paid before dividends are paid to common shareholders. Preferred stock can also be issued with callable, redeemable, and convertible options. A *callable* option permits the corporation to repurchase the stock, a *redeemable* option allows the shareholder to sell the stock back to the corporation, and a *convertible* option allows the shareholder to exchange preferred stock for common stock at a stipulated price.

The corporation has the option, but not the obligation, of paying dividends to its common shareholders. Profitable corporations typically pay dividends to common stockholders, although the rate varies greatly. Rapidly growing corporations often retain most if not all of the earnings to finance future growth. This benefits common stockholders through the appreciation in the value of

BOX 5–2

Characteristics of Common and Preferred Stock

Characteristic	Common Stock	Preferred Stock
Dividend payment	Dividends are variable	Fixed dividends paid quarterly
Dividend priority	Dividends paid only after interest on debt and preferred stock dividends	Dividends are cumulative and paid after interest on debt but before common stock dividends
Claim on assets	Paid after debt-holders and preferred stockholders	Paid after debt-holders but before common stockholders
Voting rights	Vote for board of directors	No voting rights
Special features	Class A nonvoting and class B voting	Callable, redeemable, and convertible options

their stock. On the other hand, corporations operating in mature industries typically distribute much of their earnings through dividend payments.

An additional difference between preferred and common stock relates to their claim on the corporation's assets. In the event that the corporation should be forced to liquidate its assets, preferred shareholders have claim to the assets prior to common shareholders, but only after all of the firm's debts have been paid.

DEBT CAPITAL

Debt capital is classified based on when the debt must be repaid. Short-term debt consists of money that is owed and that will be payable during the current accounting period. As discussed in Chapter 3, short-term liabilities, or current liabilities, consist of short-term loans, accounts payable, and accrued expenses. Long-term debt consists of loans, including notes payable and mortgages, and bonds. Long-term debt is classified as such because it does not become payable during the current accounting period.

As a general rule, the length of financing should approximate the expected production life of the assets being financed. This means that current obligations should be financed with short-term debt capital, whereas noncurrent assets should be financed with long-term debt capital or equity capital. Understanding the rationale behind this principle will serve to illustrate its importance. It is expected that all investments will pay for themselves through the revenues they generate (or through a reduction in costs), and ultimately add to the firm's profits. Because a good investment "pays for itself," it is important that the terms of repayment match the productive life of the asset.

As an example, consider an investment in a new wine press for Alberto Alvarez's winery, Alavarez Winery. Assume that the estimated useful life of the press is 10 years. Financing the press over a two-year period would probably mean that Alberto would have to seek another source of financing to repay the loan because the wine produced from the press may not even be sold until the end of the second year or later. On the other hand, financing the investment over a much longer period, such as 20 years, would create another set of problems because the loan payments would continue long after the asset had been replaced. Furthermore, most creditors would not loan money for longer than the estimated useful life of the asset being financed because the value of the collateral (the remaining value of the asset) would soon be worth less than the outstanding principal due. Determining the repayment period is usually simple for long-term (noncurrent) assets. When debt financing is used, the period for repayment of the loan is usually somewhat less than the estimated life of the asset being financed. A seven- to eight-year loan might be appropriate for a wine press with a useful life of 10 years.

Long-Term Debt Capital

When long-term debt is issued, both the borrower and the lender formalize the terms of the debt. Lenders may place certain restrictions on the borrowing company in order to provide the lender with some assurance that the company will repay the debt. For example, a *note* is a written promise made by a firm that borrows funds. It describes how the borrower will repay the lender for the use of the funds. We will use the example of a note to illustrate the terms used in long-term debt instruments and the implications of these terms for borrowers and lenders.

The *face value* of the note indicates the amount that the borrower will ultimately pay the lender when the note is due. The amount of cash raised from the issuance of the note is called the *proceeds*. Each note carries a rate of interest (called the *coupon* or *face rate*) that will be paid on the debt. The actual interest rate charged on the debt is called the *market rate*, or effective rate. The market rate is negotiated by the borrower and lender, and may be different from the face rate. The length of the loan is its *maturity*.

Companies issue three different types of long-term debt, and each differs in how the lender is paid. A *periodic payment note* (or installment note) is a promise to pay the lender a series of equal payments consisting of interest and principal at equal time periods over the life of the note. Assume that Alvarez Vineyards proposes to borrow $40,000 at 10 percent interest and agrees to pay the lender back over the next 36 months. The amount of the monthly payment would be $1290.69. (We will explain the details of calculating principal and interest payments on loans in the next chapter.) Each payment would consist of interest for the last installment period and a portion of the principal. A monthly car payment is another example of an installment loan.

A *lump-sum payment note*, or noninterest-bearing note, represents a promise to pay a specified amount of money at the end of a specific period of time. The note specifies the amount the borrower promises to pay back at the end of the life of the note. The amount of money the lender will give for the note depends on the market interest rate that the borrower and lender negotiate. For example, in the case of Alvarez Vineyards, the lender would loan the business $29,669.59 today in return for a promise to be paid $40,000 in three years. In this case the lender would earn a 10 percent return on the money. The difference between the proceeds of the note ($29,669.59) and the face value of the note ($40,000) depends on the market interest rate and the amount of time before the note is paid back. A higher market interest rate or a longer repayment period would result in lower proceeds to the borrower.

A *balloon payment note* combines features of the periodic payment note and lump-sum payment note. Periodic payments are made over the life of the note with a lump-sum payment at the due date. The periodic payments are calculated based on the face value and coupon of the note. The proceeds of the note depend on the market interest rate. In the Alvarez Vineyards example, if the market interest rate were negotiated to be the same as the coupon, the proceeds of the note would be the face value of $40,000. On the other hand, if the negotiated market interest rate were higher than the coupon, the proceeds of the note would be less than $40,000. In either case, the lender would receive $40,000 from the borrower at maturity.

Loans. A *loan* occurs when a firm enters into an agreement with a person or an institution to borrow funds. For most businesses, loans are the most common form of debt financing. Firms use the types of notes discussed in the previous section to borrow funds. The type of note used depends on the firm's cash flow and the agreement reached with the lender. When a firm borrows from a lender, the terms of the note include the value of the note, the interest rate, the length of the borrowing period, and covenants, or restrictions on the borrowing company, that protect the lender.

A typical mechanism that lenders use to protect their claim is to require the borrower to use some assets as collateral for the note. (This is also referred to as a *lien*.) *Collateral* is an asset that is specifically named in the debt agreement. Should the borrower *default* on (not repay) the loan, the lender has the option of claiming the secured asset. While the debt is outstanding, the borrower cannot sell the asset without the lender's permission. The lender will often retain title to the secured asset to ensure that it is not sold. For example, to get a loan on a home the buyer has to put up the home as collateral. Collateral real estate loans, such as for land and buildings, are referred to as *mortgages*. The distinction between a loan and a mortgage is a minor one; often the terms are interchangeable. Whereas a mortgage is always secured by some property, a loan need not be.

Historically, most long-term loans and mortgages were installment loans with fixed payments and interest rates. Because of this practice, the profitability of many lenders was squeezed when their cost of acquiring money to lend increased dramatically, while most of the outstanding long-term loans they had issued were at low interest rates. Many financial institutions experienced this situation during the financial crisis of the 1980s. Now, graduated payments and variable interest rates, that is, interest rates that fluctuate with market interest rates, have become common. *Variable interest rates* are typically tied to a base interest rate, such as the interest rate on federal government treasury bills. The interest rate on variable rate loans is periodically adjusted to reflect current interest-rate levels. The rate may be adjusted annually, on the anniversary of the loan, or more frequently.

A *participation loan* involves financing whereby the lender takes a financial position in a project in addition to providing debt capital. A lender may be willing to make a substantial loan but require a percentage share of the ownership of the business in addition to the interest rate charged. This is often used as a partial hedge against inflated values of the invested assets. Recently, many such arrangements have been used to finance the purchase of agricultural lands. Whereas the lender expects to profit as the value of the asset increases, it does not generally have a controlling interest and therefore it has no managerial responsibility. Under certain situations, participation loans may offer financing opportunities that would not otherwise be available.

The terms of loan agreements follow a strict set of lending practices put into place after the banking crisis of the 1980s. Lenders generally require that the borrower pay a sufficient down payment on the asset so that the salvage (resale) value of the asset will be sufficient to repay the loan if the loan cannot be repaid as originally planned. Lenders are generally willing to loan from 50 percent to 80 percent of the total appraised value of long-term assets. This range and the interest rate charged depend on the particular circumstances involved, primarily as it relates to the expected risks. Long-term lending institutions generally expect the loan to be paid off over time from the profits or cost savings of the firm. The *term* or length of the loan (payback schedule) is generally related to the useful life of the asset being acquired. For loans exceeding five years, lenders place a heavy emphasis on the current and expected future market value of the asset. The long time period involved increases the likelihood that business conditions may change and makes accurate future income projections difficult. Lenders generally give consideration to the possibility of having to foreclose or force the sale of the asset in order to repay the loan. Therefore, *progress payments* are generally scheduled so that the expected force-sale market value of the asset is sufficient to repay the loan at any time throughout the term of the loan, if necessary. Again, it is important to stress that the lender intends for the loan to be repaid out of profits or savings resulting from the act of acquiring the asset.

Bonds. *Bonds* are long-term debt instruments issued by corporations. Bond obligations appear on the balance sheet as a long-term liability (see Box 5–1). Corporations of all sizes can issue bonds, but they are generally issued by firms needing large sums of money. Bonds usually take the form of periodic-payment, lump-sum notes. Bonds are issued in units of $1,000, with fixed interest payments made every six months. A bond with a maturity of less than two years is known as a *short-term note*; *medium-term bonds* mature between two and 10 years; and *long-term bonds* have maturity rates longer than 10 years.

Bondholders do not have an ownership interest in the firm. If the firm gets in financial trouble and needs to dissolve, bondholders must be paid off in full before stockholders receive anything. If the corporation defaults on any bond issue, any bondholder can go into bankruptcy court and request that the corporation be placed in bankruptcy.

The bond contract is called a *bond indenture*, which specifies the amount of the bond issue, the maturity of the bond, the face value of the bond, and the face (coupon) rate. The bond indenture may also include covenants on the issuing corporation as well as repayment and security provisions. *Covenants* usually restrict the amount of debt the corporation may carry. Repayment provisions may include a *callable* feature that allow the corporation to redeem the bond prior to the maturity date and a *convertible* feature that allows bondholders to exchange their bonds for common or preferred stock. With secured bonds, the corporation provides some form of collateral to serve as security for the loan. If the assets are real estate (buildings or land), they are called *mortgage bonds*. Unsecured bonds (usually called *debenture bonds*) are backed by the general assets of the firm. *Subordinated bonds* are so named because holders of these bonds have rights to repayment ranked behind holders of secured bonds, unsecured bonds, and leases (discussed in the section on Leasing), but ahead of owners' claims.

The sale of a bond is usually handled by an investment banking firm, which will either underwrite (buy) the entire bond issue from the corporation and then resell the bonds to the public or sell the bonds for the company for a commission. In some cases, the company issuing the bonds may sell them directly to the public as a private placement. A *private placement* involves selling securities to a few individuals or institutions as opposed to making a broad-based public offering. The selling of bonds takes place in the *initial issue market*. Individual bonds are issued as certificates that have the face value, coupon rate, maturity, and an identification number printed on them. When the bonds are sold, the identification number and the bondholders' names and addresses are recorded in a *bond register*. After the bonds are initially sold, they may be resold prior to maturity in the secondary bond market. The *secondary market* is composed of a network of bond brokers and, for large bond issues, the New York Bond Exchange. If the bond changes ownership, it is endorsed by the seller and sent back to the company, where the ownership is recorded and the bond is reissued.

Bonds are redeemed at their face value at maturity. Prior to maturity, the price of the bond depends on the amount of time remaining to maturity and the financial strength of the corporation. As interest rates rise, the price of the bond declines because that particular bond is less attractive compared to current offerings. Likewise, as interest rates decline, the price of the bond increases because that particular bond becomes more attractive when compared to current offerings. Long bonds (those that have a longer time before maturity) generally command higher market interest rates, as do bonds issued by corporations with greater perceived risk.

Short-Term Debt Capital

The principal sources of short-term capital are loans from banks or other lending institutions, credit cards, and credit provided by the firm's suppliers. Frequently, businesses establish a line of credit with a bank, lending institution, or supplier. A *line of credit* allows a business to borrow money or receive supplies up to a specified level. When a supplier extends a line of credit, this is commonly referred to as *commercial* or *trade credit*. Many suppliers offer some type of financing. Some suppliers expect to be repaid quickly, within 30 days, for example. Other suppliers are willing to finance the purchasers of their goods for a longer period.

The need for short-term credit is largely determined by a company's short-term working capital needs. Working capital is defined as the excess of current assets over current liabilities. It is important because it is needed to operate the business. Working capital allows the firm to pay for the inventories, supplies, accounts receivable, and operating expenses, such as payroll and utilities, needed in the day-to-day operations. Determining the need for working capital is more difficult than determining the need for long-term financing. The need for working capital will fluctuate from one business to the next depending on inventory turnover rates, the volume of sales, and business activity. In many agribusinesses, the need for working capital will be seasonal. For example, at Alvarez Winery it will take approximately nine months from the time the white wine grapes are crushed until the wine is bottled, racked, and ready for sale. During this time, Alberto will need a source of short-term capital to finance the firm's capital tied up in the year's production of wine. Because of the variable need for working capital, it is helpful to classify working capital as permanent or temporary.

Permanent working capital is the minimum level of working capital or the smallest amount of current assets over current liabilities that the firm expects to maintain. Permanent working capital needs should be met through long-term financing, because providing for permanent working capital through short-term financing would mean constant refinancing. Equity capital or unsecured debt capital is often the source of financing because no specific asset is required for collateral.

Temporary working capital is defined as working capital that is not needed on an ongoing basis, such as to meet the seasonal needs of a food or agribusiness firm. Temporary working capital needs are typically financed through short-term loans and trade credit from suppliers. Short-term financing for temporary

BOX 5–3

Characteristics of Equity and Debt Capital

Characteristic	Equity Capital	Debt Capital
Payments	Dividends paid only when a profit is made	Interest payments required on a periodic basis
Tax implications	Dividends are paid out of after-tax profits	Interest is a tax deductible expense
Control	Sale of new common stock may dilute control of existing common stockholders	No affect on control of management but lenders may add covenants and can force bankruptcy if an interest payment is missed
Length of financing	Long-term	Short- and long-term
Special features	Returns depend on the firm's profitability	Returns are fixed but bonds can be callable and convertible

needs ensures that the firm uses the funds only when it needs them and that the funds do not remain idle for long periods of time. A summary of equity and debt financing is presented is Box 5–3.

Leasing

In recent years leasing has become an increasingly common alternative to purchasing an asset. In the past, leasing was used primarily for real estate—land and buildings. Today, almost any kind of fixed asset can be leased, including land, buildings, vehicles, office equipment, farm equipment, and processing equipment.

The decision to purchase or lease an asset is complex because it involves many factors. In this section we describe the different types of leases and discuss some of the most important considerations in determining whether an asset should be purchased or leased.

There are three basic types of leases, commonly referred to as operating leases, capital leases, and sale-leaseback arrangements. These terms refer to the underlying arrangements and responsibilities of the parties to the lease. All leases have a *lessor*, or owner of the asset, and a *lessee*, or user of the asset.

Operating Lease. A principal characteristic of an *operating lease* is that it is of relatively short duration. Equipment that is needed only temporarily is often leased under such an arrangement and is commonly referred to as a rental. The lease payment (or rent) is typically a fixed rate based on the duration of the lease (per hour, per day, per week, or per month, for example). The lessor is responsible for maintaining the asset. Farm machinery, construction equipment, processing equipment, and office equipment are often leased in this manner. For example, a farmer might lease a backhoe from a construction equipment firm for a period of a week, paying the owner of the asset the weekly rate for the use of the equipment.

Capital Lease. A *capital lease,* also referred to as a financial lease, is distinguished from an operating lease by its duration. A capital lease involves a long-term contract giving the lessee exclusive use of the asset for the term of the lease. The lease payment is designed to cover the full cost of the asset and a return to the lessor. This arrangement is similar to a credit-financed purchase except that the lessor maintains title to the asset. The lessee is responsible for paying the cost of maintenance, property taxes, and insurance on the asset. Capital leases usually appear on the balance sheet as a long-term liability because they involve a fixed commitment for future cash payments (see Box 5–1).

A typical capital lease would be initiated by the user who would select the equipment needed, choose a buyer, and negotiate the terms of the purchase. The user would then arrange for a leasing company to purchase the asset at the same time it enters into a lease with the leasing company.

Sale-Leaseback Arrangement. The sale-leaseback arrangement is very similar to the capital lease. The major difference is that the lessor buys the asset from the lessee. As with the capital lease the lease payments are set up so that the full purchase price and a return are paid to the lessor. Likewise, maintenance costs, insurance, and property taxes are the responsibility of the lessee.

Tax Considerations. Generally speaking, the full amount of the lease payments is a tax-deductible business expense to the lessee. The lessor is able to deduct those benefits associated with the ownership of property, including depreciation and the investment tax credit (not currently available based on current tax laws). It is critical that the lease be a written contract that complies with the requirements for a lease as established by the Internal Revenue Service (IRS). The rationale underlying the IRS guidelines is simple—the government wants to prevent firms from using leases as a way to write off the cost of capital items much more quickly than they could if they owned the assets and depreciated them.

The Lease-Buy Decision. The decision to purchase or lease an asset is typically determined based on which alternative is least costly. Equipment that is used for only short periods of time is usually leased with an operating lease. The decision is more complicated for assets that are needed on a permanent basis. In this case, the tax considerations are usually of prime importance. The benefits of leasing are most apparent when there is a large difference in the marginal tax rate of the lessor and lessee with leases being of the greatest benefit to those firms in low tax brackets. A typical case is where a lessor in a high tax bracket leases an asset to a lessee in a low tax bracket. The lessor derives the greatest benefit from being able to depreciate the asset based on the firm's higher tax bracket. The lessee benefits because his or her firm is able to share in the tax savings to the lessor, which are reflected in a lower lease payment by the lessee, and which are fully tax deductible. Similarly, when the investment tax credit is available, leases are often arranged by those in high tax brackets who act as lessors and can best take advantage of the tax credit, to those in low tax brackets who act as lessees.

Leasing might also be less expensive than purchasing an asset when the lessee is running a special sales promotion. The sales promotion might involve a large discount available only through a lease agreement.

Whereas leases may sometimes be used to reduce a firm's capital requirement, this is not typically a key factor in the decision to lease or buy an asset. As a matter of fact, the credit rating required to obtain a capital lease or sale-leaseback contract may be as high or higher than that required by most lenders when financing a purchase for the simple reason that the lease agreement is very similar to a purchase financed by a loan.

Because of the complexity involved in arranging and evaluating lease agreements, it is advisable to consult with a tax accountant to determine whether to lease or purchase an asset and to ensure that the lease agreement is properly written.

LEVERAGE

Leverage refers to the use of borrowed money with the objective of enhancing the rate of return on owners' equity. Put another way, leverage refers to the proportion of debt in a firm's total capitalization structure. The greater a firm's reliance on debt, as a proportion of its total financing, the more highly it is said to be leveraged. An example of the use of leverage can be seen by looking at Alvarez Vineyard's capitalization structure (Box 5–1). The debt-to-equity ratio is nearly 1 to 1. By incurring additional debt, and more highly leveraging his firm, Alberto may be able to increase the amount of funds available to his company in order to increase its profitability.

The use of leverage can work both for and against a firm. To justify using debt as leverage, the expected after-tax average return on total capital (total liabilities plus owners' equity) must exceed the after-tax cost of the debt capital. This is demonstrated by examining a situation in which Alberto might consider expanding his proposed winery. Suppose that after several years of operating the winery, it has generated a 12 percent after-tax return on total capital—and that Alberto expects this to continue in the near future. If he can borrow money at a 9 percent after-tax rate, he should do so. For every dollar borrowed, he would have to pay the lender nine cents after taxes. He would earn 12 cents after taxes on every dollar and have three cents left over in profits. The opposite would be true if the interest rate were 15 percent after taxes. In after-tax terms, for each dollar he borrows he would still earn only 12 cents while paying out 15 cents.

BOX 5–4

After-tax Return on Investment with Different Capitalization Structures

	Case 1		Case 2	
Total Capital	$100,000		$100,000	
Amount Borrowed	10,000		50,000	
Interest Rate	10%		10%	
Investor Capital	90,000		50,000	
Return on Capital	5%	15%	5%	15%
Income	$ 5,000	$15,000	$ 5,000	$15,000
Interest Paid	1,000	1,000	5,000	5,000
Net Income	4,000	14,000	0	10,000
Return on Invested Capital	4.4%	15.6%	0.0%	20.0%

Note: All income and interest paid figures represent after-tax dollars.

Another aspect of leveraging is the increased variability in profit. Assume that Alberto is considering a $100,000 expansion by borrowing either $10,000 or $50,000. The after-tax interest rate is 10 percent. He is evaluating the investment at two different after-tax rates of return on total investment, 5 percent and 15 percent. In Case 1, in which he finances 90 percent of the investment himself and the return on total investment is 5 percent, the return on his investment is 4.4 percent (Box 5–4), whereas at a 15 percent return on total investment he would earn 15.6 percent on his personal investment. In Case 2, in which he finances 50 percent of the investment with debt capital, the return on his investment is 0.0 percent and 20.0 percent, for returns on the total investment of 5 percent and 15 percent, respectively. The spread of 20 percent with 50 percent debt financing (0.0 to 20.0 percent) is much greater than the approximately 11 percent spread (4.4 percent to 15.6 percent) with 10 percent debt financing.

An analysis of the two scenarios demonstrates why an increase in the proportion of debt financing will result in greater variability in the return on invested capital. In the first case, a return of only $5,000 (or 5 percent) yields a return on invested capital of 4.4 percent. However, a total return of 15 percent increases the return on the firm's invested capital to only 15.6 percent because of the high ratio of invested capital to borrowed capital (9 to 1). Borrowing a greater proportion of the capital needs illustrates the power of leverage. When the firms borrows one half of its capital needs for the investment, a 5 percent return yields a 0.0 percent return. However, a 15 percent return on the investment yields a return on invested capital of 20.0 percent. The high level of debt increases the return needed to make the interest payment on the debt, but once the payment is made, the return on invested capital increases very rapidly because of the low rate of invested capital to borrowed capital (1 to 1).

The flip side of the coin is that the use of leverage increases the risk associated with the investment. In the first case, where only $10,000 was borrowed, a relatively small return of 1.0 percent was needed to cover the cost of interest on the loan (10 percent of $10,000). However, a return of 5.0 percent (10 percent of $50,000) would be needed to cover the interest cost of the money borrowed in the second case.

DETERMINING THE APPROPRIATE FINANCIAL STRUCTURE

How do owners and managers determine the optimal financial structure of a food or agribusiness firm? First of all, there is no one optimal financial structure for most food and agribusinesses firms. Most existing agribusinesses tend to live with their existing financial structure, making incremental changes over time. Most new businesses have to take what is available to them, within reason, and financially ailing firms take any financing they can get.

Firms that fit into one or more of the following categories should expect to have a relatively high percentage of equity relative to total assets, probably ranging between 75 percent and 100 percent. This is especially true for new or rapidly expanding firms because most lending institutions, for good reasons, tend to limit the amount of credit given to these firms. The categories include:

- Business activities that involve a high level of risk
- Businesses that have few hard assets that are easily liquidated
- New types of business or businesses new to a particular geographic area
- Firms with inexperienced management
- Businesses that require a long period before generating a positive cash flow (for example, a fruit or nut orchard may require eight years or longer before realizing a positive cash flow)

Food or agribusiness firms that have an established history of stable earnings can safely take on a greater debt load than firms fitting any of the five listed categories. Such firms may be able to obtain 50 percent or more of their financing needs from debt sources. However, firms providing debt capital are often reluctant to be in the position of having a greater stake in the firm than do the owners. A notable exception to this rule is a firm whose major assets are highly liquid and secure. For example, a diversified ranch with a stable earnings history can often safely borrow up to 75 percent of its capital needs. Because a high proportion of a ranch's assets are typically comprised of land and cattle, which can easily be sold if necessary, lenders are often willing to finance a high percentage of the assets.

SOURCES OF FINANCING

Equity Financing

Equity financing through the issuance of stock has already been described at length earlier in this chapter. Although this avenue has been successfully used by many new and ongoing businesses, it is not an option available to most food and agribusiness firms. The small size of most firms and the amount of funds required usually make this type of transaction prohibitively expensive for small food and agribusiness firms, even if an investment banking firm could be persuaded to handle the proposed transaction. The stock-purchasing public tends to buy well-known stocks or new, glamorous issues. The glamour issues are those that, if successful, offer the potential of a large increase in the stock price. Internet companies and technology companies that have new patents or other highly valuable discoveries are examples of the kinds of smaller, relatively unknown firms that have successfully sold stock to the general public. Unless a company can be classified as a potential glamour firm, it is essential that it provide liquidity for buyers and sellers by having a large volume of stock in the hands of the general public; otherwise, there will not be much interest in a stock by an investment banking firm.

Most food and agribusiness firms do not meet any of the above criteria. They are neither large nor well known to the investing public. They do not have a large volume of stock in the hands of the general investing public and they can rarely be classified as potential glamour issues. Hence, as a group they are of little or no interest to investment banking firms.

The amount of equity capital needed is generally sufficient to limit the opportunity of smaller firms to sell stock to the general public. Most national investment banking firms (such as Merrill Lynch or Goldman Sachs) generally require an issue of at least $10,000,000 before they are interested in handling a new issue of stock. Smaller regional or local investment banking firms may be willing to handle an issue as small as $1,000,000, but the costs of this service are very high. Investment banking firms typically charge in the neighborhood of 12 percent for selling a small issue. Legal fees, accountant fees, development, printing fees for the prospectus, and other miscellaneous expenses will likely amount to well over $100,000. Therefore, in order to raise a net amount of $1,000,000 for the business, $1,300,000 in stock would have to be sold. Other external sources of equity financing for businesses are venture capitalists and "angels." *Venture capitalists* are groups of individuals or firms, acting through an association or "club," that seek new business opportunities with the potential for a substantial, quick payoff if the company is successful. *"Angels"* are individuals with high net worths who invest in companies. Both of these sources look for investments that provide higher returns than alternative investments, such as stocks and bonds. In return for their money they take seats on board's of directors, giving them an active voice in major company decisions. Most food and agribusiness companies seeking financing do not generally appeal to venture capitalists and angels, who are typically interested in financing technology development companies, which if successful can yield an extremely high return on their investment.

Two categories of food and agribusiness firms have attracted the interest of investment bankers, venture capitalists, and angels. These are biotechnology firms and firms at the intersection of the food industry and electronic commerce, firms such as WebVan, an on-line supermarket that provides home delivery of groceries.

If selling stock to the general public and obtaining funds from venture capitalists and angels is not an option, how can agribusinesses raise equity capital? Equity capital for most agribusinesses comes from the owner's own personal savings. It is common for many business owners to start their businesses after first acquiring experience working for someone else. Over time they build up equity in their homes, retirement accounts, and stock and bond investments, which they use to fund their business. Owners that cannot generate sufficient equity for their business from their personal savings often turn to family and friends for additional capital.

Such arrangements, although highly practical, are fraught with potential pitfalls. Many relationships between family members and friends have been ruined because of problems arising out of a common business venture. Two problems are extremely common among people when family and friends jointly enter a business venture. First, a person may get along fine with his or her family and friends when the relationship is a social one, but as soon as money enters into the equation it may change the relationship. One way to reduce the chances of this problem occurring is to attempt to keep the business relationship separate from the social relationship. The second problem is that unexpected difficulties with the business often blindside the outside investors. It is a good idea for the owner-manager(s) to communicate regularly with those owners not involved in the business so as to keep them apprised of the status of the business and not surprise them with bad news.

As the firm grows, the source and amount of additional financing will depend largely on the venture's profitability. New financing might be needed because of a planned expansion or change in ownership. The need to build a new facility in order to sell a new product is a common example of a financing need derived from a planned expansion. The desire to buy out family members or friends is another example of a need to have new financing for the business. A successful, ongoing food or agribusiness company has many sources available to finance an expansion, particularly when compared to new or failing firms. Firms that are generating good profits may rely on retained earnings to finance expansion. Moreover, a firm with a solid track record of profitable performance will be attractive to investors, particularly if its business environment remains positive.

Additional equity usually, but not always, comes from someone in the local area. Generally, new investors have prior knowledge of the firm and know the present owners or managers. New investors may be friends or family members of the owners, customers, suppliers, or other people that have an association with the business. An investor (or investment group) may be interested in putting equity funds in the firm, provided that:

- The potential return appears to be sufficient.
- There are opportunities for future growth.
- The firm is well managed and has a good image.
- The investors are satisfied that the current owners and managers are honest and can get along with other people.

Another method of finding additional investors is through a *private placement*. Individual or investment groups are often interested in providing equity capital to successful firms. Investment banking firms are often willing to structure a private placement for successful firms wanting to expand. The size of the private placement can range from $500,000 to several million dollars, depending on the firm's needs and the personal interest of investors. However, potential investors will want assurances that the owners are capable of managing the expanded business. Historical balance sheets, income statements, and statements of cash flow, as well as pro forma statements and a detailed business plan, are essential.

Debt Financing

Commercial lending institutions are the largest source of loans to food and agribusiness firms. In the past, commercial lenders did business primarily with those firms that had the lowest level of risk, and then only in business areas familiar to them. However, competition among financial lenders has lead them to relax these standards. Moreover, institutions now package loans and sell them on the securities markets, where investors shoulder the risks.

Obtaining debt financing is not a problem for most agribusinesses; in fact, debt is so easy to obtain that personal bankruptcy from business failures is relatively high, based on historical records. When money is relatively easy to borrow, the cost of borrowing (the interest rate charged) is the primary issue. Interest rates charged to food and agribusiness firms vary substantially depending on the collateral available, the borrower's character and *creditworthiness* (debt repayment history), and the borrower's capacity (cash flow) to repay the loan.

Sources of Long-Term Debt Financing. Businesses borrow long-term funds from relatives and friends, banking institutions, the Farm Credit System, and insurance and mortgage companies. Relatives and friends, particularly parents, are important sources of credit for long-term assets for many business owners. If the parents can afford it, such transactions are very desirable in that they provide funds (interest payments) for the parents to live on while the son or daughter gets the use of the asset. A business owner borrowing money from family or friends must follow guidelines established by the Internal Revenue Service in order for the transaction to be treated as a business loan. First, the owner should draft a promissory note (an IOU) to document the details of the loan and establish a legal commitment to pay back the loan. Second, the owner should consult with a certified public accountant to ensure that the interest rate passes the Internal Revenue Service (IRS) minimum interest rate test. Otherwise the Internal Revenue Service might rule that the funds are a gift, which could result in a gift tax if the amount of the loan is over $10,000. If the loan is handled properly, the interest paid on the loan is a deductible business expense.

The most common sources of financing for equipment or machinery, sometimes called intermediate assets, are commercial banks, the Farm Credit System banks, and equipment dealers. Most equipment dealers operate credit departments as a means of increasing sales. They generally borrow funds by selling bonds in the public credit markets. Receipts from the sale of bonds are used in the equipment-financing program. The cost of borrowed funds to the equipment dealer is generally higher than for commercial banks. Hence, they either have to operate their financing business more efficiently than banks or charge higher rates. Many equipment suppliers provide financing to customers who either have been turned down or have been discouraged from borrowing from a bank. The customers are considered to be poor credit risks; thus, a higher rate of loan defaults is expected, which increases the overall cost of financing. On the other hand, the equipment supply firms are better able to resell repossessed assets than are banking institutions, which reduces the risk the equipment firms face should the borrower default on the loan.

Dealers occasionally offer financing at interest rates at or below those of other lenders. This may be the case when the manufacturer subsidizes the financing as a means of increasing sales or encouraging early orders. Such promotions are often advertised as "sales," with lower financing offered as a part of the deal. This practice is common for consumer products such as cars but typically occurs with producer goods when an equipment manufacturer has excess inventory. On balance, it is usually advantageous to shop around before selecting a source for financing machinery or equipment.

Seller financing (usually by individuals) is a common method of acquiring land assets. Sellers are often willing to finance land sales at attractive interest rates as a means of disposing of property and reducing the burden of taxes on the capital gain realized on the asset. A common practice is to make a down payment ranging from 20 percent to 30 percent with the balance amortized (paid off) over 20 to 30 years at an interest rate tied to the rate charged by the Farm Credit System banks. The Farm Credit System is the single largest lending institution for agricultural land. In recent years almost all of the loans made by the Farm Credit System have been variable interest rate loans. For this reason many noncommercial lenders use the Farm Credit System rate to establish an interest rate for loans.

Sources of Short-Term Debt Financing. Short-term loans, or money to finance ongoing operations, for food and agribusiness companies are usually obtained from commercial banks, the Farm Credit System (for farms, ranches, and many agricultural processing firms), vendors, and credit card companies. Commercial banks and the Farm Credit System provide favorable credit terms for firms with a successful track record. They require firms to provide them with balance sheets, income statements, statements of cash flow, and a business plan and they do an extensive check of the borrower's creditworthiness. They may also require the borrower to sign personal financial responsibility agreements or pledge collateral to secure the loan. These institutions generally operate on a margin that ranges from 2 percent to 6 percent over the cost of the funds they borrow. Because of their relatively low margins, they cannot afford large losses on their loaned-out funds. These lenders expect to be repaid in cash as the loans mature. They operate under strict government regulations and have to satisfy bank auditors by showing that their loans are secure.

Understanding what a lender is looking for can be helpful in obtaining a loan. Banking institutions expect loans to be repaid out of the profits and cash flow from the business operations rather than from the sale of an asset pledged as security for the commercial loan. For example, a bank that loans an agricultural supply firm funds to finance an inventory buildup expects the loan to be repaid out of operating profits received when the inventory is sold. They do not expect to have to take ownership of the inventory or, possibly, buildings that were also pledged as security for the loan. The banker is in the business of providing credit for safe, legitimate business needs.

In order to obtain a loan, it is necessary that a business be able to demonstrate the economic need or logic of the loan, as well as how and when the loan will be repaid. Business firms demonstrate how and when the loan will be repaid by providing the lending institution with cash flow projections (cash budgets) that show the amount of money needed, when it will be needed, and when it will be repaid. Bankers will be interested in historical and pro forma income statements because, in general, they do not want to lend funds to firms that are not profitable, even though it may appear that the firm will have the cash flow available to repay the loan. Lenders will also be interested in the balance sheet to ensure that firms have sufficient unpledged assets available that can be sold in case the investment does perform as expected and the loan cannot be repaid out of the operating cash flow.

Commercial banks and Farm Credit System banks provide a revolving line of credit to agribusinesses to meet their short-term credit needs. A *revolving line of credit* is an annual loan to the agribusiness based on the firm's projected working capital needs estimated from the cash flow budget. Collateral for the loan is often the firm's accounts receivable or inventory. As an example, assume that a tomato paste processor will need $300,000 during the next fiscal year to cover short-term expenses occurring between the time when the tomatoes are purchased until the finished product is sold. Prior to purchasing the raw tomatoes, the bank will place in the firm's account the $300,000 to be used to finance the processing and storage of the tomato paste. The firm will only be charged interest on the funds it withdraws. The firm can then pay back the loan when it generates sufficient income to do so. The bank expects the firm to pay back the loan by the end of the year, even when it is apparent that the business will need a similar loan the following year. If the firm performs as expected and repays the loan, the loan will be renewed for a subsequent time period as needed. This is why the process is often referred to as a "revolving" credit line.

Another important source of short-term funding for most food and agribusiness companies is trade credit. Suppliers of inventory or other suppliers (vendors) often allow their business customers to make purchases on credit. This allows businesses to finance inventory and serves as a means for the supplier to encourage sales. Common credit terms are for 30, 60, or 90 days. Vendors charge no interest—and some offer a discount—if the amount financed is paid within 30 days of the invoice date. After 30 days the company will owe the vendor the entire amount on the invoice plus a monthly interest penalty. Most vendors expect the entire invoice to be paid within 90 days.

Credit cards have become a major source of short-term financing for many startup companies and for those companies that lack the financial clout to get a line of credit from a bank or trade credit from vendors. American Express, MasterCard, and Visa all have small-business credit card divisions. In addition to credit, these divisions offer concierge services, technology assistance, and legal hotlines. Traditionally, credit cards have been used by small businesses as a convenience to charge food, entertainment, and travel purchases. Today, many businesses use a credit card as a means of short-term financing for the purchase of goods and services. Whereas American Express requires businesses to pay off the monthly balance within 30 days, MasterCard and Visa allow revolving balances. The benefit to businesses is that they get a two to six week "float" period between the time they receive the merchandise and the time they have to pay the credit card company, depending on the timing of the purchases. However, this can be an expensive way to obtain short-term financing because most credit card companies charge interest of 18 percent or more on unpaid balances.

Recently, credit card companies have begun signing up vendors selling supplies and inventories to businesses. Vendors pay the credit card companies a 1.5 percent to 4 percent commission when businesses use their cards to make purchases. However, the vendors receive payment within three days of the purchase from credit card companies as compared to the 45 to 60 days that is common when they invoice customers.

Government-Backed Financing. To increase economic development, local, state, and federal governments have developed programs to assist small business startups. Large companies usually seek help from state governments and small companies are often assisted by local governments and chambers of commerce. State governments have economic development agencies that make loans directly to businesses or provide loan guarantees through conventional lenders. Local and state governments provide tax incentives and reduced land costs to new firms willing to locate in their area. This is especially true in rural areas. Tax incentives usually take the form of property- and sales-tax abatements for a period of years. Many communities purchase and develop parcels of land in industrial parks that are resold to new businesses at a reduced cost. Local business incubators and community colleges offer new businesses guidance in developing business and marketing plans, as well as assistance in obtaining government-backed loans. In some cases local governments help new businesses find and train employees.

At the federal level, the Rural Business-Cooperative Service of the United States Department of Agriculture (USDA) has established programs to fund businesses in rural areas. The Business and Industry Direct Loan Program provides loans to business owners who cannot obtain credit from conventional sources. Loans can be used to improve, develop, or finance business and industry; create jobs; and improve the economic and environmental climate in rural communities. A rural community includes all areas other than cities or unincorporated areas with a population of more than 50,000 people and their adjacent areas. The maximum loan amount to any borrower is $10 million.

The Business and Industry Guaranteed Loan Program provides financial backing for rural businesses by providing loan guarantees to commercial lenders. These programs guarantee up to 90 percent of the loan. Loans can be used to fund working capital, machinery and equipment, buildings and real estate, and certain types of debt refinancing. Assistance under this program is available to any form of business organization and to public entities. The maximum loan amount to any borrower is $25 million.

The Farm Service Agency (FSA) of the USDA offers direct and guaranteed ownership and operating loan programs to farmers who are unable to obtain debt financing from commercial sources. The FSA provides loan guarantees to commercial lenders for up to 95 percent of the loan. Applicants unable to qualify for a guaranteed loan may be eligible for a direct loan from the FSA. The maximum amount to any borrower is $200,000 in the direct loan program and $700,000 in the guaranteed loan program.

The Small Business Administration (SBA) provides loan guarantees to conventional lending institutions to encourage and promote loans to small businesses. Food and agribusiness companies qualify for this assistance; however, those located in rural areas typically work through the USDA programs. The SBA guarantees loans to food and agribusiness firms that have annual receipts between $500,000 and $3.5 million, depending on the industry. To receive a SBA loan, the business owner must have applied for and been denied a conventional loan. The SBA guarantees 75 percent of the loan up to $750,000 under the Regular 7(a) Loan Guarantee Program. The SBA's MicroLoan Program makes funds available to nonprofit intermediaries, who in turn make loans to eligible business owners. The maximum loan to any borrower is $25,000. Loan eligibility is based primarily on the cash flow of the business; however, the SBA also considers good character, management capability, collateral, and owners' equity contributions. Owners of 20 percent or more of the business typically must sign personal guarantees.

FINANCING NEW FOOD AND AGRIBUSINESS FIRMS

The subject of how to successfully finance a new business is complex and beyond the scope of this book. In the following paragraphs we discuss a few basic guidelines for financing new businesses.

New food and agribusiness firms have more difficulty than established firms in obtaining financing because they do not have proven track records. Their options are much more limited than those of successful firms because new firms depend largely on the owners' personal assets and their ability to convince other potential investors and lenders that the business will succeed. Thus, the owners and the owners' families and friends are the primary source of funds for most new businesses because many creditors are unwilling to lend funds to a new business.

Lending institutions are particularly reluctant to make loans to new businesses if they have no experience with the type of firm requesting the financing. An example might be a vegetable processing facility in a geographic area where there are no vegetable processing plants. Even if the bank or other lenders agree to make a loan, they almost always require that the owners pledge their personal assets in addition to the firm's assets as a condition for receiving the loan.

FINANCING COOPERATIVES

Equity Financing

Cooperatives finance their operations by using equity and debt capital just like other businesses; however, there are important differences in the way equity and debt capital is obtained. One of the key features of cooperatives is that the users of the cooperative also provide financing, thereby guaranteeing that control will remain in the hands of members. Cooperatives obtain equity capital from members in three basic ways: (1) through direct investment, (2) by retaining a portion of net income distributed as patronage refunds, and (3) by retaining a portion of the proceeds from the sale of members' products as per-unit retains.

Figure 5–1 illustrates the sources of equity capital for the largest 100 agricultural cooperatives. Direct investment of equity is usually obtained through the purchase of common and preferred stock. Common stock is an important source of initial investment because it is usually associated with members' voting rights. However, some cooperatives issue common stock as notices of allocation of patronage refunds. Preferred stock is a second source of initial capital and is frequently allocated as a patronage issue. Preferred stock carries no voting rights but usually pays a dividend. Together these sources represent 21 percent of the equity in the cooperatives represented in Figure 5–1.

FIGURE 5–1

Sources of Equity, Top 100 Cooperatives, 1998

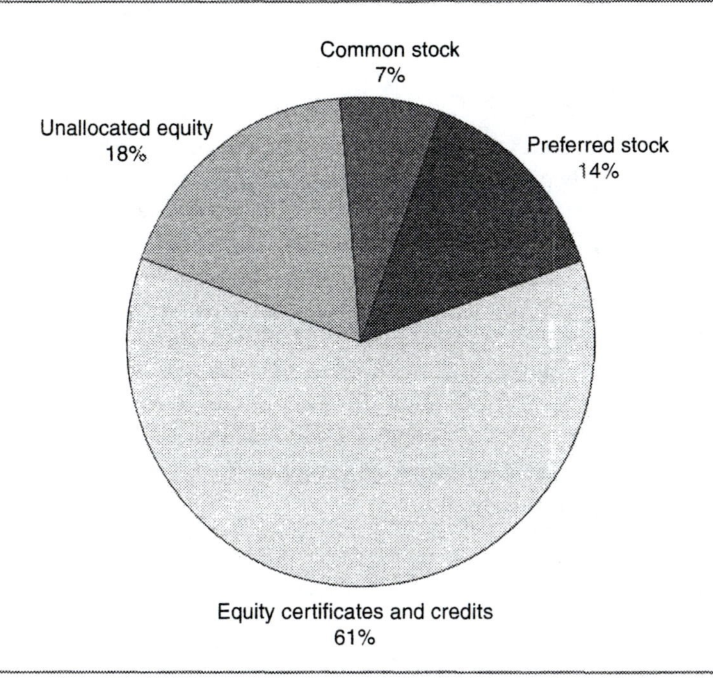

Source: "Asset Growth for Largest Co-ops Shows Resilience to Declining Revenues," ***Rural Cooperative Magazine***, Rural Business-Cooperative Service, Jan./Feb., 2000.

Unallocated equity is capital not allocated to specific member accounts. It is usually generated from the net margins of nonmember business. This form of equity is usually retained by the cooperative to protect member equity in bad times. For the cooperatives represented in Figure 5–1, unallocated equity represents 18 percent of the capital.

The majority of equity financing is in equity certificates and credits, which are another form of ownership in a cooperative. These equity sources represent net margins from patronage retained by the cooperative to provide equity capital. A net margin results when a farm supply cooperative prices products or services slightly above cost or when a marketing cooperative deducts a portion of the sales proceeds from the price it pays members. Net margins are returned to patrons at the end of the year. They take the form of patronage refunds in farm supply cooperatives and per-unit retains in marketing cooperatives.

The net margins are distributed in proportion to the patrons' volume of business as either cash, stock, or equity certificates and credits. At least 20 percent is customarily paid to members in cash, with the remainder held as retained patronage refunds or per-unit retains. These refunds are accumulated until sufficient capital is available, at which time the board of directors may decide to redeem a portion of the equity capital held by its members.

Equity Redemption

Stock in a cooperative is not bought and sold on an exchange, and most cooperatives do not allow one member to purchase the stock of another. In addition, unlike public corporations, where the value of a stock is determined on a public exchange, cooperative stock is normally redeemed at book or par value, whichever is less. Most of the equity in a cooperative comes from patronage and is therefore temporary because cooperatives have an implied obligation to redeem it. The board of directors decides when and if it will redeem equity based on the equity needs of the cooperative. This has been an area of concern for many cooperative members who want to redeem their stock in a cooperative but are unable to do so. Cooperatives have addressed the problem of equity accumulation and redemption by adopting two strategies: revolving fund financing and base capital plans.

Revolving Fund Financing. The objective of revolving fund financing is that current users finance the cooperative. To do this, the cooperative retires or buys back the oldest outstanding equity capital, as well as the capital of members who have died. By issuing new stock or certificates each year and retiring the oldest stock, the cooperative ensures that the most recent beneficiaries of the cooperative finance its operation.

Base Capital Plan. The base capital plan is a modification of revolving fund financing. Members are required to invest capital in proportion to their use of the cooperative over a base period of years, usually from one to 10 years. Each year the board of directors determines the equity needs of the cooperative and each member's proportional membership investment. This is then compared with the member's current capital investment. Each member's capital investment is then adjusted so that members maintain an investment level in the cooperative in proportion to their use. Members who have overinvested receive a larger portion of the patronage refund than they would otherwise.

Debt Financing

Figure 5–2 illustrates the source of long-term debt financing for the largest agricultural cooperatives. Cooperative banks, with 41 percent of outstanding debt, are the largest source of debt financing to agricultural cooperatives. These banks operate much like other commercial banks in making and servicing loans but are restricted to financing cooperatives. A major advantage for agricultural cooperatives is that these banks make funds available to cooperatives in situations where many commercial banks would not.

FIGURE 5–2

Sources of Long-Term Debt, Top 100 Cooperatives, 1998

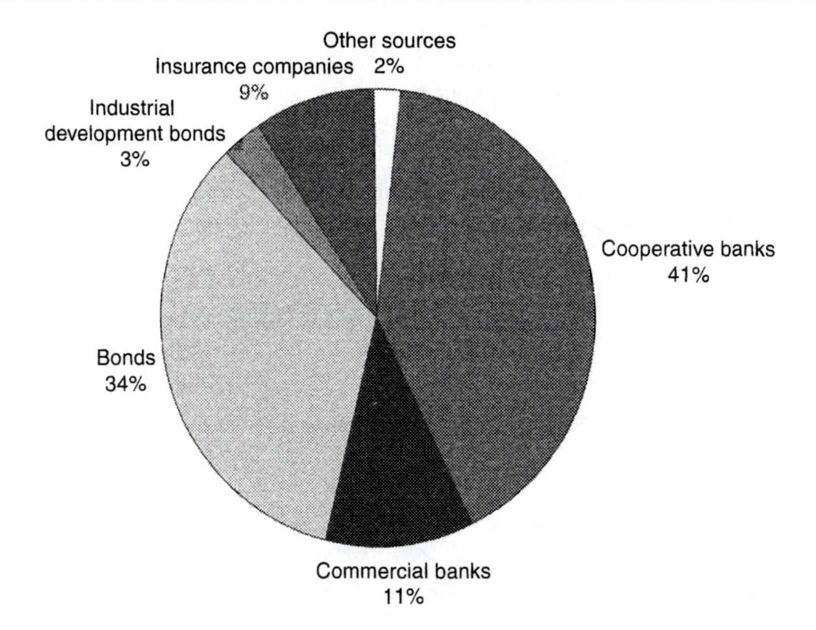

Source: "Asset Growth for Largest Co-ops Shows Resilience to Declining revenues," ***Rural Cooperative Magazine***, Rural Business-Cooperative Service, Jan./Feb., 2000.

Large cooperatives also use the services of investment banking firms to issue corporate bonds. Cooperatives are increasingly relying on the issuance of bonds as a source of debt capital. Commercial banks and insurance companies provide 11 percent and 9 percent of long-term debt financing, respectively. Other sources of long-term debt include leases, government loan programs, and other nonfinancial institutions. In addition to leasing companies, cooperatives have access to special institutions that lease only to cooperatives. The remaining funds (3 percent of total long-term debt) come from industrial development bonds. These bonds are used to construct facilities and must be repaid out of fees generated from the use of the facility.

SUMMARY

Financing is necessary to secure the funds required to operate a business. The critical finance decisions include whether to use debt or equity capital, the specific source of financing, and the length of time the funds will be committed.

The length of financing depends primarily on the type of asset. Nonconcurrent assets are financed with long-term funds, as is permanent working capital. Seasonal or temporary operating capital needs are financed with short-term funds.

Equity capital includes funds invested in the firm by the owners, called contributed capital in sole proprietorships and partnerships, membership interest in limited liability companies, and stock in corporations, as well as retained earnings (undistributed profits). While large, well-known agribusinesses may raise equity capital by selling stock, most small- and medium-sized firms, which constitute the majority of food and agribusiness companies, raise capital through the addition of partners or through the retention of the firm's profits.

Individuals or organizations that do not have an ownership interest in the firm provide debt capital. To obtain debt financing, it is essential that the agribusiness demonstrate how the debt will be repaid. The most common sources of long-term debt financing are commercial banks, Farm Credit System banks, equipment dealers, individuals, and insurance and mortgage companies. Short-term credit may be obtained from commercial banks, Farm Credit System banks, and vendors.

Several new practices that affect the financing of food and agribusiness firms have emerged in recent years. Two new lending practices are variable-rate loans, in which the interest rate fluctuates with the market rate, and participation loans, whereby the lender takes an ownership interest in the business. Leasing has become commonplace and serves as an alternative form of financing a long-term asset. Credit cards have become an important source of short-term funds for businesses that cannot obtain a line of credit from a bank or vendor.

Cooperatives generate equity capital from members through direct investment, retaining a portion of net income distributed as patronage income, and retaining a portion of the proceeds from the sale of members' products as per-unit retains. Debt financing is obtained primarily from cooperative banks and through the issuance of bonds.

CASE QUESTIONS

1. What factors should Alberto consider in determining how much equity capital versus debt capital he will use in fudning the vineyard and winery? What mix would you recommend?
2. Based on your answer to Question 1, what sources of debt financing would you recommend that he pursue? Why?
3. Based on your answers to the first two questions, what sources of equity financing would you recommend that he pursue? Why?

REVIEW QUESTIONS

1. Give two examples of each of the major types of assets.
2. Explain how the length of financing of assets is determined.
3. What are the principal features of common stock and preferred stock?
4. Construct a chart outlining the advantages and disadvantages of debt financing and equity financing.
5. Why might leasing be preferable to purchasing an asset?
6. Explain how using leverage can increase the profitability of a firm.
7. What are the principal factors that determine a firm's capital structure (i.e., debt and equity financing)?
8. Explain why capital contributed by the owners and not a public stock offering is the principal means used by most food and agribusiness firms to raise equity capital.
9. List the principal sources of short-term and long-term debt financing.
10. How do cooperatives obtain equity financing and maintain an adequate capital structure?

CHAPTER 6

Capital Investment Analysis

Learning Objectives

In this chapter we will cover the following topics:

- Factors important in making investment decisions
- Methods for analyzing investment decisions
- The concept of discounting future expenses and revenues
- Factors that determine the firm's cost of capital
- The capital budgeting process

CASE STUDY

Sarah Henderson operates a farm supply store in Newburgh, a rural town located in southern Indiana. She is considering buying new, point-of-sale electronic cash registers that read bar-coded information on most of the products that the business sells. The main purpose is to automate the marking of sales prices on products. The company would avoid the labor cost of marking initial sales prices and sales price changes on its products, which takes many hours each time products are stocked and prices are changed. The new system would reduce the costs of reordering inventory and conducting the annual inventory audit. Furthermore, the new registers would reduce the number of errors that the cashiers make when ringing up sales.

The investment in the new cash registers would generate future labor cost savings. The company's future annual outlays for wages would decrease if the new cash registers were installed. Avoiding a cash outlay is as good as a cash inflow because they both increase the cash balance of the business. The company adopts a five-year planning horizon for this capital investment. The new cash registers cost $3600. This would be paid immediately. The total after-tax cost savings (mostly in labor) are estimated to be $1000 per year for five years. This chapter will provide you with the tools necessary to analyze whether the new cash registers are a good investment.

INTRODUCTION

Investing is the commitment of money in order to earn a financial return in the future. Making wise investment decisions is a key factor in achieving and maintaining high returns on the capital invested in the food and agribusiness firm. Investment decisions often require the commitment of large sums of money over long periods of time. Therefore they warrant a thorough and careful analysis. This chapter focuses on methods to evaluate capital investments. A *capital investment* is an outlay that is expected to result in benefits extending more than one year into the future. *Capital budgeting* is the process of evaluating capital investments.

Most investment decisions involve durable, physical assets, but the concepts involved in capital budgeting apply to nonphysical assets also. *Physical assets* are tangible assets such as buildings, equipment, vehicles, and computer software. *Durable assets* usually have a useful life of at least three years and are listed under the noncurrent asset section of the balance sheet. *Nonphysical assets* include monetary assets and intangible investments. A *monetary asset* is a claim against some other party for monetary payment. Examples of monetary assets include stocks, bonds, and bank accounts. *Intangible investments* include assets that have no tangible or physical presence, such as patents, copyrights, franchises, and goodwill. These latter items are capital investments because benefits are expected over a number of years.

Some examples of common capital investment decisions encountered by businesses include:

- Deciding on whether to repair or replace an existing major piece of equipment
- Determining if the expansion of the existing activities of the firm is a wise decision
- Considering research and development projects

Before discussing how the firm makes investment decisions, it is important to understand the fundamental concepts underlying financial and investment decisions. We begin the chapter with a definition of some of the terms common to all methods of investment analysis. We then introduce the payback method, a simple method for making capital investment decisions. This is followed by a discussion of the time value of money and discounted cash flow analysis, concepts that provide the foundation that all businesses use in making informed investment decisions. In addition to being useful for evaluating capital investment opportunities like those described above, these concepts are useful in making personal financial decisions such as determining how much money to save for college or an automobile. We then introduce the primary methods that modern businesses use to evaluate capital investments, net present value and internal rate of return, and discuss how to choose a method for analyzing investments. We end the chapter with a discussion of the capital budgeting process.

Cash Flows

The typical investment requires an initial payment of cash and generates returns (or cost savings) in several future periods. The cash payments, or *outflows,* are recorded as negative cash flows. Cash returns, or *inflows,* including cost savings, are recorded as positive cash flows. Investments that generate cost savings reduce the outflow of cash and hence it is proper to consider the savings as a positive cash flow. The cash outflows and cash inflows are used by all of the methods of investment analysis to analyze investments.

For investments that require inputs of cash over several time periods, there will be some periods when there are both positive and negative cash flows. The positive and negative cash flows are summed to arrive at a single *net cash flow* for the period that may be either positive or negative. An example would be a machine that has to be periodically overhauled or repaired. In the period where the machine has to be overhauled there would be both a cash outflow, due to the repair, and a cash inflow, due to the cost savings.

Although it is expected that a successful investment will increase profits, it is the cash flow and not the accounting profit that must be used in evaluating the investment. As explained in Chapter 3, in accrual accounting revenues are reported when they are earned and expenses when they are incurred. However, an investment must be paid for up front, even though it will be expensed over many accounting periods as it is depreciated. Likewise, revenue resulting from the investment is of use to the firm, to reinvest or to pay off debt, only when cash is received or savings are realized. All changes in the cash flow of the firm, that is, increases or decreases in cash receipts or payments resulting from the investment, should be used to arrive at the net cash flow. The cash generated from increased sales resulting from an investment is an example of a positive cash flow. The increased depreciation due to the investment is not a cash payment and therefore would not be recorded as a cash flow.

It is necessary to account for income taxes to obtain the true impact of the investment on the firm's cash flow. Unless this is done, cash flows (both inflows and outflows) would be overstated. Cash inflows that represent taxable income should be reduced by the amount of the income tax that must be paid on the inflow. Similarly, cash outflows that are tax deductible should be reduced by income tax savings resulting from the deduction. For example, even though depreciation is not a cash flow item, it is a tax-deductible expense that reduces income taxes paid. Therefore tax savings resulting from depreciation are included in calculating the cash flows.

Investment Returns

The terminology used in discussing investment returns can be confusing because there is no uniformity in its usage. The following guidelines will help clarify the meaning of some commonly used terms. It is useful to distinguish between return *of* investment and return *on* investment. The return *of* investment refers to the repayment of the invested capital. The payback method, discussed in the next section, is a measure of the return *of* investment. Return *on* investment, sometimes referred to as the rate of return on investment, refers to measures that express the benefits generated by an investment relative to the amount of capital invested, typically in percentage terms. Measures of return on investment use some sort of annual or average return expressed as a percentage of a measure of the invested capital, such as the amount of the initial investment, or average amount invested. The internal rate of return is an example of a sophisticated measure of return on investment.

PAYBACK METHOD

The *payback method* is one of the simplest and least sophisticated methods for evaluating an investment. The *payback period* is the length of time required for the initial cost of an investment to be recouped (paid back). In order to calculate the payback period, the net after-tax returns (or savings) for each period are summed until they are greater than or equal to the initial investment. The number of periods is the payback period. If the entire amount of the returns for a period is not needed to pay back the investment, the fraction of the period required to cover the initial cost is calculated.

The following example illustrates the use of the payback method. A Kansas wheat farmer invests \$140,000 in a new combine. The new combine will reduce the farmer's cash expenses each year because a custom harvester will no longer have to be hired. It is anticipated that the net after-tax savings will be \$28,000 per year for the next ten years. The payback period for the combine is calculated as:

$$\text{Payback Period} = \frac{\$140{,}000}{\$28{,}000} = 5 \text{ years}$$

Typically, the calculation of the payback period is not so simple, primarily because cash flows are uneven. To illustrate this case we assume that, as in the previous example, a farmer purchases a combine for \$140,000, but that the net after-tax cash flows are as depicted in Box 6–1. The reduced cash flow in year four is due to a major overhaul that is expected after three years of heavy use. As the cumulative cash flow column shows, all of the cash flows through the third year and part of the cash flows from the fourth year are required to recover the \$140,000 initial investment. Specifically, \$20,000 of the estimated \$30,000 cash flow in the fourth year is needed to reach \$140,000. Thus the payback period for the combine is 3.67 years and is calculated as:

$$\text{Payback Period} = 3 + \frac{20{,}000}{30{,}000} = 3.67 \text{ years}$$

The payback method was the principal capital-budgeting criterion used for many years. A major advantage of the payback method in evaluating investment choices is its simplicity. Many firms establish a maximum acceptable period of time, such as five years, in which investments must be paid back. When alternative investments are being compared, the one with the shorter payback period is usually preferred. The payback method works best for investments that have identical lives and are characterized by one initial outlay followed by relatively level net after-tax cash flows over their useful lives.

The payback method is also used as a measure of the risk associated with an investment. Managers believe that the longer it takes to recover the original investment, the more likely that the cash flows may be lower than expected.

A high level of net cash flow from an investment is also important for liquidity purposes. A short payback period indicates that the investment makes a greater contribution to profits each year, thereby freeing cash flow for debt repayment or for other purposes. For this reason, the payback method is popular with new firms that are concerned primarily with maximizing the cash inflow from investments in the firm's early years. Likewise, firms that have cash flow problems often use the payback method to analyze investments.

BOX 6–1

Net Cash Flows Resulting from the Purchase of a $140,000 Combine

Year	Net Cash Flow	Cumulative Cash Flow
1	$40,000	$ 40,000
2	$40,000	$ 80,000
3	$40,000	$120,000
4	$30,000	$150,000

The payback method has two major disadvantages. First, it ignores the timing of cash flows and the fact that a dollar received now is worth more than a dollar received later. With the payback method, an investment with a $30,000 outlay and a net cash flow of $10,000 in the first year and $20,000 in the second year is just as good of an investment as one with a net cash flow of $20,000 in the first year and $10,000 in the second year. Both investments have a two-year payback period. Second, the payback method ignores all net cash flows that occur after the payback period. A $100,000 project that provides net cash flows of $20,000 a year will have the same payback period whether it has a life of five years or ten years. For these reasons, the use of the payback method as an investment analysis tool is declining in favor of the more sophisticated methods. Furthermore, the widespread use of financial calculators and computerized spreadsheets have made the more sophisticated methods of investment analysis extremely easy to use.

TIME VALUE OF MONEY

The *time value of money* is the tool used by investors and businesses to account for cash flows that occur at different points in time. As the name of the concept implies, the value of money depends on the time it becomes available. Returns received today are preferred over returns received in the future. Investors and businesses use the time value of money as a tool to determine the cash equivalent today of cash flows that occur in future time periods.

Simple and Compound Interest

The use of money is not free. Just as a business must pay to rent a car for a specific period of time, it must pay to "rent" or borrow money. The "price" of borrowing money is called the *interest rate.* The interest rate is expressed as a percentage for a specific period of time. The payments are called *interest,* which can be calculated on either a simple or compound basis.

Simple Interest. *Simple interest* is calculated only on the amount borrowed. The amount of interest paid depends on the size of the loan (the *principal*), the

BOX 6–2

Simple Interest Computation

Period	Principal	Interest Rate	Interest
1	$1,000	0.10	$100
2	$1,000	0.10	$100
3	$1,000	0.10	$100
Total			$300

interest rate, and the length of time the loan is held. The formula for calculating the amount of interest due each period is the principal multiplied by the interest rate. For example, assume that you invest $1,000 in a certificate of deposit for three years at a 10 percent annual rate of interest. Using the simple interest formula, the amount of interest would be computed each year using the original $1,000. The calculation of simple interest is illustrated in Box 6–2. The investor would receive $100 per year in simple interest for a total of $300 in interest plus the return of the principal amount of $1,000 at the end of the term.

Compound Interest. *Compound interest* is interest that is based on a principal amount that includes interest from previous time periods. In the first period, compound interest is calculated just like simple interest. However, at the start of the second period, the interest from the first period is added to the principal and becomes part of the principal. Interest in the second period is calculated on the new principal. This process of adding the interest earned in the previous period to the previous period's principal is repeated at the start of each period. We refer to this process of adding the interest to the principal amount as *compounding.* The calculation of interest compounded annually on a $1,000 certificate of deposit earning 10 percent interest for three years is illustrated in Box 6–3. The total amount that the investor would receive at the end of the three-year period is $1,331 or $1,000 in original principal and $331 in interest.

Compounding refers to earning "interest on interest." In the above example, if you invest $1,000 at 10 percent interest a year, you will receive $100 in the first year. If you reinvest that $100, you will receive a greater dollar return the following year. You will earn $100 in interest on your original $1,000 investment, plus $10 in interest earned on the $100 you reinvested. If you reinvest the interest as well as the principal earned during the second year, the total interest earned during the third year is $121.

Another way to calculate the value of the investment using the compound interest assumption is by using a simple formula. It may be expressed as: $\$1,000 \times 1.10 \times 1.10 \times 1.10 = \$1,331$. This formula illustrates an initial investment of $1,000 that increases in value by 10 percent per year. The implicit assumption is that the interest earned each year is reinvested and earns the same 10 percent return as the initial principal. Note that we could rewrite the above formula more simply as $\$1,000 \times (1.10)^3 = \$1,331$.

BOX 6–3

Compound Interest Computation

Period	Principal	Interest Rate	Interest
1	$1,000	0.10	$100
2	$1,100[1]	0.10	$110
3	$1,210[2]	0.10	$121
Total			$331

[1] $1,000 principal in year 1 + $100 interest in year 1 = $1,100 principal in year 2
[2] $1,100 principal in year 2 + $110 interest in year 2 = $1,210 principal in year 3

Notice that under the assumption of simple interest, $300 in interest is earned on a $1,000 investment. This amounts to a 30.0 percent total return over three years ($300 return / $1,000 beginning principal = 0.300 or 30.0%). However, under the assumption of interest compounded annually, $331 in interest is earned on the $1,000 investment. This results in a 33.1 percent total return over three years ($331 return / $1,000 beginning principal = 0.331 or 33.1%). The power of compounding results in an additional $31 or 3.31 percent higher return over the life of the investment.

Over a short period of time, compounding produces a slightly higher return, roughly 3 percent over the three-year period in this example. However, over long periods of time, compounding produces incredible results. Assume that you invested $1,000 in a retirement account starting immediately after graduation. At the end of 40 years with interest compounding annually at 10 percent, you will accumulate $45,259.26 ($\$1{,}000 \times (1.10)^{40}$). Simple interest will generate a much smaller value. An investment earning simple interest will earn a total of $100 each year for 40 years. The value of the account at the end of 40 years will earn a total of only $5,000 ($1,000 in original principal and $4,000 in interest).

Two additional examples illustrate the power of compounding. Because interest accumulates rapidly when it is compounded, relatively small differences in the number of years or in the interest rate can result in large differences in the total value of the investment. Assume that in the previous example you had waited five years before you started investing in your retirement account. You will then have 35 years instead of 40 until retirement and you would have accumulated $28,102.44 ($\$1{,}000 \times (1.10)^{35}$). The difference in portfolio values is $17,156.83 ($45,259.26 – $28,102.43)!

Now, let's assume that you make 9 percent on your $1,000 instead of 10 percent over 40 years. At 9 percent interest compounded annually you will save $31,409.42 ($\$1{,}000 \times (1.09)^{40}$). The difference in portfolio values is $13,849.84! Relatively small differences in the rate of return translate into large differences in the end values of portfolios compounded over many years.

BOX 6–4

Semiannual Compound Interest Computation

Period	Principal	Interest Rate	Interest
1	$1,000.00	0.05	$ 50.00
2	$1,050.00	0.05	$ 52.50
3	$1,102.50	0.05	$ 55.13
4	$1,157.63	0.05	$ 57.88
5	$1,215.51	0.05	$ 60.78
6	$1,276.29	0.05	$ 63.81
Total			$340.10

The above examples all used interest compounded annually. It is possible, and more common, to compound interest more frequently. Bonds pay interest semiannually (twice a year), stocks pay dividends quarterly, loans charge interest monthly, and many investments provide benefits on a weekly, daily, hourly, and even continuous basis.

As an example, assume that 10 percent interest on a $1,000 certificate of deposit is compounded semiannually for three years. Thus, the interest on the certificate is added to the principal every six months. The amount of interest earned is calculated by adjusting the interest rate. Since the 10% rate is an annual rate, we calculate the interest rate for the six-month period to be ½ of the annual rate, or 5%. The calculation of the total amount of interest is illustrated in Box 6–4. When semiannual compounding is used, the amount in the account after three years will be: $\$1,000 \times 1.05 \times 1.05 \times 1.05 \times 1.05 \times 1.05 \times 1.05 = \$1,340.10$. Note that this equation could also be expressed using exponents: $\$1,000 \times (1.05)^6 = \$1,340.10$.

You will recall that in the earlier example a $1,000 certificate of deposit paying 10 percent interest compounded annually grew to $1,331.00. Compounding semiannually increases the return by approximately $9. The account effectively grew at a rate of 10.25 percent per year rather than 10 percent per year because of semiannual compounding. By increasing the frequency of the compounding, more interest accumulates on the principal. The annual rate, before considering the effect of compounding more than once per period (10% in this case), is called the *nominal interest rate* or the *annual percentage rate.* The effective annual growth rate with compounding more than once per period is called the *effective interest rate.*

Future Value

We have shown how to calculate the total amount received on an investment at some point in the future. The combination of principal and interest at some future date is the investment's *future value.* The basis for all time value analysis is that a dollar invested today will grow to a greater amount in the future. By using the following formula, you can quickly calculate the future value of an amount that is invested at a fixed interest rate where:

$$FV = PV(1 + i)^n$$

where:

FV = future value,
PV = present value,
i = interest rate per period, and
n = number of periods.

The equation may be used to calculate the future value of a sum invested today (the *present value*) when the interest rate and number of periods are known. If you are compounding annually, the interest rate (i) is the same as the annual interest rate and the number of times interest compounds (n), or the number of periods, is the same as the number of years the investment is held. As an example, the future value of $100 earning 10% interest compounded annually for two years may be calculated as $FV = \$100(1 + 0.10)^2 = \$121.00$.

The formula may also be used to calculate future values when the period for which the interest rate stated is different than the compounding period. In this case the interest rate must be restated so that it corresponds to the period over which compounding occurs. This adjustment is made by multiplying the interest rate by an adjustment factor that is calculated by dividing the period over which compounding occurs by the period for which the interest rate is stated.

An example illustrates this calculation. Assume that, as in the previous example, $100 is invested at a 10% rate of interest for two years, but that interest compounds quarterly. The interest rate is for a period of one year, or 12 months, but interest compounds quarterly, or every three months. Therefore, the interest rate is recalculated on a quarterly basis by dividing the time period over which compounding occurs (three months) into the time period for which the interest rate is stated (12 months). This adjustment factor, 3 divided by 12, or 0.25, is then multiplied by the stated interest rate of 10% to arrive at the interest rate we use in the equation ($10\% \times .25 = 2.5\%$). The present value of the investment is $100 and the number of periods is eight (four quarters per year for two years). The future value of this investment may be calculated as $\$100(1 + 0.025)^8$ or $121.84.

Future values may be calculated using a calculator or a spreadsheet program. Table A–1 in the appendix contains values for numerous combinations of i and n based on this equation. The table shows the future value of $1 for various interest rates and periods. The table is used by selecting the factor that corresponds to the appropriate interest rate and number of periods and multiplying the factor by the present value.

By calculating the future value of an amount, you are determining the amount of money in the future that is equivalent to an amount today. Individuals use future value calculations to determine if they will have sufficient funds to send their children to college, make the down payment on a house, or be able to retire at a given age. Businesses use future value calculations to determine if they are retaining sufficient capital to redeem their bonds when they mature, to replace buildings and equipment, and to fund retirement plans for employees.

Present Value

The *present value* is the cash equivalent today that you would accept in exchange for a specified amount of cash at some future date. It is equivalent to the future amount less the interest that would have accumulated over the intervening period. The present value is the reciprocal of the future value. By using the following formula, you can quickly calculate present value of future sums using the compound interest assumption:

$$PV = FV\left(\frac{1}{1 + i}\right)^n$$

Two examples will serve to illustrate the application of this formula. Assume that someone is willing to pay you \$500 three years from now assuming an 8% interest rate compounded annually. What would be the present value of this amount? The formula would be applied as follows:

$$PV = \$500\left(\frac{1}{1 + .08}\right)^3 = \$396.92$$

The present value formula may also be applied to situations where the compounding period is different than the period for which the interest rate is stated. The adjustment is made in exactly the same manner as it was in the previous section on Future Value. Now let us recalculate the present value of \$500 to be paid three years from now assuming an interest rate of 8% and semiannual compounding. In this case, the interest rate must be stated on a semiannual basis (8% / 2 = 4%) and the number of periods must be adjusted accordingly (2 semiannual periods × 3 years = 6). The present value calculation would be:

$$PV = \$500\left(\frac{1}{1 + .04}\right)^6 = \$395.16$$

As you can see, the size of the present value is affected by the number of times the rate of return is compounded. However, the present value decreases as the number of compound periods increase. Thus, the present value of \$500 to be received in three years at an interest rate of 8% is \$396.92 when compounded annually, compared with \$395.16 when compounded semiannually.

As with future value calculations, a calculator or spreadsheet program can be used to calculate present values. Table A–2 in the appendix contains values for numerous combinations of i and n based on this equation.

Individuals and businesses use present value calculations to resolve investment questions similar to the examples given for future value calculations. Present value calculations are made when the future value is known and a present value equivalent must be found. For example, an individual may want to know how much to save today to accumulate a lump sum of money in the future. Similarly, a business owner may want to know how much to invest today to meet some future obligation.

Annuities

An annuity is a series of equal cash payments made at equal intervals. Automobile and house payments are two common examples of annuities. Understanding the future and present value of annuities is important in making these and numerous other financial decisions. For example, let's assume that you won your state's lottery. The state offers to pay you a lump sum of $3 million immediately. Alternatively you may choose an annuity that will pay you $200,000 a year over the next 20 years. Which should you choose? The information in the following sections will help answer this question by examining the future and present values of annuities.

Future Value of an Annuity

The future value of an annuity is the amount of money that accumulates at some future date from making equal payments over equal intervals at a specified interest rate. Individuals and businesses use the future value of an annuity to calculate the value of saving on a regular basis. The future value of an annuity is particularly useful when a target level of savings exists as might be the case with a major asset purchase or when a large debt must be repaid. For example, as part of its capital investment plan, a firm may want to know how much money may accumulate if it sets aside an equal amount of money each year for the replacement of equipment. Although it is technically possible to answer this question using the future value formula, the process would be extremely tedious.

To illustrate the process for finding the future value of an annuity, consider the amount of money you would have accumulated at the end of three years if you had deposited $1,000 in a bank account at 10 percent interest at the end of each year for three years. Box 6–5 shows the steps followed in calculating the future value of this ordinary annuity. Notice that an ordinary annuity has a level payment that occurs at the end of each period. Thus, with an ordinary annuity it is assumed that the final payment is made on the future value date; consequently, there is one less interest period than number of payments. In this example, because payments are made at the end of the year, the first annual payment will have had two years to earn interest, the second annual payment will have had one year to earn interest, and the third payment will earn no interest. The future value of this annuity is $3,310.

BOX 6–5

Future Value of an Ordinary Annuity

Payment Number	Payment	Years of Interest	Interest Earned	Compound Value
1	\$1,000	2	\$210	\$1,210 = \$1,000 $(1 + 0.10)^2$
2	\$1,000	1	\$100	\$1,100 = \$1,000 $(1 + 0.10)^1$
3	\$1,000	0	\$ 0	\$1,000 = \$1,000 $(1 + 0.10)^0$
Total	\$3,000		\$310	\$3,310

The general formula for the future value of an annuity is:

$$FVA = PMT\left[\frac{(1 + i)^n - 1}{i}\right]$$

where:

FVA = future value of an ordinary annuity, and
PMT = annuity payment.

The future value of an annuity formula can be solved by using a calculator, a spreadsheet program, or by using Table A–3 in the appendix. When using Table A–3, simply look up the appropriate interest rate (i) and number of periods (n) and multiply the future value factor in the table by the amount of the payment (PMT).

To illustrate, we'll use the table to find the future value of a 10-year, \$5,000 annuity, with 10 annual payments occurring at the end of each year and earning 6% interest. To find the appropriate future value of the annuity factor, first locate the row containing the number of payments (Number of Periods = 10) on the left side of Table A–3 and the column containing the interest rate (6%) along the top of the table. At the intersection of this row and column you will find the factor 13.1808. This factor represents the amount that would accumulate at the end of the last period if 10 payments of \$1, earning 6% interest per period, were made. The future value of this annuity is calculated by multiplying the annuity factor by the payment. The future value of the annuity is \$65,904 (\$5,000 × 13.1808).

In this example, the implicit assumption is that interest is compounded once per period. As in the Future Value and Present Value sections, this assumption may be modified by making the appropriate adjustment to the number of periods and interest rate.

The future value of an annuity formula may also be used to calculate the equal periodic amounts that must be set aside to accumulate a specific sum. Individuals use this formula to determine if they are saving enough money to reach a specific investment goal. All that is required to do this is to rearrange the future value of an annuity formula so that the payment is on the left-hand side.

This formula is:

$$PMT = \frac{FVA}{\left[\frac{(1 + i)^n - 1}{i}\right]}$$

Once again this formula may be applied using a calculator or spreadsheet. If Table A–3 is used, the payment is calculated by dividing the future value of the annuity by the annuity factor. The annuity factor is taken from the table based on the number of periods and periodic interest rate.

Present Value of an Annuity

The *present value of an annuity* is that amount of money that if invested at a specified interest rate will generate a set number of payments over equal time intervals. Individuals and businesses make decisions to invest money now that will generate benefit streams in the future. The purchase of a business is an example of an investment that involves a payment now in exchange for a series of cash benefits in the future. The present value of an annuity is useful in determining whether the future stream of benefits from an investment justifies the present cost. Likewise, a typical loan involves a cash benefit now in return for an equal stream of payments in the future. The present value of an annuity may be used to determine the loan payments for a fixed interest rate loan.

To illustrate the process for finding the present value of an annuity, consider a loan that will be paid back in the amount of $1,000 at the end of each year for three years. If the interest rate on the loan is 8% compounded annually, the total that will be loaned is $2,577.10. Box 6–6 presents the steps used in calculating the present value of this ordinary annuity. An alternative interpretation of this table is that a three-year loan in the amount of $2,577.10, compounded annually at an 8% rate of interest, would be repaid in three annual payments of $1,000 each. Each $1,000 payment consists of interest on the present value amount at the beginning of the period and the return of a portion of the present value (principal). The first $1,000 payment includes $206.17 in interest on the initial loan amount ($2,577.10 × 0.08 = $206.17). The remaining $793.83 reduces the principal on the loan. The principal paid may also be calculated by using the present value formula. $1,000 paid one year from now has a present value of $793.83. The amount due after the first payment has a present value of $1,783.27 ($2,577.10 − $793.83). Each subsequent $1,000 payment reduces the principal in the same manner until the loan balance reaches zero at the end of the third year.

The present value of an annuity procedure in Box 6–6 could be used for loans requiring equal periodic payments or investments that generate equal periodic cash flows for a number of years; however, the calculations would become tedious. For this reason it is easier to use the general formula for the present value of an annuity:

BOX 6–6

Present Value of an Ordinary Annuity

Year	Payment	Interest Paid[1]	Principal Paid = Payment − Interest Paid	Present Value[2]
0				$2,577.10[3]
1	$1,000	$206.17	$ 793.83 = $1,000[1/(1+.08)^3]	$1,783.27
2	$1,000	$142.66	$ 857.34 = $1,000[1/(1+.08)^2]	$ 925.93
3	$1,000	$ 74.07	$ 925.93 = $1,000[1/(1+.08)^1]	$ 0.00
Total	$3,000	$422.90	$2,577.10 = Present Value	

[1]Interest Paid = Present Value at the end of the previous year × 8%
[2]Present Value = Present Value at the end of the previous year − Principal Paid
[3]Present Value in Year 0 is equal to the loan amount

$$PVA = PMT \times \left[\frac{1 - \frac{1}{(1+i)^n}}{i}\right]$$

where:

PVA = present value of an ordinary annuity.

As with other time value formulas, the present value of an annuity can be solved using a calculator, spreadsheet, or a table. Table A–4 in the appendix contains present value factors of $1 ordinary annuities that were generated from the formula. The present value of an annuity is found by locating the factor that corresponds to the specified interest rate and number of periods and multiplying it by the annuity payment. As with the future value annuity table, the assumption is that interest is compounded once per period. If this is not the case, then the appropriate adjustment must be made in the interest rate and number of periods.

As we did with the future value of an annuity formula, we can also rearrange the present value of an annuity formula so that the payment of an annuity may be calculated when the present value is known. This is particularly useful in determining the loan payment when the loan amount, interest rate, and number of payments is known. Rearranging the present value of an annuity formula so that the payment is on the left-hand side yields the following formula:

$$PMT = \frac{PVA}{\left[\frac{1 - \frac{1}{(1+i)^n}}{i}\right]}$$

This formula may be used directly to calculate an annuity payment with a calculator or spreadsheet or by using Table A–4. If the table is used, the payment is calculated by dividing the present value of the annuity by the annuity factor. The annuity factor is taken from the table based on the number of periods and periodic interest rate.

Factors Affecting the Market Interest Rate

Several factors affect the *market interest rate,* which firms must pay for the use of funds. Until now we have simply stated that current dollars are preferred to future dollars. In this section we will discuss those factors that affect interest rates in general. They include the real, risk-free rate of interest, an inflation premium, a risk of default premium, a liquidity risk premium, and an interest rate risk premium. Market rates for different types of securities are based on the real, risk-free rate of interest and are adjusted based on the expectation of inflation over the life of the security, as well as the risks associated with the security. The factors that determine a firm's actual cost of capital are covered in the Capital Budgeting section of this chapter.

Real, Risk-Free Rate of Interest. The *real, risk-free rate of interest* is that rate of interest that would exist on an investment that entailed absolutely no risk in an inflation-free world. It is largely determined by people's preference for consuming goods today versus deferring consumption until the future. It is the true time preference for money, because it is not affected by other factors. It is also unobservable, because no investment is absolutely free of risk and the risk of inflation is always present, even when the expectation of inflation is relatively low. The real rate of interest is generally considered to be 2 to 4 percent per year.

Inflation Premium. The second factor influencing interest rates is the *rate of inflation,* or the rise in the general price level. Because inflation erodes the purchasing power of money, an adjustment for inflation is necessary in order for the real value of the borrowed money to remain constant. For example, suppose that you deposited $100 in a savings account, earning 5 percent interest, at the beginning of the year and that with that $100 you could have bought 50 pounds of ground beef at $2.00 per pound. Furthermore, assume that all goods rose in price by 10 percent that year. If you were to withdraw your money, you would have $105, your original deposit plus 5 percent interest, but you would be able to purchase less than 48 pounds of ground beef at the inflation-adjusted price of $2.20 per pound. Inflation would have wiped out any real gain you had received on your money and you would still be worse off because your money would purchase fewer goods at the end of the year than it could have at the beginning. For this reason investors and lenders demand an *inflation premium* based on their expectation of inflation over the life of the security.

Nominal, Risk-Free Rate of Interest. We noted that the *real,* risk-free rate of interest is not observable. However, there is a rate that closely approximates the *nominal*, risk-free rate of interest. It is the interest rate on short-term U.S. securities, U.S. Treasury Bills, commonly referred to as T-bills. The word nominal refers to the quoted or market rate for the security, which includes an adjustment for the expected rate of inflation. The U.S. Treasury securities are generally considered to be risk free, because the possibility that the U.S. Treasury will default on its obligations is considered to be extremely low. Thus, the rate on the U.S. T-bill includes both of the first two interest rate determinants, the risk-free rate of interest and an inflation premium.

Risk of Default Premium. The *risk of default premium* is the risk that a borrower will not meet its obligation to the lender, either to pay the interest on the security or to repay the principal. The risk on U.S. Treasury obligations is considered to be virtually zero. The securities of blue chip corporations are also considered to be relatively low risk. On the other hand, the rate of interest on the debt obligations of many new companies, or those that are struggling financially, is much higher. This reflects the higher risk of default by these firms.

Liquidity or Marketability Risk Premium. Investments that can be quickly turned into cash are considered to be liquid. *Liquidity risk* is the chance that an investment cannot be readily converted into cash. Cash is, of course, the most liquid of all assets, followed by financial assets, particularly those that are openly traded. For example, an investor who holds 100 shares of John Deere stock could quickly sell the shares through a broker and receive cash for the transaction at the end of the trading day. For this reason, such an investment is considered liquid. Real estate, on the other hand, is much less liquid. A farmer who wants to sell 80 acres of land would have to wait to find a buyer, and in most cases, for the prospective buyer to get a loan. It could reasonably take weeks or months for the farmer to get a fair price for the land. If the farmer needs cash immediately, he or she might be forced to reduce the price of the land in order to attract a buyer. For this reason, a liquidity risk premium is included in the market interest rate of investments to compensate investors for the lack of liquidity.

Interest Rate Risk Premium. The rate of return that investors require from an investment is directly tied to the risk-free rate. When interest rates rise, the value of an investment paying a fixed interest rate declines. Conversely, when interest rates fall, the value of an investment paying a fixed interest rate rises. The premium that investors require in order to be compensated for this risk is known as the *interest rate risk premium*. There is also another factor that influences the interest rate risk—the maturity of the security. The opportunity for an adverse move in interest rates is greater for an investment with a long time to maturity than it is for an investment that will mature in a short period of time. Moreover, the longer the term of the security, the larger is the impact that a change in the interest rate will have on the security value. Therefore, long-term securities include an interest rate risk premium that increases with the time to maturity.

DISCOUNTED CASH FLOW ANALYSIS

Discounted cash flow analysis is a method used to evaluate investments that incorporate the time value of money in the analysis. It addresses the major deficiency of the payback method, which does not distinguish between returns earned early in an investment's life versus those earned in later years. In this section we will first introduce the concept of the firm's cost of capital, followed by the primary methods businesses use to evaluate capital investments, net present value and internal rate of return.

Cost of Capital

A firm's *cost of capital* is the weighted-average cost of a firm's debt and equity financing. Knowing the firm's cost of capital is critical in conducting sound investment analysis because only investments with returns that meet or exceed the firm's cost of capital will be pursued.

There are many methods for estimating a firm's cost of capital. For example, the cost of debt capital may be estimated based on past borrowings, the purpose of the loan, the length of the borrowing period, and current interest rates. On the other hand, the owners determine the cost of equity capital. It is the rate of return required by the owners for the use of their money. Equity capital is nearly always more expensive than debt capital because the owners have more at risk than creditors, which typically require collateral. Therefore suppliers of equity capital require a higher return for the use of their money.

One method for calculating the firm's cost of capital for a specific investment is to determine how the investment will be financed and to estimate a cost of capital based on the proportion of debt and equity capital. A major disadvantage of this method is that the investment decision may be different depending on whether debt or equity capital is used. Furthermore, the proportion of debt and equity capital chosen is often arbitrary, because the firm may be financing or refinancing several investments at the same time, using both equity and debt capital.

A compromise, which is widely used throughout industry, is to use the firm's weighted cost of capital. The weighted cost of capital is calculated by estimating the cost of debt and equity capital each weighted by their respective proportion of the firm's total capital. Box 6–7 shows how the weighted cost of capital is calculated for a firm whose cost of short-term debt, long-term debt, and equity capital is 8, 10, and 14 percent, respectively. The firm's financing consists of approximately 15 percent short-term debt, 25 percent long-term debt, and 60 percent equity capital. Ignoring the income tax implications, the weighted cost of capital would be 12.10 percent.

Because interest is a tax-deductible expense, the cost of debt capital should be reduced by the *marginal income tax rate* (the rate charged on the last dollar of income). This is accomplished by multiplying the cost of debt capital by one

BOX 6–7

The Weighted Cost of Capital

Type of Capital	Cost of Capital	Weight	Weighted Cost
Short-term Debt	8.00%	0.15	1.20%
Long-term Debt	10.00%	0.25	2.50%
Equity	14.00%	0.60	8.40%
All Capital		1.00	12.10%

BOX 6–8

The Weighted After-Tax Cost of Capital

Type of Capital	Cost of Capital	1 – Marginal Tax Rate[1]	Weight	Weighted After-Tax Cost[2]
Short-term Debt	8.00%	0.70	0.15	0.84%
Long-term Debt	10.00%	0.70	0.25	1.75%
Equity	14.00%	1.00	0.60	8.40%
All Capital			1.00	10.99%

[1]Only the cost of debt capital is reduced by the marginal tax rate.
[2]Cost of capital × (1 – marginal tax rate) × weight = weighted after-tax cost

minus the marginal tax rate (Box 6–8). In this example the firm's marginal income tax rate is 30 percent. After accounting for income taxes, the weighted after-tax cost of capital is 10.99 percent.

Net Present Value

The *net present value* (NPV) is a discounted cash flow analysis whereby the present value of all expected future cash flows are calculated and summed. Discounting allows us to value future cash flows on an equivalent basis, the time of the initial investment. The interest rate used to determine present values in this analysis is called the *discount rate.* The discount rate may be viewed as the exchange rate between future dollars and present dollars. The discount rate, which we will use for investment analysis, is the firm's weighted after-tax cost of capital.

The NPV of an investment is determined by summing the present values of all of the net cash flows generated by the investment. The NPV is the value of the investment over and above what is required to cover the initial cost and the specified return. A zero or positive NPV indicates that the NPV of the future cash flows is at least as great as the cost of the asset and that the investment should be accepted. This means that the investment's return is at least as great as the firm's cost of capital. An investment with a negative NPV should

BOX 6–9

Net Present Value—Feed Mixer

Year	After-tax Cash Flow	Present Value Factor	Present Value[1]
0	($10,000)	$[1/(1+.11)^0] = 1.0000$	($10,000.00)
1	$ 3,000	$[1/(1+.11)^1] = 0.9009$	$ 2,702.70
2	$ 3,000	$[1/(1+.11)^2] = 0.8116$	$ 2,434.80
3	$ 4,000	$[1/(1+.11)^3] = 0.7312$	$ 2,924.80
4	$ 4,000	$[1/(1+.11)^4] = 0.6587$	$ 2,634.80
5	$ 5,000	$[1/(1+.11)^5] = 0.5935$	$ 2,967.50
Total			$ 3,664.60[2]

[1]Present value = after-tax cash flow × present value factor
[2] The net present value ($3,664.60) is the sum of the annual present values (the initial capital outlay in year 0 and the after-tax returns in years 1 through 5).

be rejected. When alternative investments are being compared, the investment with the highest NPV should be accepted.

Box 6–9 illustrates the calculation of the NPV for a potential investment in a feed mixer for an agribusiness. First, we must identify the timing and the amount of all after-tax cash inflows and outflows associated with the feed mixer over its five-year life span. The current after-tax cash outflow of −$10,000 is the cost of acquiring the feed mixer in the current period, year 0. The related after-tax cash inflows generated by the feed mixer are expected to be $3,000 in years 1 and 2, $4,000 in years 3 and 4, and $5,000 in year 5. Second, we calculate the present value of the future cash flows using a discount rate of 11 percent, which is the firm's weighted after-tax cost of capital. The discount factors are calculated using the present value formula, presented earlier in this chapter. Third, we compute the NPV by summing the present values of all of the cash flows. This includes the initial cash outflow used to acquire the feed mixer (a negative number) as well as the present values of all of the future cash flows (all positive in this case). The present values are calculated by multiplying each net after-tax cash flow by the appropriate discount factor. Last, we calculate the NPV by summing the present values for all the periods. We invest in the new feed mixer if the NPV is zero or positive, and we reject the investment if the NPV is negative. The NPV of the feed mixer is $3,664.60. If the cash flows occur as projected, the new feed mixer's rate of return will be greater than the firm's 11 percent cost of capital, and the firm should purchase the feed mixer.

The NPV does not represent the amount of profit or loss that the asset will realize. The positive NPV of $3,664.60 in this example does not mean that the firm will make $3,664.60 on this investment. Rather, it means that if the re-

BOX 6–10

Net Present Value of an Investment when the Timing of the Cash Flows Varies

		Case 1		Case 2	
Year	Present Value Factor, i = 0.10	After-tax Cash Flow	Present Value	After-tax Cash Flow	Present Value
0	1.0000	($11,000)	($11,000.00)	($11,000)	($11,000.00)
1	0.9091	$ 1,000	$ 909.10	$ 5,000	$ 4,545.50
2	0.8264	$ 2,000	$ 1,652.80	$ 4,000	$ 3,305.60
3	0.7513	$ 3,000	$ 2,253.90	$ 3,000	$ 2,253.90
4	0.6830	$ 4,000	$ 2,732.00	$ 2,000	$ 1,366.00
5	0.6209	$ 5,000	$ 3,104.50	$ 1,000	$ 620.90
Total			$15,000	($ 347./U)	$15,000

turns occur as expected, the net value of the returns is $3,664.60 in today's dollars. Similarly, the NPV does not reveal what the expected rate of return is. The NPV method only reveals that the rate of return on the feed mixer is greater than the cost of capital. For this reason, a positive NPV indicates that an investment should be pursued. Put another way, if the firm pays $10,000.00 for the feed mixer and receives the cash inflows as projected in the example, then the firm will recover its investment as well as the cost of the capital employed and still have a positive return on its investment in the mixer.

Let's look at another example to see what impact varying the timing of the cash flows will have on the investment decision. In Box 6–10 the initial cost of the investment is $11,000 and the discount rate is 10 percent. The total net cash flows for Cases 1 and 2 are the same, but the timing of the net cash flows varies. Case 1 has a NPV of −$347.70 and consequently would be rejected. This does not necessarily mean that the investment would lose money from an accounting perspective, but rather that the firm should not make the investment if it wants to recover its initial outlay and make a 10 percent return on its capital. Case 2 has a NPV of $1,091.90 and consequently would be accepted. Notice that the total cash flows for both Case 1 and 2 sum to $15,000. However, because of the timing of the cash flows in Case 2, where the higher returns are earned in the early years, this investment yields a higher NPV than Case 1, where the returns are lower in the early years. The higher NPV in Case 2 reflects the fact that a positive cash flow early in the life of the investment is preferred to one later on, because it can be reinvested or used to pay off debt.

The impact of using different discount rates can be seen in Box 6–11. In this example, we want to purchase an $8,000 asset with expected annual net cash flows over five years of $2,000 assuming two different discount rates, 6 and 9 percent. Notice that as the discount rate is increased, the present value of the future cash flows is decreased. At a 6 percent return the investment is acceptable. At a 9 percent return the investment is rejected since the NPV is negative.

BOX 6–11

Net Present Value of an Investment when the Discount Rate Varies

		Discount Rate = 6 Percent		Discount Rate = 9 Percent	
Year	After-tax Cash Flow	Present Value Factor	Present Value	Present Value Factor	Present Value
0	($8,000)	1.0000	($8,000.00)	1.0000	($8,000.00)
1	$2,000	0.9434	$1,886.80	0.9174	$1,834.80
2	$2,000	0.8900	$1,780.00	0.8417	$1,683.40
3	$2,000	0.8396	$1,679.20	0.7722	$1,544.40
4	$2,000	0.7921	$1,584.20	0.7084	$1,416.80
5	$2,000	0.7473	$1,494.60	0.6499	$1,299.80
Total			$ 424.80		($ 220.80)

One last consideration is the life of the investment. Investments have different useful lives, and when comparing investments it is better if the same number of periods is used. One method of accomplishing this is to choose the life of the shortest investment as the basis for evaluation. For investments with a longer useful life, a salvage value must be determined based on the estimated value of the investment at the end of the evaluation period. The salvage value is then added to the cash flow for the last period. By employing this approach, all investments are evaluated over the same time period.

Internal Rate of Return

The *internal rate of return* (IRR) is another discounted cash flow method that is used to determine whether a firm should acquire a long-term asset. The method uses the present values of the projected after-tax cash flows, including the initial outlay, in order to determine the rate of return of the proposed investment. In other words, the IRR is the discount rate that will give an NPV of $0. We can apply the present value equations defined earlier in the chapter to find the IRR for an investment that has even cash flows. In the example shown in Box 6–11, we analyzed an $8,000 asset that generated annual net cash flows of $2,000 over five years and found that the discount rate would be somewhere between 6 and 9 percent. To find the exact discount rate, we apply the present value for an ordinary annuity formula:

$$\text{PVA} = \text{PMT} \times \left[\frac{1 - \frac{1}{(1 + i)^n}}{r} \right]$$

We may restate this formula as:

$$PVA = PMT \times PVAF_{n,i}$$

where $PVAF_{n,i}$ stands for the present value of annuity factor, given n periods and an interest rate of i. Because the initial outlay is $8,000, we would then estimate the interest rate that would result in an annuity with a present value of $8,000, given five annual payments of $2,000. In other words, we would solve the following formula:

$$\$8{,}000 = \$2{,}000 \times PV_{5,?\%}$$
$$PV_{5,?\%} = \$8{,}000 / \$2{,}000$$
$$PV_{5,?\%} = 4.0000$$

Dividing the initial cost of the asset ($8,000) by the annual cash flow ($2,000) produces the present value factor of an ordinary annuity (4.000). To find the IRR, we then go to Table A–4 and find the number of periods or annual cash flow payments (5) at the left of the table, and then move across the row until we find the factor closest to 4.000, which is 3.9927 under the 8 percent column. Thus, the investment's IRR is approximately 8 percent.

Finding the IRR with Uneven Cash Flows

Finding the IRR is tedious if cash flows are uneven over the asset's life. There is no formula for solving the IRR for uneven cash flows. Instead, an iterative process is used to find the point where NPV equals $0. To solve for the IRR with uneven cash flows, you select a discount rate that you think will generate a NPV of $0. In the example shown in Box 6–10, we know that the IRR for Case 1 is below 10 percent, because at this discount rate the asset has a negative NPV. Conversely, we know that the IRR for Case 2 is above 10 percent, because at this discount rate the asset has a positive NPV. This hit and miss procedure is continued until the discount rate, which gives an NPV of $0, is found. The IRR for Case 1 is 9.00 percent and the IRR for Case 2 is 14.93 percent.

Finding the IRR is a time-consuming process when done by hand, especially when an investment generates returns over many years. However, you can easily solve complex IRR problems using a financial calculator or a computer spreadsheet program.

Use of the Internal Rate of Return

The decision to acquire a new asset using the IRR is straightforward and identical to the accept-reject criterion of the NPV. If the IRR is above the firm's cost of capital, then the NPV would be positive, and the firm should acquire the asset. If the IRR is below the cost of capital, then the NPV would be negative, and the firm should reject the asset.

The use of the IRR method has two principal advantages. First, because the IRR is stated in terms of a rate of return, it is easy to explain, and it is easy to compare to the firm's cost of capital. Second, because a higher IRR is more desirable, the IRR can be used to rank competing assets when those assets are similar in terms of costs, lives, risk, and cash flow patterns.

Unfortunately, there is also a problem associated with the IRR. An implicit assumption with the IRR method is that cash flows are reinvested at the IRR. This assumption can be unrealistic because increased competition in the marketplace will usually drive down the returns on capital investments over time to some average rate of return. In the investment world this phenomenon is called *"reversion to the mean."* In some cases, analyzing two identical investments may lead to different rankings of the investments depending on whether the NPV or IRR method is used.

Choosing a Method of Investment Analysis

Both the NPV and IRR methods are far superior to the payback method because they take into consideration the timing and value of the cash flows. The principal advantage of the payback method is its simplicity, both in terms of ease of calculation and ease of communication. However, the use of the payback method can lead to inappropriate investment decisions, because of the naive assumptions on which it is based. If it is used at all, it is most appropriately used to compare investments with similar lives and similar patterns of cash flow. When the returns from potential investments are perceived to be risky, the payback method may also be used as a measure of risk, with the shortest payback period being preferred. However, because most investment decisions involve thousands of dollars, it is well worth the time and effort necessary to employ a more sophisticated technique to ensure that the correct decision is made.

The choice between the NPV and IRR methods is more complex. The difference between the two methods lies in the assumption about how the cash flows are reinvested. The IRR method is based on the assumption that returns are reinvested at the IRR, whereas the NPV method is based on the assumption that the returns are reinvested at the chosen discount rate. The NPV assumption is probably more realistic than the IRR assumption, especially when the IRR is substantially higher than the cost of capital. Most financial experts also recommend using the NPV method over the IRR method for most investments. However, the IRR is often preferred for communication purposes, because it is expressed as a percentage, which is more easily understood than the NPV of an investment.

Accounting for Risk in Investment Analysis

Up to this point we have assumed that the costs and returns of an investment are known with certainty. In reality they are not; investments are risky. Risk may arise from several sources. For example, the actual cost of an investment is often not known. This is particularly true for investments that involve the construction of buildings or equipment. Furthermore, the actual level of cost reductions or increased sales is not typically known with certainty until the investment is in place, and even then the true costs and returns may not be known for years. One method for reducing the risk associated with an investment is to make the most accurate estimates possible of the costs and returns for an investment and to fix as many of the costs as possible. For example, some construction contracts can be written on a fixed cost basis, thereby transferring some of the costs of overruns to the contractor.

When the costs and returns associated with an investment are not known with certainty, it is necessary to account for risk in investment analysis. Most people are risk averse, that is, they prefer to avoid risk when possible. However, investors in a company usually only make money by assuming some risks. Given two investments, which are otherwise equal, the less risky investment should be chosen. Unfortunately, most investment decisions are not that simple. The more risky investments usually offer the prospect of higher returns. Alternatively, investors require a higher return on their money if they are to accept a higher risk. Therefore, a means for assessing the tradeoffs is needed.

When the NPV method is used, risk may be handled by adding a premium to the discount rate based on a subjective evaluation of the risk involved with each investment. For example, investors may decide that one investment is more risky because they have some doubts that it will reduce their labor requirements. They may decide to require a return on their investment of 18 percent as compared to their usual 15 percent. In this case, the NPV would be recalculated for the investment based on the higher discount rate.

When the IRR method is used, it is not necessary to make any additional calculations. The risk of an investment can be considered when choosing between investments. For example, a "safe" investment with an IRR of 20 percent would probably be chosen over a "risky" investment with an IRR of 21 percent.

THE CAPITAL BUDGETING PROCESS

The *capital budget* is a formal plan for meeting the long-term capital needs of the firm. Capital budgeting is a four-step process that includes:

- Identifying long-term investment opportunities for the firm
- Evaluating investment opportunities
- Determining the means by which capital expenditures will be financed
- Monitoring these investments to ensure that they meet the firm's long-term goals

We address each of these tasks next.

Identifying Long-Term Investment Opportunities

The capital budgeting process involves investment decisions regarding fixed assets. There are many reasons why firms make capital investments. These include replacing or repairing worn out or unproductive assets, expanding existing capacity, expanding into new products or markets, and complying with government regulations that require a safe, accessible, and environmentally-friendly workplace. Some decisions are fairly routine, such as the major overhaul of equipment on a regular basis. Other decisions are much more complex and involve a high degree of risk, such as developing a new product line. Capital budgeting is important because it represents a large, long-term commitment by the firm. In this way, the capital budget is an expression of the firm's strategic direction. Careful planning is required if the firm is to grow, operate efficiently, and remain competitive.

The capital budget is developed several years in advance of implementation because long lead times are required to properly plan and arrange financing for large projects. The capital budget is usually reviewed and revised annually as the previous year is dropped out and a new year is included. Box 6–12 presents the capital budget for a hypothetical firm over the next five years. It shows that capital expenditures of $10,000 are planned in Year 1 for the purchase of a new computer system and $3,500 for the repayment of a long-term loan. The firm will also have proceeds of $8,000 from a new three-year loan. The capital budget reveals a deficit of $5,500 for the year. This difference will have to be covered through some other source of financing or through the generation of internal funds.

BOX 6–12

Example Capital Budget

	Year 1	Year 2	Year 3	Year 4	Year 5
Capital Expenditures					
Computer equipment	$10,000	—	—	—	—
Repayment of long-term loan	$ 3,500	$20,000	$10,000	—	—
Purchase of a new truck	—	—	—	—	$35,000
Total Capital Expenditures	$13,500	$20,000	$10,000	$0	$35,000
Capital Proceeds					
3-year bank loan	$ 8,000	—	—	—	—
10-year bank loan	—	$15,000	—	—	—
Sale of preferred stock	—	—	—	—	$40,000
Total Capital Proceeds	$ 8,000	$15,000	$ 0	$ 0	$40,000
Capital Surplus (Deficit)	($ 5,500)	($ 5,000)	($10,000)	$ 0	($ 5,000)

Evaluating Investment Opportunities

Capital expenditure projects should be evaluated using the methods discussed earlier in this chapter, the payback, NPV, and IRR methods. Many capital investment decisions are relatively simple, such as the routine repair of a major piece of equipment. Some decisions are mutually exclusive, such as whether to repair or replace a piece of equipment. However, evaluating all of the investment opportunities a firm may have is often very complex, particularly when managers have been actively encouraged to identify potential capital projects. For this reason, it is useful to rank all of the investment opportunities, a task that can be accomplished with all three of the methods listed previously.

The ranking of projects is useful for two reasons. First, it is unlikely that the firm could pursue all of its potential capital projects because it could probably not finance all of the projects simultaneously. Second, implementing capital projects can require a great deal of managerial planning and oversight. By ranking the alternative capital investments the firm can choose those that have the potential for the greatest net returns.

Financing Capital Expenditures

Once management identifies those capital expenditures that it intends to pursue, the plan must be put into operation. However, before this can be done, financing must be arranged, particularly for large expenditures. Sources of financing for food and agribusiness firms were discussed in detail in Chapter 5.

Monitoring Capital Expenditures

Monitoring capital expenditures occurs in two phases. In the first phase, the *acquisition phase,* the costs incurred in conjunction with the asset's acquisition are monitored. The purpose of this phase is to control the cost of the project to make sure that the costs do not exceed the budgeted amount. For example, the managers in charge of the construction of a building should be held accountable for any construction cost overruns that exceed the construction budget. Because cost overruns are common with capital projects, many firms build in a contingency factor to account for unexpected costs.

The second evaluation phase is called the post-audit. The *post-audit* involves comparing the cash flow projections made in the preacquisition analysis with the actual cash flows generated by the asset. The post-audit is used to evaluate the accuracy of the original investment analysis. It is particularly useful in determining ways of improving future estimates. A capital project that is not living up to expectations should be investigated to determine the source of the project's shortcoming and whether corrective action is warranted. Occasionally, a project performs so poorly that the project must be abandoned.

SUMMARY

Investment analysis is a decision aid used in evaluating decisions concerning the purchase of capital items such as machinery, buildings, or land. It is used to determine whether or not a particular investment decision should be made and for ranking alternative investments.

Investment analysis requires the estimation of the net, after-tax cash flows resulting from an investment for each period of the investment's useful life. The net, after-tax cash flow for a period is calculated by summing the after-tax positive cash flows such as increased sales or reduced expenses, and the after-tax negative cash flows such as the initial investment cost or increased expenses.

The simplest method of investment analysis is the payback method. With this method the payback period is determined by calculating the number of periods necessary for the cost of the initial investment to be paid back. Although this technique has several drawbacks, the most important one being that the timing of the cash flows is ignored, it is widely used in industry because of its simplicity and the information it provides concerning a project's liquidity and riskiness.

In order to account for the different timing of cash flows, the concept of discounting, which allows for the valuation of cash flows in different periods on an equivalent basis, is employed. When the firm's cost of capital can be estimated, it may be incorporated into an NPV or IRR analysis to give a more accurate ranking of investment alternatives than that provided by the less sophisticated payback method.

The capital budget should be used to plan, evaluate, finance, and monitor capital expenditures for items such as buildings, equipment, land, and the repayment of debt or equity capital.

CASE QUESTIONS

In addition to the information presented at the beginning of the chapter, the new cash registers are estimated to have no salvage value at the end of five years. Sarah has a 10 percent required return.

1. Explain how you would incorporate this information into the investment analysis.
2. Calculate the payback period for this investment.
3. Calculate the NPV for this investment.
4. Calculate the IRR for this investment to the nearest percent.

REVIEW QUESTIONS

1. Why are net cash flows and not accounting profits used in analyzing investment decisions?
2. Explain why it is more appropriate to use after-tax cash flows than before-tax cash flows in analyzing investment decisions.
3. What are the advantages and drawbacks to using the payback method for evaluating investment decisions?
4. Explain the concept of discounting.
5. Give a brief definition of the factors that affect market interest rates.
6. Give an interpretation of a firm's weighted after-tax cost of capital.
7. What does it mean to say that the NPV of one investment is higher than the NPV of another?
8. What factors should be considered in choosing between a simple method of investment analysis such as the payback method and more sophisticated methods such as NPV or IRR?
9. What factors should be considered in choosing between the NPV and the IRR methods?
10. How are risky investments handled in NPV and IRR analyses?

CHAPTER 7

Strategic Marketing

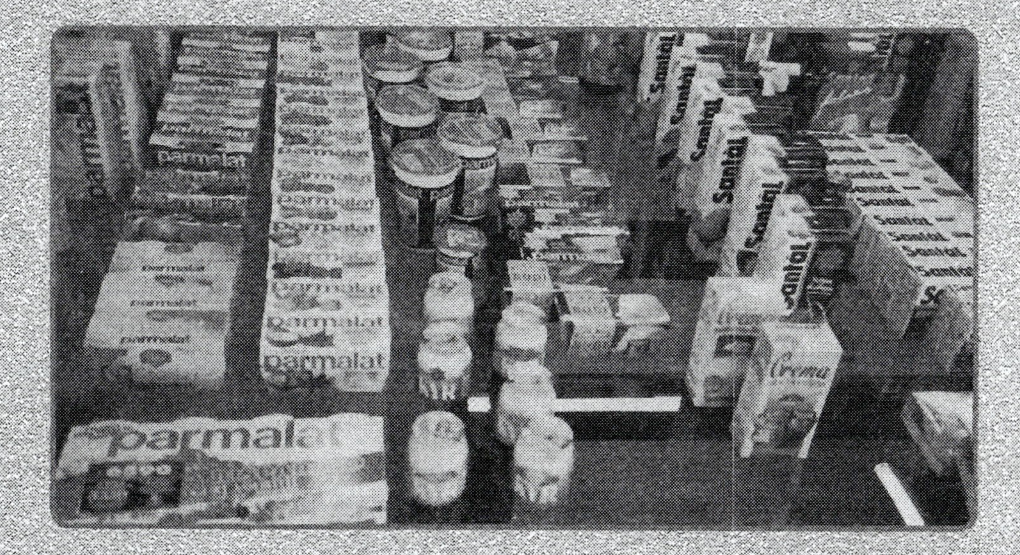

Learning Objectives

In this chapter we will cover the following topics:

- A definition of marketing
- Marketing management
- Target marketing
- Product positioning
- Marketing opportunities
- Branding
- Consumer buying behavior
- Business buying behavior
- Marketing to wholesalers and retailers

CASE STUDY

Jack Anderson is the owner and operator of Anderson's Farm Store, located in Garfield, Nebraska, a town of about 300. The business was started by Jack's father, and Jack took it over seven years ago when his father retired. It has grown from a small feed store to a business that sells feed, seed, fertilizer, farm chemicals, and some commonly requested tools and parts. Although the business was successful under his father's management, most of the growth has occurred since Jack took it over. Jack's father devoted most of his time to his ranch, and the business was more like a hobby to him. Jack, who is 35, decided to manage the business full-time. Since then, both sales and profits have increased substantially.

Although the store is located in Garfield, most of the customers live on farms or ranches in the outlying areas. The store is the local gathering place for farmers and ranchers, as well as for many townspeople. Consequently, Jack has gotten to know many of his customers very well. Many of them have complained to him about having to make the trip to Newton, about 30 miles away, to purchase food and other items. Jack, who has built his business by being sensitive to his customers' needs, had already toyed with the idea of building a supermarket for several years. He had, however, concluded that the amount of business generated would be far too small to operate at a profit. He reached the conclusion that a convenience store had a much greater potential for profitability.

After gathering some figures, Jack was convinced that such a store could be quite profitable. He went forward with his plans and built the store, and is now faced with some major decisions in choosing products. He realizes that the key to the new venture's profitability is the proper selection of merchandise. When he requested information on items to sell, he was startled by the number of choices he had. For toothpaste alone, he received information on more than 100 different items, including many different brands, flavors, and package sizes. He also received information on ice cream alternatives, such as sherbets, sorbets, and frozen yogurts. He is leaning toward stocking only a few traditional flavors of a popular brand. However, his wife has suggested that people purchase ice cream for many different reasons and that he should provide customers with more variety to meet their diverse needs. Jack is not convinced—he strongly believes that ice cream is ice cream, and that other than flavor differences there are few real differences between types or brands of ice cream.

This chapter will help you understand why people and firms buy the things they do and how producers of goods and services exploit this information. After reading the chapter, you should be able to analyze Jack's problem and make a decision on the types of products he should stock.

INTRODUCTION

Companies need to grow if they are to attract good employees, satisfy stockholders, and compete more effectively. However, a company must be careful not to grow simply by pursuing every possible customer—the objective is to achieve profitable revenue growth. To succeed, companies must carefully identify, evaluate, and select marketing opportunities. Managers engage in marketing to assess customers' needs and determine whether a profitable opportunity exists. (We will use "needs" generically to include both needs and wants.)

There is a general misunderstanding of what marketing is and what it can do for a company. This misunderstanding is primarily based on the most visible aspects of marketing, for example, selling and advertising.

One common misunderstanding is that marketing and selling are the same thing. Selling is part of marketing, but marketing includes more than just selling. For example, marketing starts long before the company has a product, whereas selling cannot start until the product is manufactured. Marketing continues throughout the product's life as the company attempts to create repeat sales, find new customers, and improve the product.

Another misconception is that marketing is mostly advertising. Although advertising is critical in the marketing of products, it is only one of the functions of marketing, albeit one of the most visible.

A DEFINITION OF MARKETING

Marketing is the task of finding, developing, and profiting from business opportunities by fulfilling customers' needs. Our approach to marketing, in this and the following chapter, focuses primarily on products that may be differentiated. Commodity products, which are characterized by their lack of differentiation, are sold primarily on the basis of price. Commodity products include grains, oilseeds, and livestock products that are important at the farm and processor level. Increasingly, food products are differentiated and branded, particularly at the wholesale and retail levels. Successfully competing in this market for differentiated food products requires an in-depth understanding of the marketing opportunities, sources of competitive advantage, and marketing strategies that successful companies employ.

MARKETING MANAGEMENT

A key decision every company must make is to determine how homogeneously to treat the market. Companies that produce one product or service for the entire market practice *mass marketing*. Companies that produce products or services for one or more specific market segments rather than the whole market practice *target marketing*. Finally, companies that produce products or services for each individual customer practice *customer-level marketing*. The appropriateness and characteristics of each of these marketing approaches is addressed in the following sections.

Mass Marketing

Mass marketers produce one product for the entire market. Mass marketing had its origins in the Industrial Revolution, as food and agribusiness firms used new technologies to mass produce, mass distribute, and mass promote common food products. Although food products were originally sold in bulk as commodities, today most food products are packaged for sale to the consumer and carry brand names. The use of brand names has allowed food processors to advertise to persuade customers to request their brands, thus encouraging supermarkets to stock these brands on their shelves. Food processors have also promoted their products by offering grocers incentives to display their brands prominently to assist customers in finding their brands among the huge array of available products. Thus, by heavily advertising their brands to consumers, food processors "pull" their products through the market by encouraging supermarkets to carry their brands. Conversely, by offering incentives to supermarkets, food processors "push" their products through the market by securing prime shelf space.

A principle advantage of mass marketing is the low cost of these products, which their manufacturers achieve by realizing *economies of scale*. Economies of scale occur when the average per-unit cost of production declines with the volume of production. This is a common characteristic of industries where there is a high fixed cost associated with production or marketing. In order for its product to be attractive to consumers, a company that produces an undifferentiated product for a mass market must price it sufficiently lower than the differentiated products offered by its competitors. By doing so, the company hopes to achieve a volume that is high enough so that the margin between the product's price and its cost will yield the company an attractive profit. Over time the importance of mass marketing has diminished, as marketers have developed a cost-effective appeal to smaller consumer segments with distinct tastes. As a result, target marketing has become the preferred approach for most products because of its effectiveness in identifying new marketing opportunities.

Target Marketing

There is a well-known saying that goes something like this—you can please some of the people all of the time and all of the people some of the time, but you can't please all of the people all of the time. It's worth repeating—you can't please all of the people all of the time. This statement could serve as the basis for the principle of target marketing. Because consumers have so many different needs, it is difficult for one product to satisfy all consumers. By grouping consumers with similar needs into market segments, it is possible to target a product to consumers in the chosen market segment (the target market).

Market Segments. A market segment is defined as a group of consumers with similar needs. Many markets can be broken down into a number of broad segments using various approaches. *Benefit segmentation* consists of grouping consumers together who are seeking a similar benefit. Some of the more common benefits associated with food products include taste, nutrition, safety, appearance, and convenience. *Demographic segmentation* means grouping customers who share a common demographic makeup, such as age, income, education, sex, or nationality. *Occasion segmentation* means grouping customers according to the occasions for their use of a product. For example, customers who eat out do so for business, pleasure, or to save time. *Usage-level segmentation* means grouping customers according to whether they are users or nonusers of a product, or whether they are light, medium, or heavy users of a product. *Lifestyle or cohort segmentation* means grouping consumers by their lifestyle, such as "yuppies" or "soccer moms."

Although markets may be segmented in many ways, it is important for the company to design market segments that include consumers with similar product needs. The company hopes to profit by identifying a need that is not being met by existing products. Then the company will market differentiated products designed to meet the well-defined needs of the market segments it has identified. However, developing different products for each group is achieved at a cost. The measure of success for a company practicing target marketing is whether it can attract a sufficient number of customers who are willing to pay the price for its differentiated products that are designed to meet the needs of each distinct customer group. This is in sharp contrast to mass marketers, who produce a relatively undifferentiated product that are designed to meet, at a low cost, the common needs of consumers across all market segments.

Another key decision facing the agribusiness is the number of market segments to pursue. A company pursuing a single segment runs the risk that the segment will decrease in size or attract too many competitors. Focusing on multiple segments reduces these risks and often offers an opportunity to lower costs because of the opportunity to share facilities and personnel in the production and marketing of the various products. The major disadvantage in marketing to multiple segments is that the firm may lose market share to more focused competitors. For an example of how market segmentation has been used in the United States brewing industry, see Industry Profile 7–1.

Niches. Niches describe smaller consumer segments that have more narrowly defined needs or unique combinations of needs. Examples of products serving niche markets include specialty teas, gourmet coffees, and exotic fruits and vegetables. Focusing on serving customers in a niche has several advantages, including a better knowledge of the customer group and the potential to

INDUSTRY PROFILE 7–1

THE U.S. BREWING INDUSTRY

The brewing industry in the United States has undergone a major transformation since World War II. In the period immediately following the war most brewers were regional in scale, and there were numerous brewers serving local and regional markets. There were only a few national brewers. Segmenting the market was relatively easy in this post-war period. Beers fell into one of three segments: low cost, premium, or import.

Starting in the 1950s, the cost savings associated with operating larger plants and national advertising spurred consolidation in the industry. By the 1990s the U.S. industry was dominated by two giants, Anheuser-Busch and Miller Brewing Company. There were numerous market segments and niche markets.

Possibly the greatest innovations in the industry resulted from the 1970 purchase of Miller Brewing Company by Phillip Morris, the tobacco giant. After the acquisition Miller adopted many of the aggressive marketing tactics of its parent company, segmenting the market, advertising heavily, and introducing new products. Miller introduced the first successful low-calorie beer, Miller Lite. Through its successful advertising campaigns, Miller effectively differentiated its products, even though most beer drinkers could not tell one major brand from another in blind taste tests. Although the initial effect of Miller's heavy spending on advertising was to reduce Anheuser-Busch's market share, the end result was that many smaller brewers were driven out of business.

Today, the beer market is extremely diverse. Anheuser-Busch has brands for every pocketbook: Busch is a popular priced beer, Budweiser is a premium priced beer, and Michelob is a super-premium priced beer. Every major brewer also has a low-calorie beer as well as a no-alcohol beer, such as Miller's Lite and Sharps. Miller also has the Lowenbrau brand, which is positioned as an import. Coor's has the Zima brand, which is targeted toward women, and Anheuser-Busch has many specialty beers targeted at niche markets including Red Wolf, Tequiza, and AmberBock. Microbreweries, such as Seattle's Redhook Ale Brewery and Boulder Brewing Company, also produce an astounding array of products filling market niches too numerous to mention.

earn a greater profit, which results from the firm's ability to meet the niche customers' needs more effectively than more broadly focused competitors. However, niche marketers face the same risks as single-segment marketers should the niche decline. Pursuing a multiniche strategy offers advantages and disadvantages similar to those of pursuing a multisegment strategy.

Customer-Level Marketing

Companies may want to identify even smaller groups of customers who share some characteristics that provide a market opportunity. Today, many companies build customer databases containing information about their customers' demographics, past purchases, and other characteristics. The intersection of computers, customer databases, and manufacturing flexibility gives companies the possibility of offering customized products and personalized services. Customized marketing takes place when the manufacturer produces a new product from scratch for the customer, for example, when an equipment manufacturer builds a piece of specialized equipment for a food processor. Mass customization occurs when the company has established basic modules that can be combined in different ways for the customer, for example, when an equipment manufacturer includes optional equipment to meet the customer's needs.

Product Positioning

Once the food or agribusiness firm has determined which market segments it will pursue, it must decide how its products will be positioned within those segments. The goal of *product positioning* is to utilize the company's strengths so that its products will have a competitive advantage relative to competitors' products. Food products are often positioned based on product quality, as defined by taste, nutrition, price, or the image associated with consumption of the product. A key factor in successfully positioning a product is to ensure that consumers know the product's key benefits and that they are able to differentiate the firm's product from its competitors' offerings.

After positioning a product, the company then establishes a marketing mix that will support the product's positioning. The *marketing mix*, also known as the *four Ps*, includes various decisions regarding the product, price, place, and promotion. The marketing mix will be covered in the next chapter, where tactical marketing is discussed.

MARKETING OPPORTUNITIES

You will recall that marketing is the task of finding, developing, and profiting from business opportunities by fulfilling consumers' needs. A marketing opportunity exists when a company identifies a group of potential customers whose needs are not being met, or whose needs it believes it could meet better than its competitors' existing products. There are several situations that may create a marketing opportunity. It is the task of the marketing manager to research the opportunity and determine if it will be profitable. This is done by examining factors such as the market segment's size, the customers' needs, and the ability of the company to deliver a superior product.

Market Research

Market research is a method employed by many successful food and agribusiness firms to identify, explore, and exploit market opportunities. It is used to reveal the different market segments based on distinct customer needs and preferences, to understand the needs of consumers in a segment, and to develop a marketing strategy. Most importantly, firms use market research to determine if the firm can profitably take advantage of the marketing opportunities it identifies. There are three situations that create marketing opportunities for a firm. These situations include (1) supplying a good or service in short supply, (2) supplying an existing product or service in a superior way, and (3) supplying a new product or service.

The first situation exists when a company can supply a good or service in short supply, a condition often caused by a rapidly growing market. Succeeding in this situation requires the least amount of marketing talent. Because buyers are lining up to buy the product, no market development is needed. Such marketing opportunities seem very attractive because they appear to be obvious profit opportunities stemming from the high prices firms charge. The high visibility of this profit opportunity may also be the greatest disadvantage because other firms will be likely to notice and respond to the same signals. The most successful firms are typically those that are first to enter the market in this situation and those that establish either a strong brand or low cost position. The remaining firms will suffer from low market share and low profits and eventually be forced to exit the market.

The second type of opportunity occurs when a company can enter a market by supplying an improved product or service. One way to identify such opportunities is to recognize areas where existing products fall short of consumer expectations. Customer surveys are used to determine if consumers are dissatisfied with or have any suggestions for improving an existing product or service. The firm may also ask customers to describe their ideal product or follow a product through the consumption channel to determine if there are areas for improvement. For example, in the last decade supermarkets started processing chuck roasts into ground chuck because customers no longer purchased as many roasts as they once did. When consumers were asked why they were not purchasing chuck roasts, they indicated that they liked to serve roasts for dinner but that they did not like the amount of preparation time required. Several meatpackers seized this opportunity and began to offer precooked pot roasts in microwavable packaging, substantially shortening the amount of preparation time required. Supermarkets now offer many precooked meat items that meet the needs of customers who want a traditional dinner but do not have the time to cook it themselves.

Finally, a firm can break into a market by offering a new product or service. The two other methods of market entry depend primarily on identifying market or product deficiencies. These methods typically produce product or service improvements rather than product or service innovations because most consumers are limited in their ability to imagine new products or services derived from new technologies or creative breakthroughs. Therefore, new product innovations normally come from a firm's research and development (R&D) section.

During the last two decades, rapid technological changes and increases in the standard of living have resulted in great opportunities for product development in the food industry. The goal of product development research is to invent and develop new products that customers will buy. Product development starts with a marketing strategy to determine what the firm is trying to achieve and how the product fits into the firm's product offerings. The marketing strategy guides the selection of research and development projects and serves to inform the research team of the firm's marketing goals.

The first step is to create and screen ideas. Two key components common to many innovative firms are fostering a creative environment and ensuring that ideas flow to key decision makers where they can be collected, reviewed, and evaluated. One method of accomplishing this is to appoint a senior member of management to be the idea manager. The manager creates a multidisciplinary committee consisting of a scientist, an engineer, and employees from purchasing, manufacturing, finance, marketing, and sales. This committee meets regularly to evaluate proposed new products and services. Everyone involved with the firm is then encouraged to submit ideas to the committee for review. Frequently, firms reward individuals who contribute good ideas that become successful products or services.

The committee sifts though the ideas and identifies those that appear to be promising. The ideas that survive the initial screening process are then assigned to members of the committee to determine if the product can be manufactured and successfully marketed. If the product appears feasible and marketable, the committee will pass the information to the R&D section, where the product will be designed and product prototypes will be developed. A *focus group,* which is a select group of potential customers representative of the target market, is then asked to evaluate the product. If the prototype is successful, the firm will then manufacture the product in a pilot plant or in small runs in the standard plant to find out if the product can be easily, consistently, and economically manufactured to the firm's quality standards. The product is then test marketed in regions throughout the country to determine the most effective promotion, advertising, pricing, and distribution strategies. Once the test marketing is completed, a financial analysis is conducted that includes the costs of launching the product and production, along with predictions for revenue, costs, profits, and the ultimate success of the product. At this point, if the product still looks promising, production and marketing plans are finalized and the product is launched into the market. During the first year of its launch, the product will be carefully monitored to determine its success and to make any production and quality adjustments or changes in distribution and marketing methods to ensure product sales growth.

Branding

A fundamental marketing decision is whether to brand a product. A brand gives a product identity in the marketplace, connotes a level of quality or service, and can be of great value to both consumers and the company that owns it. A *brand* is defined as a name, term, or symbol identifying the maker of a product. Ultimately, brands are valuable to consumers because they provide a guarantee of consistent quality and reduce the search costs involved in shopping. Brands are valuable to producers because they encourage consumer loyalty to the brand and encourage repeat purchases.

Brands, however, are costly to develop and support. They also add more value to some types of products than others. For this reason not all types of products are branded. A typical manufacturer will have to decide not only whether to brand the product but also the level at which it should be branded.

Food products are most commonly sold as unbranded, a retail brand, a foodservice brand, or a private label brand. Tomato paste provides a good example of how a product is marketed using all four alternatives. Tomato paste manufacturers produce paste from ripe red tomatoes using an elaborate process of cleaning the tomatoes and then evaporating off the water to condense the solids into a thick paste. Some processors sell the tomato paste in bulk in 55-gallon drums. The product is described in terms of standard quality attributes, but it is not branded. It is used by processors all over the world as an ingredient in other products, such as pizza sauce or ketchup. Tomato paste is also branded. Companies like Heinz produce tomato paste that they market under their own retail label (sometimes referred to as a *national label* or a *manufacturer's label*). Other companies produce tomato paste for the foodservice industry with a label used exclusively for foodservice. The last alternative used by tomato paste manufacturers is to produce the product but package it under another company's label. Supermarket chains often contract with tomato paste producers to produce tomato paste to their specifications and put the supermarket's label (referred to as a *private label*) on the can.

This overview of the tomato paste industry also provides insight as to why each of the branding options may be selected. Producing in bulk and not branding is a low-cost option that is used when the product is produced in great quantities and it entails no investment in branding. The foodservice label and private label alternatives require only a modest investment in brand development. In the private label case, the investment in the private label is made by the label's owner. Private label manufacturers typically focus on low-cost production and may produce for several firms. Tomato paste manufacturers selling under their own foodservice label must promote their label to the foodservice industry. However, this is much less expensive than promoting a national brand, which involves a large advertising and promotion budget. Developing a national label is the most expensive of the branding options. However, it also yields the greatest profit potential because it involves the greatest profit margins as well as the possibility of developing long-term consumer loyalty.

The following section describes how brands are developed, maintained, and grown. The focus is on national manufacturer's brands; however, much of what is discussed may also be applied to private label and foodservice brands.

Building Brand Equity

Once a company has introduced a product into the marketplace, it is then faced with the task of getting customers, keeping customers, and growing the customer base. Brands provide the vehicle for accomplishing this objective through the development of *brand equity* (the brand's monetary value). Establishing a valuable brand is more of an art than a science. Nonetheless, there are several general guidelines for developing a brand.

First, the company must determine how its product will be positioned relative to its competition. For example, some companies position their products by focusing on product differentiation. These firms use market research, new technology, and innovation to become the product leaders in their industry. They appeal to customers who may be looking for the newest, highest quality, or most advanced products in the industry. Other companies concentrate on being the most operationally efficient. They offer products that appeal to customers who are looking for a good-quality, low-cost version of the product. Finally, some companies position their products in a market niche. These companies cater to consumers who require products designed specifically to meet the distinct needs of the market niche.

Companies usually go beyond broad positioning to give customers a more definitive reason to buy their products. Most companies market their products to provide customers with a specific benefit. For example, soft drink manufacturers market diet soft drinks to appeal to weight-conscious consumers, whereas regular soft drinks are targeted at customers who prefer the taste of soft drinks sweetened with sugar. Food manufacturers also use quality or price positioning to market packaged food products at a certain quality and price level. For example, "gourmet" dinners appeal to consumers who seek high quality and are willing to pay a high price, whereas dinners featuring large portion sizes, such as the Hungry Man label, appeal to customers who want large portions and a good value. Food manufacturers also position products in other ways, including use or application (Hamburger Helper), category (Pace picante sauce), or attribute (Coors made from mountain spring water). Once the company establishes the product positioning, it then begins the important task of building brand equity.

Choosing a Brand Name. In naming a product or service, a company has many choices. Some examples of well-known food and agribusiness brand choices include the name of a person as a brand name (John Deere, Tyson), location (Kentucky Fried Chicken, Milwaukee's Best), quality (Safeway, Sunkist), lifestyle (Weight Watchers, Healthy Choice), and an appealing artificial name (Haagen Daz, Blue Diamond).

Companies also differ in how they use their brand name. For example, John Deere uses its brand name on all of its products, including farm machinery, industrial equipment, and lawn mowers. This is referred to as *family branding.*

In contrast, Campbell's uses a *line branding* strategy. In addition to putting its corporate name on its soups, Campbell's uses the brand names Pace, Prego, V8, and Pepperidge Farm to denote different product lines. Procter & Gamble, on the other hand, prefers to use an *individual branding* strategy, with brand names such as Pringles, Jif, Crisco, Folgers, and Sunny Delight representing different products. Although companies create and use brand names in many different ways, they generally follow these basic guidelines when choosing a brand name:

- The brand name should be easy to pronounce, recognize, and remember.
- It should suggest something about the product's benefits or special attributes.
- It should be distinctive and not carry a double meaning.

Brand Associations. The best-known brand names carry positive associations. There are five basic dimensions to building *brand associations:* attributes, benefits, values, personality, and users. As an example, let's look at how Coca-Cola, the most recognized brand in the world, uses these dimensions to build brand associations. First, a brand should trigger in the consumer's mind certain attributes. Coca-Cola is cool and refreshing. Second, a brand should suggest benefits. Drinking Coca-Cola is fun and satisfying—it picks you up! Third, a brand should connote values and associate the product with a positive image. Based on the famous song used in advertising the product many years ago, the Coca-Cola Company wants to foster peace, love, and understanding among the people of the world. Fourth, a brand should exhibit personality traits. Of course, Coke is the "real thing." Last, a brand should reflect positively on the kinds of customers who buy the product. Coca-Cola portrays those who drink Coke as individuals who strive to create a world community.

Brand Identity. It is the job of marketing to build *brand identity* using positive associations based on the five dimensions just discussed. When mentioned, a brand name should trigger a favorable response from customers in the target market. The Coca-Cola brand is so powerful that consumers frequently use the word "Coke" to refer to "colas" made by different companies. Another technique is to use a slogan repeatedly to reinforce the brand image the company is trying to project. Hearing (or seeing) a slogan over and over creates a strong association between the product and the message associated with the brand image. Coca-Cola's message that Coke is the "real thing" attempts to convince consumers there is no cola substitute for Coke. Companies often use color to aid in brand recognition; for example, Coca-Cola uses a distinctive red color in its packaging. Companies may also adopt symbols in their communications. Symbols may include the use of well-known spokespersons, such as actors, recording artists, or athletes; characters, such as Ronald McDonald or the Pillsbury Doughboy; and logos, such as the Nabisco symbol or Farmland's double circles.

CONSUMER BUYER BEHAVIOR

Understanding what motivates buyers is critical to marketers of food and agribusiness products. The last several decades have proved very fruitful in the development of models to explain the purchasing decisions of both individual consumers and businesses. Since a detailed presentation of even a representative sampling of these models is outside the scope of this textbook, we have chosen to present a few models that are very insightful to understanding purchasing behavior.

The Adoption Process

The adoption process is useful in understanding how consumers decide to buy a product, particularly a new one. The decision to buy a product may be broken down into the logical sequence of events described in Box 7–1. In purchasing a new product, a consumer must: (1) find out about the product (awareness); (2) become interested in the product; (3) evaluate how well the product fits his or her needs by comparing it to other products, evaluating its price, and weighing other factors; (4) experiment with the product, if possible (trial); (5) make a decision to purchase or reject the product; and (6) evaluate the decision every time a purchase is made (reinforcement or rejection).

The actual adoption process is often more complicated than this sequencing, and it varies considerably with the person and type of product purchased. For example, the adoption process for purchasing a package of chewing gum is very simple and may take only a few seconds, because the cost of making a mistake is minimal. However, the decision to purchase an expensive item, such as a car, may take several months because it is an important financial decision for most people.

The adoption process may also become routine, especially for low-cost, frequently purchased items. For example, satisfaction with a brand of mayonnaise may lead a consumer to simply repurchase the brand each time the need arises without becoming interested in or evaluating other brands. On the other hand,

BOX 7–1

The Adoption Process

Awareness	Customer learns of the product
Interest	Customer's interest is stimulated
Evaluation	Customer evaluates information about the product
Trial	Customer experiments with the product
Decision	Customer makes a decision on buying the product
Reinforcement or Rejection	Customer evaluates the buying decision

dissatisfaction may lead the consumer to reject the brand and initiate a previous step in the adoption process, such as reevaluating the product or initiating a search for alternative products.

An understanding of the adoption process is particularly important in promoting a product. Producers of new products typically try to hurry consumers through the adoption process, often by helping them complete several steps in the process simultaneously. Sending a product sample in the mail or passing out free samples in the supermarket is an example of this technique, which ensures that the consumer is made aware of the existence of the product. And what better way to spur interest in a product than by giving it away? Most consumers will evaluate and try the product because there is virtually no cost associated with doing so. By using this method, marketers can be virtually certain that consumers will make a decision about their product without having to coax them along at each stage. Coupons are often used in conjunction with free samples to serve as a reminder of the new product and as additional incentive to purchase the product.

Categories of Product Adopters

Understanding why some people readily adopt new products and why others are slow to do so has proved to be a useful tool in understanding and influencing buyer behavior. Marketers call it the *adoption curve.* Some people adopt a product shortly after it is introduced—to be the first on the block. Others tend to follow, and still others may never get past the awareness stage. These tendencies have as much to do with purchasing behavior as demographic or psychographic characteristics. Understanding the characteristics and motivations of each type of product adopter is extremely useful in developing marketing plans for introducing new products. People usually fall into one of five categories on the adoption curve: innovators, early adopters, early majority, late majority, and laggards (Figure 7–1).[1] Each group has a different time frame that allows them to establish a comfort level when purchasing a new product or service. Members of each group also tend to have certain demographic characteristics in common, which help marketers appeal to the individuals in each segment (see Box 7–2).

Innovators. The first group of people to adopt a product or idea is called innovators. They tend to be young, well-educated, and willing to take risks. They are often scientifically oriented and keep up with the latest technical information. They usually have many contacts outside their local community. Because of their reliance on technical information, the best way to reach innovators is through their information sources—scientific documentation in trade magazines, technical journals, public or educational television, or the Internet.

---

[1]See Everett M. Rogers, ***Diffusion of Innovations***, 4th ed., The Free Press, New York, 1995, pp. 252–280.

FIGURE 7–1

Categories of Product Adopters

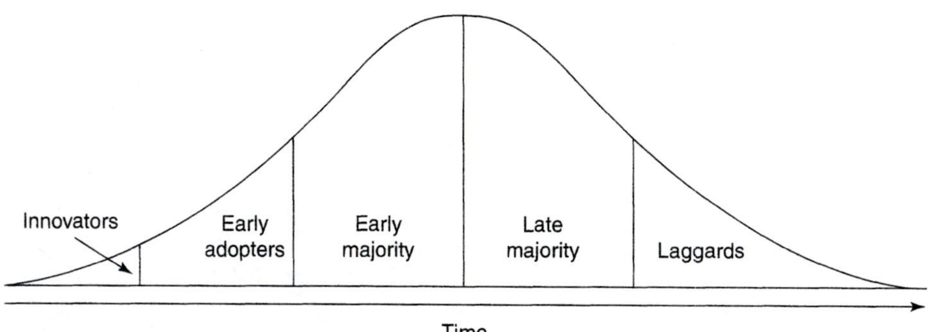

Source: Reprinted with the permission of The Free Press, a Division of Simon & Schuster, Inc., from ***Diffusion of Innovations***, Fourth Edition, by Everett M. Rogers. Copyright © 1995 by Everett M. Rogers. Copyright © 1962, 1971, 1983 by The Free Press.

BOX 7–2

Characteristics of Product Adopters

Category	Characteristics
Innovators (3% to 5% of the market)	Younger Risk takers Well-educated High social and economic status Rely on technical information Self-reliant in gathering information
Early Adopters (10% to 15% of the market)	Younger Well-educated High social and economic status Cautious in making a decision Rely on sales people and self-reliant in gathering information
Early Majority (30% to 40% of the market)	Average social and economic status Rely on media, friends, family, and early adopters in gathering information
Late Majority (25% to 35% of the market)	Below average social and economic status Susceptible to social pressure from their own peer group Wait to try product until it is widely accepted
Laggards (5% to 15% of the market)	Lowest social and economic status Tradition-bound

Early Adopters. After innovators, the next group of people to adopt a product is the early adopters. Like the innovators, they also tend to be young and well-educated, but not quite as willing as innovators to take risks. They also get much of their information from technical and scientific sources, but they rely more heavily on salespeople than do innovators. However, this alone is not enough to convince them to buy a product. Unlike innovators, they tend to have a wait-and-see attitude. They watch the innovators to see if the new product or idea is a success—and then they adopt those products they view as successful innovations. In this way they gain most of the benefits of having a new product—that is, they are still among the first owners—while assuming little risk. When an early adopter adopts a product, it usually signals to other groups that the product is a "safe bet." Consequently, the early adopters are opinion leaders; they are watched closely by other groups. Because their opinions are so well respected in the community, it is important for marketers to get the early adopters to use their product.

Early Majority. The early majority will adopt a new product only when it has been adopted by a significant number of innovators and early adopters. Members of the early majority group seek out a lot of information before making a purchase. Although they rely heavily on information from magazines, newspapers, or other popular information sources, they will usually not adopt a new product until they have seen it used successfully by family, friends, or neighbors. Before they buy a product, they want to see that it is being used by a large number of people, thereby avoiding all risks.

Late Majority. Members of the late majority group will purchase a new product only after it is no longer new. They are very cautious about adopting new ideas, tending to be less educated, conservative, and older. They are less likely than the early majority to follow the early adopters. They will, however, respond to peer pressure from their own group, and as the product becomes more widely accepted, they too will adopt it.

Laggards. The last group of people to adopt a product, if they adopt it at all, is the laggards. They don't like to try new ideas, preferring instead to do things the way they have always done them.

Understanding the classes of product adopters is useful for a firm in determining where it should concentrate its marketing efforts for new products. Marketers consider the early adopters to be the most important group because of their influence on the other categories of product adopters. Once they buy the product, others will follow. Therefore, initial advertising or personal contact is often aimed at identifying and influencing the early adopters. As more people adopt the product, the marketing strategy must change to reflect the changing characteristics of the consumers who have yet to adopt the product. For example, initial advertising may be concentrated in technical magazines read by innovators and early adopters, whereas subsequent advertising may appear in more popular publications read by the early and late majority. Advertising should also reflect the target group's responsiveness to different messages. For example, early adopters may be most impressed by the benefits of a product, whereas later adopters are more heavily influenced by peer pressure. Thus, advertising aimed at the early or late majority tends to show individuals actually using the product.

BUSINESS BUYER BEHAVIOR

Many products or services are purchased for use in the production of goods and services, which are then sold to other businesses. This is called *business-to-business marketing.* It is distinguished from consumer marketing because the customer in this case is not the end consumer but rather a business that will resell the product or use it as an input into another product. Examples of business buyers include food processors, foodservice providers, wholesalers, and retailers.

Business buyers behave differently than consumers. Whereas consumers purchase goods to satisfy their own needs, business buyers make purchases based on how the purchase will affect the company's profits.

Because the behavior of business buyers tends to be less complex than consumer behavior, relatively few models have been developed to explain the process of business buying behavior. However, one frequently used model is the problem-solving process (Box 7–3).

The initial step in the process is an awareness of the problem. A problem is defined as a difference between the actual and desired situation. This may occur when a firm's inventory is running low, when a machine breaks down too frequently, or when profits are undesirably low. A problem is a sign that something is wrong or that improvements can be made. It signals the need for action. Sales representatives and advertisers often make business buyers aware of a problem. For example, many farmers never thought that their fields were graded improperly until custom farm operators informed them of the benefits of more efficient irrigation resulting from laser planing their fields.

The second step—one that is easy to overlook—is the determination of the problem's cause. It is important to distinguish between the problem's symptoms and the problem's cause if a long-term solution to the problem is desired. For example, falling sales for a product may only be a symptom of the real problem. If the product is obsolete, more advertising may temporarily boost sales, but ultimately the problem of the outdated product must be addressed.

BOX 7–3

The Problem-Solving Process

Step 1. Awareness of the problem
Step 2. Determination of the problem's cause
Step 3. Search for alternative solutions
Step 4. Evaluation of alternative solutions
Step 5. Choice of a solution
Step 6. Evaluation of the decision

The next step is to seek out alternative solutions. Potential solutions to the problem may include alternative strategy formulations, the search for a substitute product, or simply examining different brands. The last steps involve evaluating the alternatives based on the decision maker's criteria and determining which alternative will be selected. Successful decisions will be evaluated favorably and used again, whereas unsuccessful decisions will be discarded or at least modified before they are repeated.

For first-time purchases the problem-solving process may be extensive, with each step in the process requiring a lot of time. For goods that are frequently purchased or for low-cost items, the entire process may take only a matter of seconds and be very routine. The complexity of the problem-solving process will depend on the size and frequency of the purchase and the success of previously selected solutions.

You may notice some similarities between the problem-solving model and the product adoption model. Because they are both human decision-making models, we would expect some similarities. All purchase decisions are, after all, made by people. The key to understanding a business buyer's purchasing behavior, just as it is in understanding a consumer's purchasing behavior, is knowing the buyer's motivation. By understanding the motivating force behind a buyer's actions, it is possible to determine what to market and how to market it. Consumers are driven by their needs and wants; business buyers are driven primarily by the profit motive (with the exception of nonprofit organizations and government agencies). In other words, business buyers are interested in seeing how a product will improve the economic performance of their firm. Larger firms have individuals, usually called *purchasing agents* or procurement agents, specifically assigned to the task of buying. Purchasing agents tend to be more predictable than consumers in that they have well-defined, easily understood goals. The way to successfully market to purchasing agents is to show them how the product can help them achieve the goals of their business.

The problem-solving process reflects a more extensive search for a solution to a problem than does the consumer adoption process. It is well suited to modeling behavior because businesses are constantly searching for ways to lower costs or increase income. However, the problem-solving process may also be useful in understanding consumer behavior, especially for very expensive or infrequently purchased items. By the same token, the product adoption model may in some cases be useful in understanding purchasing agent buyer behavior, because these buyers may be motivated by goals other than simply maximizing profits.

Marketing to Wholesalers and Retailers

Up to this point we have treated business buying behavior in a generic fashion. However, two categories of food businesses deserve special treatment: wholesalers and retailers.

Wholesalers may generally be defined as firms that source products from producers or processors and supply retailers, businesses, or other wholesalers.

They perform a variety of functions, including buying, transporting, storing, grading, financing, and selling. Wholesalers include firms such as brokers, grain elevator owners, vegetable packers, and food distributors.

Retailers are defined as firms that sell products to the final consumer. Food retailers include supermarkets, warehouse or wholesale clubs, online grocers, and bakeries.

Firms manufacturing a finished consumer product, such as frozen enchiladas, face a dual marketing challenge. Understanding consumer needs, wants, and behavior is essential to their survival, but they do not generally deal directly with the buying public. Their products are sold through wholesalers and retailers. For these firms to have the opportunity to sell any product to consumers, they must first convince wholesalers and/or retailers to stock their products. The "initial consumers" of their products are the purchasing agents for the wholesalers and retailers. To be successful, these food manufacturers must understand how wholesalers and retailers work and what motivates their behavior.

Buyers for wholesalers and retailers interpret what they believe to be the consumers' wants and preferences. If a particular product doesn't sell, the wholesalers and retailers will lose money. Therefore, they place a heavy emphasis on what consumers want. These consumer needs and preferences are, however, filtered by individual buyers' beliefs, views, and experiences. The buyers' perceptions may be an accurate reflection of consumers' preferences, or they may be totally incorrect. The danger for processors that have done their homework in developing a good product is that buyers may not be interested in offering the product because they have a different perception of consumer demand.

Retail and wholesale buyers must balance their views of consumer wants with their companies' policies and procedures. Manufacturers must be aware of these policies and procedures and adapt their products to comply with them or they will not succeed in making a sale. Examples of store policies and procedures include the type of packaging, display methods, pricing, and return policies. The product must be packaged so that it fits the retail shelf display space for that particular type of product. Many stores will not allow large, freestanding displays. Also, buyers have a preconceived idea of what price a particular product will sell for in their store. This is called a *price point.* If the price of the product is above their perceived price point, they most likely will not handle the product.

Retail and wholesale buyers are faced with a very difficult task. Generally, they are offered at least 10 times as many products as they have the space to stock. Food retailers typically charge a *slotting allowance,* which is a fee paid by the food manufacturer to the supermarket for access to the retailer's shelf space. This is unique to food retailing, although not all products are sold in association with slotting allowances (produce is an exception) and not all food retailers charge slotting allowances (warehouse clubs and nontraditional food retailers such as Wal-Mart are also exceptions). Slotting allowances are particularly problematic for small food manufacturers. These fees are substantial and may limit the opportunity for small firms to get their products to market through conventional retail food channels.

It is very difficult for products from new companies, and even new products from established companies, to break into the market. Buyers tend to give preference to established companies and products with proven sales records. However, each wholesaler and retailer operates differently.

The sections in this chapter on market segmentation, the problem-solving process, and categories of product adopters may be used to understand the buying behavior of both wholesalers and retailers. Some companies, either by practice or policy, may be considered early adopters of new products, whereas other companies are not. For instance, direct-mail catalogue companies tend to be early adopters. On the other hand, the major discount retailers that offer a limited selection of food products are much more conservative and typically require historical sales information before agreeing to purchase a product. This policy makes it difficult to sell new products to this segment of the market.

SUMMARY

Marketing is the task of finding, developing, and profiting from business opportunities by fulfilling customers' needs. A key marketing decision is to determine whether to produce for the mass market or for specific market segments. Companies that produce products or services for the entire market practice mass marketing. Mass-marketing companies generally produce products with a wide appeal and compete in the market by producing goods at a low cost. The downside to mass marketing is the potential vulnerability to competitors who specifically design products to meet the needs of individual customer segments.

Most companies now manufacture products for distinct target markets. By grouping customers with similar needs into homogeneous market segments, it is possible to design products that will have great appeal to customers in specific market segments. Companies segment the market in many different ways, including benefits derived, demographics, occasion, usage, and lifestyle. Some companies engage in niche marketing, which can be described as targeting relatively small consumer segments that have well-defined or unique combinations of needs. Companies focusing on the narrowest market segment, the individual customer, practice customer-level marketing.

Once the market is segmented, the company must decide how to position its product or service in a particular market segment by establishing a marketing mix that will support and deliver the product's positioning. The key to success is to position the product in the target market so that it will be favorably received relative to its competitors' products. Successful product positioning starts with market research. A company uses market research to determine if there is a marketing opportunity. Marketing opportunities occur in situations where a product or a service is in short supply, where the firm can supply a product or service in a superior way, or where a company can supply a new product or service.

Once a company has introduced a product or service into the marketplace, it is then faced with the task of getting customers, keeping customers, and growing the customer base. Branding is used by many food and agribusiness firms to differentiate their products and build long-term consumer loyalty. Building brand equity is more of an art than a science and requires a company to focus on several tasks. To be successful, companies must give customers a persuasive reason to buy their brand. They must build the brand by choosing a strong brand name and making brand associations that reflect the brand's attributes, benefits, values, personality, and users. If the company is successful in establishing positive associations among consumers of its brand, it will develop a brand identity and consumer loyalty.

Several models explain the buying process. The commonly used product adoption process explains how consumers go about buying a product. The adoption process comprises six steps: awareness, interest, evaluation, trial, decision, and reinforcement or rejection. The product adopter categories represent groups of consumers that are classified according to how quickly they adopt new products. These classes are innovators, early adopters, early majority, late majority, and laggards. The early adopters are typically the most sought-after group because they are often opinion leaders, whose opinions are highly regarded by members of other groups.

Business buyers and the purchasing agents who represent them, constitute another important buying group. They tend to use a problem-solving approach in making purchases. This approach consists of the following steps: awareness of the problem, determination of the problem's cause, search for alternative solutions, evaluation of alternative solutions, choice of a solution, and evaluation of the decision.

Wholesalers and retailers represent a special category of business buyers. They primarily purchase products for resale and perform little or no processing. Food manufacturers selling to these buyers must recognize that although they may be producing a product destined for the final consumer, the preferences of consumers are filtered by the wholesaler's or retailer's perceptions of consumer desires. These perceptions combined with store policies are important in understanding buyer behavior.

CASE QUESTIONS

1. Use the market-segmenting process described in this chapter to identify what you believe to be the important market segments for ice cream and ice cream-like products (such as frozen yogurt). Describe in detail what you have done in each step.

2. Of the market segments you have defined in Question 1, which ones do you think Jack should target for his convenience store? Why should he target them? Name a product that falls into each of the market segments Jack should target and explain why you believe it falls into that segment.

REVIEW QUESTIONS

1. Describe a product that is mass marketed. Why has the product been successful (or unsuccessful)?
2. What are the principal advantages and disadvantages associated with mass marketing and target marketing?
3. Pick a new food product that has reached the market in the last year. Describe the company's target market segment or segments for this product.
4. Describe the three different types of marketing opportunities.
5. Pick an existing food product that has been on the market for a long time. Describe how a company can use market research to reposition the product in the marketplace. How would you recommend that the company reposition the product?
6. Pick an existing food product and describe how its manufacturer has attempted to build the product's brand equity.
7. Describe how you could apply the adoption process to a recent purchase you made.
8. Which class of product adopters is currently adopting global positioning to map out their farm and ranch land? What strategy would you use to get agricultural producers in this class to adopt the product?
9. What are the similarities and differences between the adoption process for consumers and the problem-solving process for wholesalers and retailers? Why are there differences in buying behaviors between the two groups?
10. What are the special characteristics of wholesale and retail buyers?

CHAPTER 8

The Marketing Mix

Learning Objectives

In this chapter we will cover the following topics:

- The four areas that comprise the marketing mix
- Key aspects of the product decision
- The stages of the product life cycle and their impact on the marketing mix
- Commonly used pricing strategies and when they are most appropriately used
- How discounts and allowances are used in pricing
- The promotion decision
- The process of personal selling
- The place decision
- Illegal marketing activities

CASE STUDY

Wendy Heathcott is the owner and manager of a snack food company in Fairfield, New Jersey. Originally started as a bakery in 1953, Wendy began expanding her business by packaging some of the more popular products such as cakes and cookies and distributing them in local supermarkets. The success of this venture encouraged her to include potato chips and candy bars as well. She later changed the name of the company from Wendy's Bakery to SnackTyme and expanded the distribution of products to include most of the greater New York City area.

Wendy believes strongly that her employees are her firm's most valuable resource. Her policy of rewarding hardworking employees with above-average wages has paid off many times over through high productivity and low turnover and absenteeism. In the past she was often seen on the production floor discussing new ideas with employees and has found their suggestions very useful in controlling costs. In recent years with the growth in her business and the increase in the number of employees, she has found that she no longer has enough time to spend with her employees. About five years ago she installed a suggestion box and began rewarding employees whose suggestions were implemented. The following suggestion came from a 43-year-old mother who has worked for the company for 13 years.

SUGGESTION CARD

Suggestion: Have you ever considered making a candy bar entirely out of fruit and fiber? My children are always snacking on candy bars loaded with chocolate and sugar, and I think a nutritious candy bar would be quite popular.

Name: Lisa James

Department: Packaging

Phone: 555-4569

Wendy liked the idea and consulted with the product development people who thought such a product would be feasible although it would take a while to develop. She asked them to start working on the product immediately and requested that a market study be done on the product's potential.

The marketing department's report indicated that the product had a high profit potential, not only in the New York area, but nationwide, due principally to its uniqueness. The product would be positioned as a "healthy" candy bar. The major market segment for the product was described as consisting of those consumers who are health conscious but not willing to sacrifice taste. The marketing department emphasized the importance of incorporating two principal characteristics, nutrition and flavor, into the candy bar. The product will be nutritious due to its fruit content, as well as having a "natural" sweetness—no sugar added. The fiber content will give it the crunchiness desired in candy bars as well as making an important dietary contribution. The two most important consumer groups to be targeted will be parents concerned about their families' nutrition and regular buyers of health food products. Before the product is marketed, important marketing decisions will have to be made concerning product, price, promotion, and distribution of the product. The information in this chapter prepares you to analyze these marketing decisions.

INTRODUCTION

One of the distinguishing features of modern marketing is that it is externally focused. Henry Ford is famous for insisting that consumers could have any color automobile they wanted—as long as it was black! How many cars would Ford sell today with this kind of philosophy? Today's most successful marketers are keenly aware of the market environment. This includes an in-depth understanding of customers and their needs, new laws and regulations, the international trade environment, new technological developments, and the political environment, to name a few of the most important factors. This externally focused view of the marketing system recognizes the system's dynamic nature. The ability to initiate and respond to change has never been more important than in today's rapidly changing environment. Technological innovations have made possible products that were inconceivable a decade ago. The product development period, that is, the time between the conception of an idea and the mass distribution of the product, has never been shorter. And the electronic revolution has drastically reduced the time it takes a firm to respond to the actions of its competitors.

The subject of food and agribusiness marketing is very broad. Products range from basic materials such as fertilizers, chemicals, or lumber to sophisticated farm machinery, irrigation systems, and computer software to consumer food products. They are marketed all over the world in both developing and industrialized countries. The end users may be ordinary consumers or

BOX 8–1

The Marketing Mix

Product	Price	Promotion	Place
Features	List price	Advertising	Channels
Brands	Discounts	Public relations	Locations
Packaging	Allowances	Personal selling force	Inventory
Instructions	Credit	Sales promotion	Assortments
Quality	Payment terms	Direct marketing	Transportation
Service			
Warranty			

multinational firms. Therefore, it is impossible to characterize a "typical" food or agribusiness product or market. For this reason we will discuss the principles used in marketing food and agribusiness products as opposed to describing how any single product is marketed.

Marketing decisions may be classified into four areas: product, prices, promotion, and place—and conveniently called the 4 Ps. In turn, each P covers several activities (Box 8–1). The word distribution is perhaps more descriptive of the "place" decision but has much less value as a memory device (the 3 Ps and a D just don't sound as good), so we'll stick with the 4 Ps. The 4 P framework focuses the marketer's attention on the product and its characteristics, setting the price, deciding how to distribute the product, and choosing the appropriate promotion techniques.

A related framework, the 3 Cs, is often used in preparation for making marketing decisions. The 3 Cs consist of customers, competitors, and company. A thorough analysis of all three is necessary to make informed marketing decisions.

The customer analysis should focus on developing a complete understanding of the firm's current and potential customers and should address the following factors:

- Identification of existing and potential customers
- Market description (size, potential growth, segments)
- Customer needs
- Customer profiles (demographics, lifestyles, values)
- Information sources used by customers
- How and where customers purchase products

Because the company's offerings will be compared to those of competitors, it is critical to have a comprehensive understanding of competitors, their product offerings, motivations, and resources. The competitor analysis should address the following issues:

- Identification of competitors (existing and potential competitors)
- Analysis of competitors' products
- New products being developed by competitors
- Competitors' goals (market share, growth, profitability)
- Competitors' costs
- Competitors' strengths and weaknesses
- Strategies pursued by competitors

Lastly, the company itself should be analyzed. Although this may seem superfluous, an honest assessment of the company and how well it stacks up against its competitors is essential to developing successful marketing strategies. The company analysis should include the following factors:

- The company's goals
- The company's cost position
- The company's strengths and weaknesses
- The company's source of competitive advantage

Once the firm has completed the analysis of its customers, competitors, and company, it is ready to begin with the first P—the product decision.

PRODUCT

For most businesses, marketing starts with the product offering. It is the product that satisfies customers' needs and distinguishes it from competitors. By supplying a superior product, firms strive to create value for consumers, increase market share, and ultimately increase profits.

Products differ greatly in the degree to which they can be differentiated. At one extreme are commodities that, by definition, are undifferentiated. These mainly include raw or lightly processed products such as grain, flour, fruits, vegetables, and meats. In recent years, many food companies have been successful in differentiating products that were formerly commodities, even though it is especially challenging to do so. They accomplish this feat by finding a real or psychological basis of distinguishing their products from similar products. From a marketer's point of view, all goods and services can be differentiated. Tyson, for example, has been successful in differentiating chicken although most consumers would be hard pressed to distinguish Tyson's chicken from competitors' products. Tyson has done a splendid job of convincing the consumer that its chicken is more tender and flavorful than unbranded chicken as well as competing brands.

At the other extreme of commodities are products that are highly differentiable. These include products such as farm equipment, improved seeds, and convenience foods. With these products the marketer's goal is to create products that meet customers' needs and are distinctive. The differentiation may be based on physical differences, availability, service, price, or image. When a company is successful in differentiating its products, it will usually find that the competition will introduce their own version of the product, often imitating the successful product. For this reason, firms must continually renew the basis for differentiation in order to maintain market share and profitability.

Product Life Cycle

Just as human life is characterized by a fairly predictable series of events (we are born, grow, mature, grow old, and eventually die), so are the lives of many products. This pattern is called the product life cycle. Figure 8–1 illustrates the shape of the sales and profit curves for a typical product. Although the stages are depicted as if each stage lasts an equal and determinate period of time, in reality the duration of each stage in the product life cycle is highly variable. For example, the life cycle of many mature food products may last for decades, or even centuries. On the other hand, the life cycle for a fad, such as pet rocks, will be extremely short lived.

It is important to understand that the product life cycle refers to a general product, and not a specific product of one firm. Thus, there is only one market introduction stage, although some firms may introduce their particular products well after the general product is established. Marketing activities are heavily dependent on the stage in the product life cycle just as a person's physical activities are dependent on his or her age.

FIGURE 8–1

Product Life Cycle for a Typical Product

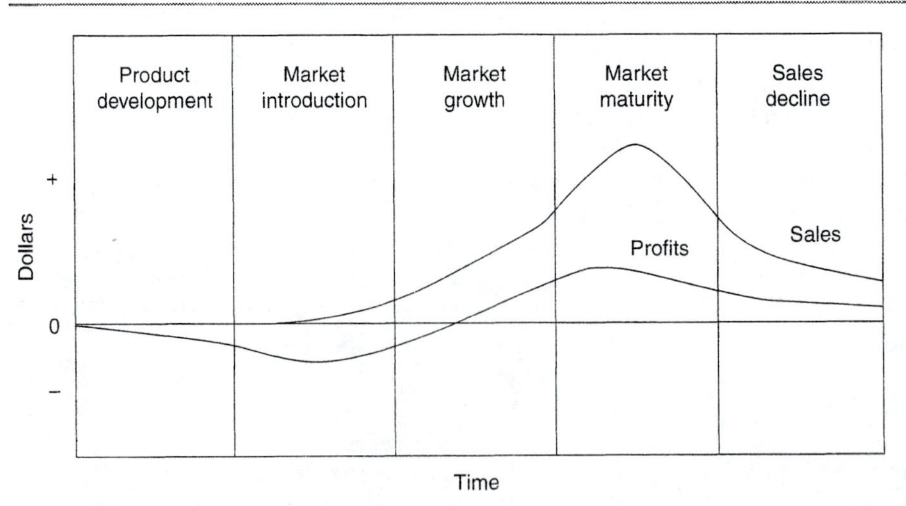

Product development, the first stage in the product life cycle, begins with an idea. Because all ideas are not created equal, it is necessary to screen the good ideas from the bad. Preliminary analyses should be conducted to determine the product's technical feasibility and market potential. Ideas that pass these two tests will be developed. It is important to conduct the market analysis in the development stage where mistakes are least costly. Researching the idea prior to marketing the product can prevent many new product failures.

In the *market introduction stage*, the new product is introduced to the market for the first time. The marketing decisions will initially focus on getting consumers to experiment with and then adopt the product. Extensive promotional campaigns are often used to launch new products and special price incentives may be used to entice the first customers to make their purchases. The number of competitors in this stage is usually small. Often a firm may have the market to itself. However, this advantage is typically only temporary for successful products. The firm that successfully introduces a new product will attempt to solidify its position in the market in anticipation of stronger competition in the future. In most cases profits are negative in this stage due to the high costs of introducing new products.

The *market growth stage* is characterized by a high growth in sales. Competitors have certainly realized by this time that the product is a success and are probably preparing to enter the market, if they have not already done so. Whether or not a firm faces competition will depend on how easily the product is copied and any relevant patents or copyrights. This may be the most profitable stage because sales are strong and the number of competitors is typically small in the beginning.

The *market maturity stage* is characterized by a leveling off of sales, as only the latest adopters have not yet purchased the product. Depending on the nature of the product, sales will consist mostly of repeat sales for products that are frequently purchased, or replacement sales for durable items. Competition is likely to be strong because the market is no longer growing and because more firms are marketing the product than in any other stage. The increased competition often leads to heavy promotional expenditures and price cutting as each firm tries to solidify its position in the market, resulting in the weaker firms being driven out. Profits typically decline in this stage due to weaker product demand and intensified competition as marketers try to maintain their market share.

The length of the product life cycle is unpredictable. Some products may pass from generation to generation virtually unchanged, whereas others may last less than a year. Most products will eventually enter the sales decline stage and be replaced by a new product that better meets consumer needs. Only the strongest firms will survive declining product sales because the small market can profitably support only a few firms. Although products become outmoded, seldom does the need behind the product disappear. Thus an obsolete product may offer an opportunity for the foresighted firm to market yet another product.

In reality, the stages of the product life cycle will not be obvious or clear cut. It may be difficult to tell whether a slowing in sales marks the end of the market growth stage or is simply due to a slowdown in the economy. However, the product life cycle does provide a useful basis for planning the marketing strategy. The transition from one stage to another is a good point to stop and evaluate the current marketing mix to determine whether it meets the changing needs of the consumers and the inevitable response by competitors.

The life cycle should not be viewed as unchangeable and uncontrollable. Products (like people) may have their life cycles extended with the proper care and attention. Some firms have even been successful at reviving products in the sales decline stage by finding new uses for their products. For example, as consumers spend less time in the kitchen baking, Arm and Hammer has encouraged consumers to use a whole box of baking soda to deodorize their refrigerator!

The Product Decision

The *product* is what the agribusiness has to offer the consumer. It may include, in addition to the primary product or service, packaging, instructions, and a warranty. These product accessories may be an integral part of the product and, more importantly, may be what distinguishes it from a competitor's product. Take, for example, farm accounting software packages. One of the factors impeding their adoption has been their complexity. Software that is user-friendly, or that comes with an understandable user guide, has a big advantage over competing packages that are less user-friendly. For many products, the decision to offer service, warranty, and instructions is easy. It is ridiculous to include instructions with an avocado, but is considered essential to include them in a box of cake mix.

Product accessories may offer an opportunity to differentiate the product by better tailoring the product to the needs of the target market. Although the trend has been towards bigger, lower-priced supermarkets, some grocery stores have maintained their clientele by providing better service, such as shorter lines, and offering to carry customers' bags to their car. At the other extreme are the "no-frills" supermarkets that offer rock-bottom prices, but require you to bag your groceries yourself. There are even some supermarkets where you have to bring your own bag, or purchase one from the store.

Some product accessories may be offered as options or bundled in alternative products, leaving the choice up to each individual consumer. For example, farm machinery product options range from a very basic piece of machinery to a machine loaded with all of the options.

The package decision is one of the most visible product decision areas, and often one of the most important. Some products have become successful more as a result of their package than because of the product in the package. Kellogg's cereal in the individual serving boxes was an instant hit, because of its convenience. Consumers bought it for occasions when it was impractical to carry a larger box, such as for camping or for packing a breakfast for their children. Similarly, bag-in-the-box wine is very popular with restaurants and for party occasions because it is convenient as a serving device as well as a storage container. Some recent innovations in product packaging are convenience foods that are sold, cooked, and served in one container, and yogurt that is eaten by squeezing it directly from a plastic pouch. Convenience has become a prime consideration in package as well as product design. This has never been more true than today when both parents work outside of the home in a majority of families.

The physical appearance of the package should not be overlooked, especially when consumers are faced with many product choices. Although an eye-catching package will not guarantee that it will be purchased, it will at least ensure that it will be noticed. For this reason, food marketers place a heavy emphasis on developing an attractive package design that attracts consumers' attention, conveys the product's principal features, is functional, and presents a consistent brand image to the consumer.

In contrast to products that are intended primarily for consumers, products marketed to business customers need not have fancy packages. Instead they should emphasize their functional nature. Business customers want the most "bang for their bucks." If it adds value, or increases their sales, it will be purchased. Business customers usually spend more time analyzing purchase decisions than do consumers because they often purchase a product in large quantities or make many repeat purchases. Therefore, an attractive package will be far less important than what is in the package. As a matter of fact, any unnecessary extras may be considered unfavorably by business customers because they understand that the cost of the packaging is passed on to them.

PRICE

The price decision is a prime determinant of the firm's revenue. Revenues, along with the firm's costs, ultimately determine the firm's profit level. Pricing is one of the most complex decisions the marketing manager must make because it is influenced by a multitude of factors. In this section we will discuss some of the key factors that must be taken into consideration, including costs, customers, and competitors. We will also present some of the more common pricing methods, including those used at the market introduction stage in the product life cycle. The section concludes with a discussion of how discounts, allowances, and credit are an integral part of the price decision.

Costs - Marketing Math

A good place to begin the discussion of price is with the costs associated with producing and marketing the product. A thorough understanding of all of the costs related to a product as well as the firm's cost structure is fundamental not only to the price decision, but to the profitable operation of the business. Understanding a product's costs will help ensure that the price is set so that the company's costs are met and that a margin for profit is included.

Contribution to overhead analysis and break-even analysis, discussed in depth in Chapter 4, are often used as a starting point in pricing a product. With these methods the first step is an accurate estimate of the variable costs associated with the product. A per unit CTO sufficient to cover fixed costs and a margin for profit is then added to the total per unit variable cost to arrive at a price. The total CTO can then be calculated based on the estimated sales at the proposed price. Alternatively, break-even analysis can be used to calculate the break-even quantity when the fixed costs associated with the product can be accurately estimated.

So far we have only considered the firm's costs in the discussion of pricing. However, the firm must consider other factors in establishing its prices. This is essential because customers evaluate a firm's price relative to the product offered, the perceived value to the customer, and competitors' products and prices. As a matter of fact, in competitive markets, including most markets for food and agribusiness products, most customers don't care what a manufacturer's costs are.

Customers - Demand Analysis

By considering consumer demand we recognize that customers are sensitive to the prices firms charge for their products, and that, in fact, it is price that to a large extent determines how much a customer will buy. In general, the higher the price for a product, the lower the quantity that will be sold. This is something that we all understand intuitively. This relationship between a product's price and the quantity that consumers are willing to buy is known as the demand curve. A typical demand curve, which is downward sloping for almost all products, is shown in Figure 8–2. The demand curve shows that as the price rises from P_1 to P_2, the quantity decreases from Q_1 to Q_2.

One of the most important uses of the demand curve is to understand how consumers react to changes in price. This concept is referred to as the *elasticity of demand* and it describes consumers' responsiveness to price changes. The *elasticity of demand* is defined as follows:

$$\text{Price elasticity of demand} = \frac{\text{percentage change in quantity}}{\text{percentage change in price}}$$

It is convenient to think of a product's elasticity in terms of a 1 percent change in price. If the quantity demanded of a product declines by more than 1 percent in response to a 1 percent price increase, demand for the product is said to be *elastic*. If the quantity demanded of a product declines by less than 1 percent in response to a 1 percent price increase, demand for the product is said to be *inelastic*. Figure 8–3 depicts both elastic and inelastic demand curves.

FIGURE 8–2

Demand Curve for a Typical Product

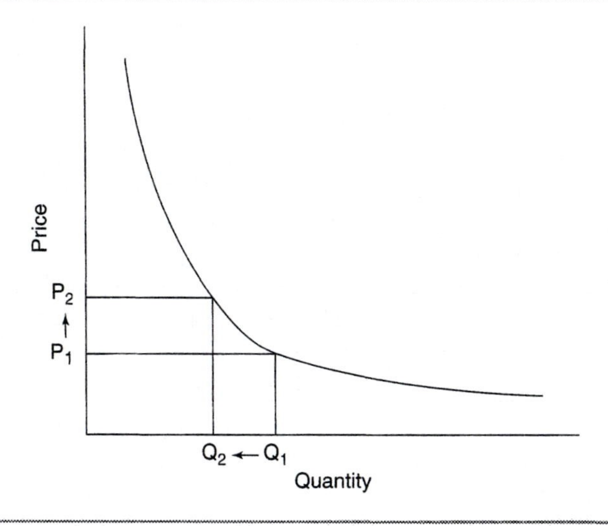

FIGURE 8–3

Elastic and Inelastic Demand Curves

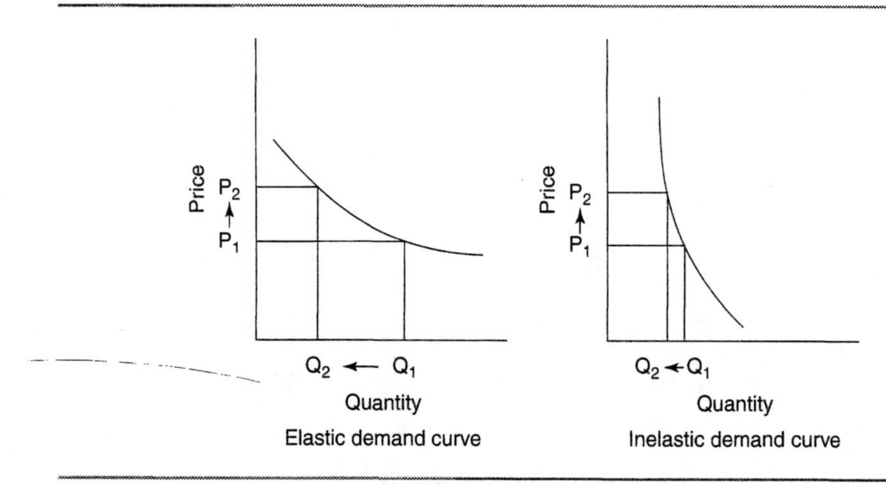

The demand for products tends to be more inelastic, that is, consumers are less responsive to price changes, when the product is a necessity, when there are no or few good substitutes, and when the price of the product is low relative to consumers' incomes. Table salt is a good example of a product with an inelastic demand curve. Consumers will reduce their consumption of table salt only slightly, if at all, in response to price increases. Table salt is a necessity, it has no good substitutes, and the average consumer spends little on table salt relative to his or her income. The mango, an exotic fruit, serves as an example of a product whose demand curve is elastic. The fruit is not a necessity and there are many alternative fruits that can substitute for mangoes. Thus consumers might be expected to substantially decrease their consumption of mangoes and substitute other fruits in response to a relatively small price increase for mangoes.

Marketers can use the concept of price elasticity to evaluate price changes. For example, producers of goods that are elastic might explore the idea of using a price decrease to stimulate sales. Alternatively, manufacturers of goods with inelastic demand curves might consider raising prices in the expectation that volume sales would decline relatively little compared to the price increase. However, this analysis would be incomplete without anticipating competitors' responses.

Competitors' Prices

Competitors' prices are probably the key factor affecting a firm's flexibility in pricing its products. Consumers typically compare products when making a purchase and seek out the best value. Higher priced products must offer more features, better quality, or a better image to justify their higher price. It is especially difficult to charge a higher price for products that are relatively undifferentiated or products that are easily compared by consumers. On the other hand, a higher price is more easily justified for unique products, those that have a well-defined image, and those whose quality attributes are not easily evaluated by consumers.

Competitors' prices are often used as a starting point for price determination. By comparing attributes such as the product's features, quality, reliability, and warranty to competitors' product offerings and prices, the product may be positioned and priced consistent with the firm's marketing strategy.

Introductory Pricing Policies

Several pricing policies are commonly used in conjunction with the introduction of new products. Because they are used during the period when the firm's product is becoming established, they are called introductory pricing policies.

Price Skimming. *Price skimming* is used to take advantage of the fact that some people are willing to pay a very high price for some new products (those with an inelastic demand curve). The price is set high initially and gradually lowered to attract new customers who were not willing to pay the higher price. This policy is often used to recapture the cost of developing a new product. It is most effectively used when the producer is the sole maker of a product, as is the case when the product is patented, subject to proprietary technology, or difficult to imitate. When too many firms sell the same type of product, competition will result in lower prices, making skimming ineffective. Price skimming is seldom used for food and agribusiness products because the majority of the products are easily imitated and such a strategy would only attract additional competitors. Price skimming is used successfully with high technology products such as genetically modified seed. Because there may be only one producer of a seed with a specific characteristic, the seed company is able to keep its prices high initially. Over time prices typically decline as other seed companies enter the market with comparable products.

Penetration Pricing. With a *penetration pricing* policy, the price is set as low as possible with the intention of capturing a large market share as soon as possible. This strategy is commonly used when a firm expects to have strong competition for the product and wants to establish itself as the market leader. A price skimming strategy might give a competitor the chance to get his or her foot in the door, because the introduction stage is prolonged and profits are high. However, a penetration pricing policy may actually discourage competitors from entering the market. Many food and agricultural products are sold on this basis. The sellers, realizing that their products can be easily copied, offer them at a low price to capture as much of the market as quickly as possible.

Introductory Price Cuts. As the name implies, a policy of *introductory price cuts* means that low prices are established as the product is introduced. However, unlike the penetration price policy, the low prices are only temporary. This strategy may be used to introduce a new product into the market, knowing that once the product is established the price cuts can be eliminated and the product will sell well at the planned permanent price level. Many new food products are introduced using this strategy in an attempt to get consumers to try the product. Coupons or other types of discounts are often used in order to indicate that the low price is only a temporary reduction off the higher permanent price. By signaling to consumers that the price cut is temporary, the firm attempts to avoid resistance when the introductory price cuts are eliminated.

The introductory pricing strategy is also commonly used when a firm is trying to enter an established market. By signaling competitors that the price cuts are only temporary, price wars, which may result in heavy losses for both the established and new firms, may be avoided. A disadvantage of this strategy is that even an attempt by a new firm to capture what it feels is a reasonable market share may be received by established competitors as a sign of hostility.

Common Pricing Methods

Ideally, a pricing policy should guide the establishment of prices. This is important for several reasons. It ensures that prices are established consistently across different products and over time. Moreover, it forces the marketing manager to consider how the pricing strategy supports the overall marketing and firm strategy. The pricing policy should also address how price adjustments (discounts and allowances) are to be used, because these affect the net price paid by customers. Lastly, the firm's credit policy should be addressed in the pricing policy, because credit terms indirectly affect price and are often an important consideration in the buying decision.

The number of pricing methods used by food and agribusiness firms is overwhelming—too many to cover in detail in an introductory text. However, several pricing methods are so commonly used that every manager should be familiar with them. Understanding the basics of the pricing methods that follow will provide a good basis for understanding the underlying principles of pricing. However, it is important to remember that any pricing method is simply a guide and not a substitute for managerial decision making. Furthermore, in using a pricing method to establish prices, the marketing manager should not forget to consider the 3 Cs discussed earlier in this chapter: the analysis of customers, competitors, and the company.

Competitive Pricing. A good starting point for beginning the discussion of price is to look at what competitors are doing. When other firms are competing in the same market, and when the products are very similar, price often becomes the determining factor for consumers. If prices are set much above the competition, the firm would garner few sales. Setting prices far below the other firms' prices would likely result in heavy losses or price wars, which would be disastrous for the firm and industry. Faced with this situation, most firms opt to set their price right at, or strategically above or below, their competitors' prices and develop a marketing strategy of trying to differentiate their product from their competitors' on some basis other than price. This is known as *competitive pricing* (also known as strategic pricing).

Cost-Based Pricing. Many companies, particularly wholesalers and retailers, use *cost-based pricing*. The price of the product is marked up by adding on to the cost a margin (usually computed as a percentage of the cost) to determine the selling price. The markup formula is:

Selling price 5 unit cost 3 (1 1 markup)

For example, a 20 percent markup on an item that costs a retailer $5.00 would produce a retail price of $6.00 or:

$5.00 3 (1 1 0.20) 5 $6.00

Supermarkets apply standard markups to each product category such as packed food products, meats, fruits and vegetables, frozen foods, dairy products, health and beauty aids, etc. These markups differ by product category to reflect both the differences in the costs of selling products within the category and the difference in supply and demand for the products. For example, the markup on frozen foods is higher in a supermarket than on nonrefrigerated, packaged food products, to reflect the increased costs of refrigeration. The price for individual food items within a product category is adjusted to reflect supply and demand conditions. Cost-based pricing is very practical because of the large number of items that most supermarkets carry.

Cost-based pricing may also be used in conjunction with break-even analysis to determine whether a given price will at least allow the firm to break even on a product or to estimate profits or losses at given prices.

Value-Based Pricing. In recent years *value-based pricing* has become a common pricing method. It owes its popularity primarily to consumers' desire for "good value" in the products they purchase. Value-based priced products are typically good quality products that are fairly priced. Often, marketers will begin by estimating what most customers will pay for a product and then charge slightly less for the product. In this way the customer will feel that they are getting a good value for their money. Examples of companies that use value-based pricing in the food and agribusiness industries include farm equipment manufacturers, seed companies, agri-chemical companies, and consumer food businesses to name a few. H.J. Heinz has positioned The Budget Gourmet as a value brand. This line of frozen entrees is of good quality but priced substantially below many of the higher-quality frozen entrees offered by competitors.

Value pricing may also be used in other situations. For example, some restaurants offer early bird specials to patrons who come early for dinner before the restaurant is crowded. And many fast-food restaurants have value menus that include items that can be purchased for less than a dollar.

Loss Leader Pricing. To attract customers, retailers often use *loss leader pricing.* By setting the prices of some carefully chosen items very low, often at or below costs, they hope to generate enough sales on other products to more than compensate for the decreased profits on these loss leader items. Products that are purchased in conjunction with other products, or in limited quantities, are ideal for this method of pricing. Supermarkets commonly use milk as a loss leader. Milk works well as a loss leader because it can't be stored over a long period of time and therefore is usually bought in small quantities. Most importantly it is often bought with many other grocery items.

Price Point Pricing. *Price points* are sometimes used as guides for product positioning as well as pricing. For example, wines are often referred to as popular premium, premium, super-premium, and ultra-premium. The popular premium wines, also referred to as the "fighting varietals," are typically priced between $5 and $7 for a 750 ml bottle. Price points are also convenient for consumers because they provide a convenient grouping mechanism that indicates quality parity among products within a group, that is, between price points.

Discounts

It is important to recognize that there is often a difference between list price and realized price. Price adjustments are used so frequently in some industries that hardly anyone pays list price. Also, it is the retailer and not the manufacturer that establishes the retail price. This is why you see the statement "manufacturer's suggested retail price" printed on some product packages. Buyers frequently receive a discount price, a price allowance, a rebate, a coupon, a free added service such as an extended warranty, or a free gift. Each of these lowers the realized price, and in some cases companies can get so carried away with these pricing strategies that they dilute their profits.

Discounts are usually used as a means of influencing customer behavior. They may be used to increase sales or to encourage customers to pay cash. Although they are often used as a part of the firm's price policy, they may also be used to make the minor adjustments necessary to be responsive to competitors.

Quantity discounts are used to encourage customers to buy in large quantities by reducing the per unit price when the order reaches a certain size. Quantity discounts may be based on individual orders or they may be cumulative, with the size of the discount dependent on the total purchases within a period. *Cumulative discounts* encourage repeat business because the size of the per unit discount increases as total purchases increase. Thus, each purchase reduces the net price of all previous purchases.

Seasonal discounts may be used to encourage buyers to stock up on the product before the major buying season begins. Producers use seasonal discounts to get a better idea of what their buyers intend to order for the season and therefore what they will need to produce. It also enables manufacturers to shift part of the storage function to other firms further along in the marketing channel by reducing the demand on their limited storage space. Seasonal discounts are sometimes given by retailers at the end of the season in order to clear their shelves and make room for the next season's products.

Trade discounts are given to participants in the marketing channel in return for the functions they perform. Manufacturers, for example, generally sell to retailers at a lower price than to final consumers (if the manufacturer sells to final consumers at all). This lower price to retailers is a payment for the task they perform. When the manufacturer sells directly to the consumer, he or she has essentially done the retailer's job and thus receives the full price for doing so—not to mention the fact that the retailer would likely be very unhappy if the manufacturer was substantially undercutting its prices.

With the extensive use of credit cards, cash discounts have become commonplace. Whether credit is offered through the food or agribusiness company itself or through a credit card company, credit is costly. Some retailers will give a discount when cash is paid for a product. Cash discounts are also common in business-to-business transactions if payment is made within a specified period of time.

Allowances

Allowances are reductions in price given because something is due the customer. A price reduction given to a customer who is not satisfied with a product, possibly because of some defect in or damage to the product, is an example of an allowance.

A *trade-in allowance* may be given when an item similar to that being purchased is traded in with a purchase. The agreed upon allowance is then deducted from the purchase price. This practice is very common in the agricultural equipment industry.

Advertising allowances are used to encourage advertising by firms further along in the marketing channel. A producing firm will typically share the cost of advertising with a wholesaler or retailer in return for them arranging the advertising in their local area. This arrangement works well for both firms because they each benefit from the increased sales but share only part of the advertising cost. In addition to being more convenient for a local wholesaler or retailer to arrange for the advertising than a distant manufacturer, the local firm will probably be more responsive to the special circumstances in their market.

Credit

The discussion of credit adds the element of time to the pricing decision. Whereas consumers have until recently used credit only to finance major purchases, such as homes, home improvements, or a car, it is now commonplace to purchase almost anything on credit. In fact, many customers will not even consider purchasing some items if they cannot be purchased on credit.

Credit is also very common in business-to-business transactions. For example, retailers and wholesalers typically expect manufacturers to offer *delayed payment terms*, called "giving terms." For example, the payment terms on an invoice may read 3/10 net 30. This means that a 3 percent discount is allowed if the bill is paid within 10 days. If it is not paid within this period, the amount listed on the invoice is due within 30 days. It is common for businesses to offer deferred payment terms of 30 to 90 days.

Not surprisingly, the widespread use of credit has introduced a whole new set of management problems and must be considered as part of the marketing mix that is offered. The decision to offer credit and credit terms must be made after careful consideration of the benefits and costs.

The chief benefit of extending credit is increased sales and, when properly managed, increased profits. Credit makes consumer purchases easier because the purchaser does not have to have the cash to buy a product at the time of purchase. This can be a powerful competitive tool, particularly if below-market interest rates or interest-free loans are offered (usually over a fairly short period of time).

The two most common ways of providing credit are to use a credit card company such as VISA, MasterCard, Diners Club, Discover, or American Express, or for the business itself to provide credit. The principal advantages of using a major credit card company are that it is very convenient and it is easy to control costs, because the major cost is a percent of the purchase price. It is virtually risk free if the rules of the credit card company are followed (i.e., checking for stolen or invalid cards and getting approval for purchases over a certain dollar amount).

The alternative to offering credit through a credit card company is to do it yourself. This method has been most successful for small businesses who know their customers well and larger businesses that can afford to have their own credit department and possibly even offer their own credit card. Many agribusinesses that sell to other businesses provide credit themselves. They are successful at doing this because their customer base is small and they have a longstanding relationship with their customers. Costco is an example of a warehouse club that offers its own credit card.

There are several costs involved in offering credit. The major cost is the interest paid for the use of borrowed funds or foregone earnings. Even if the firm does not borrow money to finance the credit, there is an opportunity cost because the cash that would have been collected could have been invested. The most difficult part of offering credit, however, lies in the other costs, which are not easily seen or measured. The most unpredictable cost is bad debt loss. The firm offering credit assumes full responsibility for collecting payment and absorbing any losses. It is very important that bad debt loss be kept at a minimum because each dollar of debt loss requires many dollars in sales to offset it. Bad debt loss can be minimized by restricting credit to only those with the best credit records, requiring sizeable down payments, or running credit checks on those applying for credit. Almost all companies restrict credit to some degree; the difficulty is in determining how to restrict it.

If the amount of credit given is sizable, it may be necessary to have a credit department. At the very minimum there should be one person responsible for administering the credit affairs of the business, even if it is not a full-time responsibility. The advantage of centralizing the credit function is that the firm's credit policy will be uniformly applied because one person administers it. Credit management can be very costly in terms of labor, record keeping, and collecting past due accounts.

The decision to offer credit is not easy. Even in the simple case when a credit card company is used and the cost as a percentage of dollar sales is known, the total cost is unknown because it depends on the amount of credit card sales. From the firm's perspective, cash purchases are preferred because they are less costly and are risk free. However, offering credit is expected of most businesses and those businesses not offering credit will likely be at a competitive disadvantage. In making the decision to offer credit many questions must be answered. How much does the availability of credit increase sales? What sales level will be generated by different credit policies? What is the true cost to the firm of offering credit? A response to these questions is difficult to quantify, and the best answers probably lie in knowing the firm's clientele well. In most cases, it will be worth the investment to seek out expert advice in evaluating the alternatives and assessing the risks.

PROMOTION

Food businesses use promotion to deliver a message to their target audience. Some form of promotion is important to all food and agribusiness firms—customers must be made aware of the firm and its products. If the importance

BOX 8–2

Promotion Activities

Advertising	Public Relations	Personal Selling	Sales Promotions	Direct Marketing
Billboards	Annual reports	Sales calls	Coupons and rebates	Catalogs
Radio and television commercials	Community relations	Trade shows	In-store displays	Direct mailing
Brochures	Events and sponsorships	Incentive programs	Contests	Telemarketing
Internet	Lobbying	Presentations	Demonstrations and samples	Television marketing
Packaging	Press kits and releases	Meetings	Exhibits	Internet
	Speeches and publications		Special financing	
			Continuity programs	
			Special offers	

of an activity can be judged by the amount of money spent, then promotion is indeed very important. Promotional activities are usually grouped into five categories: advertising, public relations, personal selling, sales promotions, and direct marketing (Box 8–2). Which kinds of promotional activities the company employs will depend primarily on the intended audience, the type of product marketed, and the kind of information to be conveyed.

Advertising

As consumers, we are all familiar with advertising. We see it on TV, in newspapers, on billboards, under the windshield wiper, and in the mailbox. Advertising is most effective at delivering a common message to large groups of people. Although it may seem expensive when compared to the other forms of promotion, it is probably the cheapest form of promotion when the cost is calculated on a per person basis.

The largest users of advertising are the sellers of consumer products. Advertising is ideal for most consumer products because the market is very large and one message is often appropriate for large groups of consumers. Advertising is commonly used to introduce new products or changes in a product, to encourage sales, to compare a product to other products, to inform consumers of a special promotion, or simply to keep the product's name in the consumer's mind. Advertising is less commonly used as a means of reaching business customers because the number of potential customers is usually small, making mass communication impractical.

An effective advertising program must efficiently target its intended audience. In other words, the goal is to get the most bang for the buck! Companies must be careful in allocating their advertising budget. The most effective approach is to establish the advertising budget on an *objective and task* basis. The company must decide how many individuals it wants to reach in the target market, the frequency with which the message will be delivered, and the desired impact. An advertising budget is then developed to deliver the desired reach, frequency, and impact.

With the advertising budget set, the company must then decide on the message. The message is shaped by whether the firm intends to inform, persuade, or remind the target market of the product. The challenge is to present the product in a manner that will create an awareness of, interest in, or repeat purchase of the product, depending on the firm's intention. Most food and agribusiness firms use an advertising agency for this purpose. The message must also be tailored to fit the media outlet.

Many *media outlets* are available for advertising, and the choice of the correct one will depend on the intended audience. Even within a media such as television, the placement of the advertising will depend on the intended receiver. For example, Saturday mornings or right after school lets out is the best time to advertise to children. Business customers may be best reached through their particular trade magazines or journals. Farm magazines, for example, are an effective means of communicating to farmers, and an industry publication, such as The Packer, is effective at reaching the produce industry. In most advertising campaigns the firm selects a combination of media outlets. Under these circumstances the execution of the message will vary depending on the media outlet that is selected.

Once the media campaign is delivered to the target market, it is important that the firm monitor and measure the effectiveness of the message and the media. An approach frequently used to measure the effectiveness of a media campaign is to conduct a consumer survey to determine recall and recognition. All too often this is not sufficient. The best measure is to tie sales directly to the impact of the advertising. This may be easier said than done, as sales can be impacted by a change in other variables not under the control of the firm. For example, while a firm is running an advertisement, a competitor could be running a sales promotion that blunts the impact of the ads. Consequently, it is imperative that the firm continually monitor and measure the effectiveness of its advertising programs, given advertising's expense and complexity.

Generic advertising has been used by agribusinesses in the milk, meat, and fruit and vegetable industries. *Generic advertising* is defined as advertising for a commodity and is contrasted to conventional advertising that promotes a firm's product or brand. It is typically used for products where branding is difficult because of the low level of differentiation between products. It would be impractical, for example, for one milk producer to advertise to promote his or her products. Not only would it be very costly, but in the absence of branding the advertising would benefit all milk producers. By banding together, a group of producers is able to share both the costs and the benefits of advertising a commodity. Because it is too easy for any one producer not to contribute but yet share in the benefits, what is commonly referred to as the "free rider" problem, participation in such programs is often mandatory. Market orders and agreements have made possible such marketing alternatives as the advertising check-off as well as research and other programs that would be impossible on an individual basis. Some of the most successful generic advertising campaigns have been the "Got Milk?" and "Dancing Raisins" promotions.

Public Relations

When used properly, public relations can be an effective form of communication. *Public relations* (PR) is one of the primary tools, in addition to advertising, that a company uses to establish its image. PR consists of a set of communication tools including publications, events, publicity, community involvement, establishing an identity, and lobbying. Because of its importance to the firm, many food and agribusiness companies hire a public relations firm to handle their PR work or supplement their PR.

Perhaps the most important form of PR is *publicity*, which can be defined as any unpaid form of mass communication. Good publicity is priceless, and bad publicity can become a firm's worst nightmare. Because it is not purchased, the firm has little direct control over it; however, when properly managed, the firm can greatly influence the publicity it receives. An item is suitable for publicity if it is of general interest to the subscribers of newspapers or magazines, recipients of newsletters, television viewers, etc. Media firms often give preferential treatment to companies that regularly purchase advertising. The grand opening of a new store, a technological breakthrough, or new product releases are typical publicity items. The preparation of a *news release* and the distribution of *press kits* will help ensure that the information is reported as accurately and completely as possible.

Bad publicity may be the result of an accident, a worker strike, or the unfair treatment of an employee. Several recent high-profile food safety incidents have shown the importance of being prepared. Odwalla was largely able to preserve its favorable public image because it responded quickly to the *E. coli* poisonings, was responsive to the media, and was perceived as handling the situation fairly.

Personal Selling

Personal selling is a form of product promotion that relies mainly on direct communication between a sales person and the potential customer. Personal selling is one of the most expensive marketing communication tools. Because personal selling is so costly, in terms of expense per customer, it is only used in select situations. It is typically used to market products to business customers, because of the large volume of business they do, or to consumers for high-priced, complex items such as automobiles.

Food and agribusiness companies commonly use personal selling to reach manufacturers, wholesalers, retailers, and farmers and ranchers. The key to selling to wholesalers and retailers is to have a product that will sell well. Retailers are not interested in using up precious shelf space for slow moving products. Likewise, wholesalers are reluctant to allocate warehouse space for products that are not likely to move. If the product is new, the salesperson's main responsibility is to convince the buyer that the product will be a success. For products with an established track record, the salesperson's main responsibilities may be to take orders, see that products are delivered, handle complaints, etc. The salesperson must ensure that wholesalers and retailers give their product as much attention as possible in terms of shelf space, advertising, special sales, etc. This entails providing supporting materials such as displays or rebate forms to those firms that handle the products.

Personal selling is an especially effective means of communicating with food processors because their individual needs and problems are best addressed on a one-to-one basis. The large size of many farms and ranches also makes them particularly good targets of personal selling efforts. In many cases the salesperson may not actually make the sale to the customer (the distributor often does that); however, the salesperson accomplishes his or her objective by convincing the customer to buy the firm's product. Because wholesalers and retailers often handle several competing products, personal selling may be the only way of ensuring that a firm's product is given the desired level of emphasis.

Personal selling is a process that may be broken down into several discrete activities. The process includes the preparation, the presentation, and the close. Each of these steps is discussed in detail below.

The Preparation. The majority of the salesperson's work is done before setting foot in the customer's door. The *preparation* includes locating customers, planning the presentation, preparing materials, and scheduling appointments. For most salespeople the actual time in direct contact with customers will be relatively small. So it is important to make this time as productive as possible by planning well.

Prospecting, or locating potential customers, is the first step in personal selling. Agribusinesses, whether they are new or established firms, must constantly search for new customers. It is particularly important for new or growing companies. Some companies will give new salespeople an established territory with regular customers, whereas other salespeople may find that they must start from scratch.

Some common sources of prospects are telephone books, trade association directories, county agents, magazines, and referrals from customers or friends. The choice of a method and source depends on the likelihood of the method to yield potential customers, and the cost of the method. For example, the telephone book would not be a very good source of prospects for a fertilizer salesperson, but it is probably an excellent source for someone selling tickets to the policemen's ball. The best source of prospects is one that contains the names of most businesses in the area of interest. For example, a county extension agent's mailing list would be a good way of locating farmers or ranchers in a county. Many industries have trade associations whose membership lists are useful in searching out firms in that industry. The mailing lists of farm or trade magazines may be used to obtain the addresses of people in these industries.

Because of the cyclical nature of agriculture, the emphasis on prospecting changes depending on the season of the year. During the peak selling season, very little time is devoted to active prospecting because most of the salesperson's efforts are devoted to securing sales from established customers. Only those new prospects with the highest potential for becoming good customers are contacted. The slower seasons of the year are used to prospect for new customers and to make the initial contact with those prospects.

Before a prospect is called on, the sales call should be planned. Information about a prospect or company should be collected when available. Personal information such as age, gender, level of education, hobbies, etc., is useful in determining how to approach the prospect and in breaking the ice in the initial meeting. Information about the prospect's company, its products, structure, problems, competition, etc., is extremely useful in planning an effective presentation.

The sales plan should be put in writing, specifically listing the objectives to be accomplished in each meeting. Large or complex sales may require several meetings. The sales plan is used to monitor the progress toward the ultimate objective of completing the sale. Written objectives also ensure that there are in fact objectives and provide a framework around which to plan the specifics of each presentation. An example of objectives for the first of several sales calls is presented in Box 8–3.

BOX 8–3

Example Sales Call Objectives

Objective 1. Introduce myself and the company, and discuss the purpose of my visit.

Objective 2. Establish a good rapport with Mr. Johnson.

Objective 3. Present our irrigation systems and discuss their major benefits.

Objective 4. Show the results of field trials comparing our systems with several competitors' systems.

Objective 5. Request permission to do a preliminary design based on Mr. Johnson's needs.

Objective 6. Make an appointment for the next sales call.

A plan should not be confused with a script. Plans are made with the best information available at the time. It will often be necessary to deviate from the plan in order to address the situation as it unfolds. In most cases the presentation should not be memorized, because it will come off as exactly that—memorized. "Dress rehearsals" or "mental rehearsals" are useful to ensure that the presentation goes off as planned.

The final step in preparing for the sales call is to make an appointment. This may not always be possible as in cases where the prospect does not have a listed telephone number or is difficult to reach. However, sellers and buyers alike usually appreciate appointments, so that both can make better use of their time. Appointments made far in advance should be confirmed before the visit is made.

The Presentation. The actual sales call is called the *presentation*. The first few minutes of the sales call are the most crucial. Most managers are very busy and view sales calls as an intrusion on their valuable time. Therefore, it is important to make a good initial impression on the prospect. The car you drive up in, the way you talk, the way you dress, your mannerisms, and the way you introduce yourself will all influence the impression you make on the prospect.

The cue as to how to start the sales call should be taken from the prospect. Some people want to chat about the weather or how the local baseball team is doing, whereas others want to get right down to business. When possible, conversation should be encouraged because it allows both the prospect and seller to get to know one another and to establish the trust necessary for a successful business relationship. Conversation is also one of the best means of learning about the prospect—of really understanding what makes him or her tick. By directing the conversation and asking the right questions throughout the sales call (probing), the salesperson may determine what is important to the prospect, how he or she makes decisions, his or her company's goals and problems, and other information that may be used later in the sales call. Good salespeople are almost without exception good listeners, who listen not only with their ears but also with their eyes. They pick up information wherever it is offered and file it away for future use.

It is essential to grab the prospect's attention at the outset and hold it. An effective way to start the presentation is to show the prospect what the product will do for him or her. A demonstration may be appropriate for some products. If the product will increase profits, or cut costs, then show how. If the buyer does not think there is anything in it for him or her, then you've probably already lost the sale. Your most exciting and impressive points don't have to be used right at the start of the presentation. However, you must give enough bait to lure the prospect in for the rest of the presentation.

The best method of ensuring that your points hit home is to emphasize the product's benefits or how it solves a problem the client has. People do not want to hear about a product's features. Features are bland. Features are rpm's, kilobytes, and amps. Instead tell the prospect about decreased costs or increased profits. For example, a feature of a new seed variety is that it has an 8-percent higher germination rate. This feature translates into a higher yield, which in turn translates into higher revenues and higher overall profits—a point that is much more interesting than a higher germination rate.

Salespeople are often faced with objections from the potential buyer. Expert salespeople differ as to how they should be handled, although it is generally agreed that they should be handled immediately. All objections are important because they are important to the prospect. In order to handle an objection, it is critical to understand why it was raised. A prospect who says that the price is too high may really be saying that he or she is loyal to a competitor's product and that any price is too high. Price is clearly not the problem and addressing this issue would probably just lead to further objections.

Objections that are not true should be tactfully refuted. Valid objections, on the other hand, may be put into perspective. Many objections concern some deficiency in the product, often relative to a competitor's product. Because customers seldom purchase a product because of just one feature, the product as a whole should be emphasized. Although some farmers object to the high cost of John Deere equipment, the company is successful because it is effective in selling the whole package. After considering the high quality of the equipment and its reliability, the annual cost of a John Deere tractor may actually be lower than many competitors' products.

The Close. For many salespeople the most difficult part of selling is *closing* (or completing) the sale. Some people never complete the sale because they simply never ask! Regardless of how or when you choose to close, you should always make an attempt. The timing of the close is probably more important than how it is done. Sometimes it is obvious that the prospect wants to purchase the product. When this happens, it is time to close, even if you are only one-third of the way through your presentation. Finish the presentation alone in your car on the way home if you have to, but don't give the prospect the chance to change his or her mind! Questions concerning credit or specifics about the product may indicate a willingness to buy.

When you are uncertain as to whether the prospect is ready to buy, a trial close may indicate whether the prospect is willing to make a decision. A *trial close* is a question whose answer indicates the prospect's level of interest without directly asking whether he or she wants to purchase the product. The question "Do you prefer the red or the green model?" is an example of a trial close. A positive response most likely indicates an interest in buying the product, whereas a negative response may be followed by returning to the presentation.

Lastly, it is important to follow up on sales calls. It may be necessary to make a return trip to seal the deal, or ensure that the product was delivered or the customer was satisfied. In many cases, developing a personal relationship is important to establishing a long-term business relationship. This is particularly true when there is very little product differentiation and the personal relationship may be the only thing that differentiates one company's products from another.

Sales Promotions

Sales promotions include a wide ranging set of incentives aimed at retailers (trade promotions) and consumers. Sales promotions make up the bulk of the promotion budget for food manufacturers—a large percentage of which goes to trade promotion. *Trade promotion* includes offering supermarkets and other retailers special allowances, discounts, and gifts. Specific forms of trade promotions include *slotting fees* (money paid to retailers to reserve shelf space for new products), advertising allowances and cooperative advertising, buying allowances, *merchandising* (point-of-purchase promotions in a supermarket, such as displays and themes), and free items given as premiums. Supermarkets depend on trade promotion revenues, particularly slotting allowances, for much of their profit margin.

Consumer promotions include in-store displays, rebates, coupons, contests, giveaways, and free samples to entice consumers to buy a certain brand. Consumer promotions are especially effective when a company has a superior brand with a low brand awareness. The promotion can be used to encourage the customer to try the product and build the customer base. New or improved products are often given away, sold at a very low price, or offered in conjunction with some other incentive. Many companies use special promotions, such as contests or rebates, to periodically increase sales or to keep the product's name fresh in consumers' minds. For example, cereal manufacturers occasionally use buy-one-get-one-free offers, known in the trade as BOGOs, to encourage sales of their brands.

Farm input suppliers have successfully used customer meetings to display their products. It is common for one or several agribusinesses (usually noncompeting firms) to sponsor a meeting focused on the introduction or use of several products. A dinner or some other attraction may be offered to encourage attendance. Field days, workshops, trade shows, or other such gatherings are an effective means of reaching business customers. They are good opportunities to present new products because they are typically attended by people representing the most progressive companies, who are most likely to be receptive to new ideas and products.

Direct Marketing

Direct marketing is defined as marketing targeted at consumers that encourages them to make a direct response. It bypasses the wholesale and retail channels and includes direct mailing, mail-order catalogs, telemarketing, television marketing, and Internet marketing.

Direct marketing has been made possible because of two relatively recent developments. The advent of computer and information technology has given manufacturers and retailers the ability to collect and manage large amounts of information that allows firms to specifically target small groups of consumers, and in some cases individual consumers. Food retailers provide a good example of how this information is collected, unknown to most consumers. Supermarkets that issue frequent purchase cards encourage customers to use the coded cards to receive discounts on the items they purchase as well as discounts based on cumulative purchases. As the customer goes through the checkout line, the card is scanned and all of the purchase information is stored. This information, along with demographic information on the customer, becomes part of a database. Similarly, many companies share information on consumer purchases made through catalogs or Internet sales. The second event spurring the growth in direct marketing has been the growth in rapid delivery services such as Federal Express and United Parcel Service, which have made it possible to provide reliable, inexpensive delivery of many products.

Some of the most common direct marketing techniques used by food and agribusiness firms include direct mailers such as those used by supermarkets, catalogs such as Harry and David's, and, more recently, Internet marketing such as companies that e-mail customers with special offers.

Industry Profile 8–1 describes how the fresh cut produce industry has approached the introduction of their product.

PLACE

A food processor may produce the best products in the world; however, if they are not made conveniently available to consumers, they will still have the best products in the world! The distribution system should be chosen and controlled to make the product available to customers in a manner consistent with the product, price, and promotion efforts of the firm. The *place* decision includes all decisions that affect how, how much, where, and when the product is made available to consumers.

A food or agribusiness company can either sell goods directly to customers or sell through an intermediary. A farmer who operates a fruit and vegetable stand and the producer of boxed cake mixes both market food products, but have chosen very different distribution systems. Let's analyze some of the differences between the two businesses to understand why they have occurred.

Farmers with roadside stands perform all of the distribution functions themselves. Why are they able to do this? A major reason is that their production is small relative to that of the boxed cake producer. Thus, they can market a significant amount of their production through just one outlet. Second, their point of production is relatively near the consumers, making it convenient for them to operate a roadside stand. They are providing a unique service that their customers appreciate. Customers receive fresh, high-quality products and friendly service.

The boxed cake maker, on the other hand, produces a large quantity of cake mixes. The operation of its own distribution system would be impractical and very costly because it would require many, many outlets over a large geographical area. And unlike fruits and vegetables, a store selling only cake

INDUSTRY PROFILE 8–1

THE FRESH CUT PRODUCE INDUSTRY

The fresh cut produce industry emerged in the 1980s in response to consumers' increased demand for healthier foods. The produce industry did not initially respond directly to consumers, but instead to the foodservice industry that wanted to provide healthier alternatives to their customers. Firms like McDonalds and Burger King wanted to offer prepackaged salads to their customers, but lacked the ingredients in the proper form.

Following the introduction of fresh cut vegetables to the foodservice industry, in the late 1980s, fresh cut vegetables were introduced at retail. They were immediately popular as they met consumers' twin demands of convenience and a healthy product. Fresh cut vegetables include a wide variety of prepared salad mixes, sometimes with a salad dressing, and other vegetables, such as broccoli florets, baby corn, and carrot sticks.

The development of the industry was made possible to a large extent by an innovation in packaging. Vegetables that were to be cut, packaged, and remain on the shelf for several days would have to remain fresh, particularly if they were to meet consumers' expectations for a convenience product. In response to this need special films were developed for packaging. These films allowed producers to change the content of the gases in the bag, thereby slowing the respiration of the vegetables and greatly extending their shelf life.

The product was introduced at retail without a lot of fanfare. Because of the publicity garnered by this new product form, and the consumer interest generated because of the product's innovativeness, it was not necessary to spend heavily on advertising. However, the product was priced substantially higher than regular fresh produce and the industry encountered resistance among many consumers unwilling to pay the price premium. In reality the price difference per unit of finished product was less than it initially appeared to consumers. The trimming and removal of defects at the processing plant meant that consumers ended up with a much higher yield than for conventional produce because very little product had to be discarded.

In order to convince reluctant consumers to try the new product, free tastings were given in supermarkets and warehouse clubs. Coupons were distributed at the tastings as well as in supermarket ads, newspapers, and magazines. By lowering the cost of the initial trial, the industry hoped to overcome any initial price resistance.

Today, the fresh cut produce industry has provided the produce industry with an avenue for growth. Fresh cut produce is a higher-margin product than traditional products. An additional benefit for the industry is that some of the product used in fresh cut products would have been otherwise unsuitable for sale, because products with some minor defects were previously discarded. Fresh cut products have also allowed firms to develop brands and some degree of consumer loyalty.

mixes would not be very convenient for consumers because consumers want to buy cake mixes with other grocery products. The distribution of boxed cake mix is handled by wholesalers who serve as a central distribution point for thousands of products. Wholesalers are an efficient means of collecting products from the many producing firms over a wide geographical area and balancing their availability with the needs of supermarkets throughout the country.

Some of the most important factors influencing the place decision are the location, size, and timing of production relative to that of consumption. From the standpoint of the consumer, the decision will be affected by where and when they wish to purchase the product.

Market Channel Management

The primary place decision is selecting a *marketing channel*. In other words, the agribusiness must choose the path by which its products will travel to the ultimate consumer. Marketing channels may be described as direct, contracted, or administered.

Under the *direct system* the food or agribusiness firm handles the entire distribution of its product. The roadside stand operated by the farmer, the nursery that grows and sells its own plants directly to consumers, and the winery tasting room are all examples of direct marketing channels. The obvious advantage of this system is that the company has complete control over how the product is displayed and priced, special promotions, inventory levels, etc. Most food and agribusiness firms, however, find that marketing all of their products through such a direct system is impractical. It would either severely limit their size or require an extensive distribution system, something that is already performed efficiently by wholesalers and retailers. Indeed many agribusinesses that use the direct system market only a small portion of their products directly.

Some food and agribusiness firms *contract* with another firm to distribute their products. Farm machinery dealerships are a good example of contracted distribution systems. Although the producing firm may have no ownership interest in the distributing firm, it may still exercise a great deal of control over it. A major advantage of the contracted system is that the local distributors are more sensitive to the needs of their customers than a manufacturer's outlet would be, because most of the decisions are made at the local level rather than at some corporate office.

The great majority of food and agribusiness products are sold through *administered channels* where no formal agreement exists between the producer and the distributor of its products. Only by the use of its economic power can the food or agribusiness firm manage the distribution of its products to ensure that they are prominently displayed, promoted, priced, etc. Some food manufacturers, because of their dominance in a product area, or because of their sheer size, may exert a great amount of pressure on distributors of their products whereas others may exert none at all.

Viewed from another perspective, the place decision represents functions that must be performed. Products must be transported, stored, and financed, buyers and sellers must be brought together, and someone must hold ownership of the product and bear the risk of loss due to damage, price fluctuations, or deterioration. Even food producers that directly market their products to consumers may contract with other firms to provide some of the marketing functions, taking advantage of the many firms that specialize in one or several of these place functions. For example, a direct marketer may contract with a warehouse to store its product and a shipper to ship the product.

Wholesalers. *Wholesalers* are firms that sell mainly to retailers and commercial users, but not in large quantities to consumers. There are many different kinds of wholesalers, usually making it possible for the agribusiness to choose a wholesaler that specifically meets its needs. Wholesalers are typically classified by whether they take ownership of the goods they market. *Merchant wholesalers* take ownership of the goods they sell; *agent wholesalers* do not. The many types of wholesalers reflect the different functions they perform. Many merchant wholesalers maintain an inventory of goods for their customers, thereby anticipating their needs. They may also provide other services such as credit, delivery, advertising, store layout, store location, accounting, and automated product reordering. The functions that wholesalers perform have evolved over time to suit the needs of their customers. Grocery wholesalers, for example, own, store, finance, and deliver the goods they handle, catering to the needs of the supermarkets they serve. On the other hand, cash-and-carry wholesalers maintain an inventory of grocery products but do not provide all of the services offered by the grocery wholesalers. They operate much like retail grocers and may even sell to consumers. They meet the needs of the small retail grocers who may not be able to justify the cost of a full-service grocery wholesaler.

Another type of wholesaler servicing supermarkets is the rack jobber. *Rack jobbers* not only deliver products to the supermarket but actually stock the products on the supermarket shelves. Rack jobbers distribute food products that have short shelf lives, are perishable, or can easily be damaged if improperly handled. Examples of products distributed by rack jobbers include bakery goods, soft drinks, beer, and snack foods.

Agent wholesalers are most often used when a long-term relationship between a buyer and seller is not practical, as is the case with products that are bought and sold infrequently, where supply or demand vary greatly or are unpredictable, and in geographical areas where sales volumes are low. They do not take title (ownership) to the products they represent and are compensated through commissions based on sales. Unlike wholesalers who frequently carry a wide assortment of products, agents typically specialize in a product line such as fruits and vegetables or frozen foods. They often represent several noncompeting manufacturers at the same time. These specialists perform a valuable function by bringing into contact the buyers and sellers of these products. Brokers, auctioneers, and commission agents are examples of agent wholesalers.

For a discussion of the difficulties Ben and Jerry's has faced in choosing a distributor for its ice cream, read Industry Profile 8–2.

INDUSTRY PROFILE 8–2

BEN AND JERRY'S HOMEMADE, INC.

Ben and Jerry's Homemade, Inc. was founded in 1978 in Burlington, Vermont by Ben Cohen and Jerry Greenfield. The two had grown up together in New York and had been friends since grade school. They started making ice cream in a renovated gas station after completing a $5 correspondence course. They claim to have gotten a perfect score because the test was open book.

Ben and Jerry produce what is known as superpremium ice cream, and more recently superpremium yogurt. The company has become at least as well known for their offbeat style and social mission as they are for their high-quality ice cream. Initially, Ben and Jerry delivered their ice cream themselves to mom-and-pop stores using an old Volkswagen Squareback wagon. By 1983, they had expanded their distribution, opening the first out-of-state franchise and using independent distributors to deliver their ice cream in Boston.

In the early 1980s, Ben and Jerry's increasing popularity got the attention of the market leader in the superpremium ice cream segment, Häagen-Dazs. Ben and Jerry's accused Häagen-Dazs of attempting to limit its distribution in the Boston area by threatening to stop supplying any store that sold Ben and Jerry's ice cream. Ben and Jerry's filed a lawsuit and mounted a campaign against Häagen-Dazs' parent company, Pillsbury, asking "What's the Pillsbury Doughboy Afraid of?" About a year later the lawsuit was settled with Häagen-Dazs agreeing to stop its anticompetitive practices. This was followed by a second lawsuit in 1987 as Ben and Jerry's once again accused Häagen-Dazs of trying to enforce exclusive distribution by its retailers.

Ben and Jerry's grew rapidly in the 1980s, using a hodgepodge of local distributors. Then in 1986, they entered into an agreement with one of the leaders of the premium ice cream segment, Dreyer's Grand Ice Cream, to deliver Ben and Jerry's ice cream using Dreyer's distribution system. By the early 1990s, Dreyer's delivered more than half of Ben and Jerry's ice cream. By the mid-1990s, Ben and Jerry's and Häagen-Dazs were in a virtual dead heat for market share leadership in the superpremium ice cream segment.

Then in 1998, Ben and Jerry's shocked the world by announcing that they were ending their 12-year-old relationship with Dreyer's to distribute their ice cream and that they would enter a new alliance with Häagen-Dazs to deliver Ben and Jerry's ice cream. This followed a takeover offer of Ben and Jerry's by Dreyer's. It was unclear what Ben and Jerry's hoped to gain by having their fiercest competitor deliver its products.

By 2000, the outlook for Ben and Jerry's had soured. Dreyer's had used the end of its relationship with Ben and Jerry's to introduce a superpremium brand of its own. While Ben and Jerry's and Häagen-Dazs were still the leaders in the segment, competition had increased substantially. Industry analysts wondered how long Ben and Jerry's would be able to control their own destiny without controlling the distribution of their product. The speculation ended that same year when it was announced that Unilever would purchase the company.

Retailers. *Retailers* are firms that sell principally to consumers. The firm's selection of retailers will determine where the product is sold, the intensity of its distribution, and the kind of stores in which it is sold.

To most effectively reach the target market, the product must be sold in the appropriate type of stores, located in the right places. Oranges are not sold in department stores or auto part stores, but in supermarkets with other food items. Likewise, expensive shrubbery is typically not sold in city centers, but in nurseries located in the affluent sections of a town or in the suburbs. By understanding where consumers shop and the characteristics of consumers that shop in the different kinds of retail outlets, food and agribusiness firms can ensure that their distribution system will be effective in achieving the desired exposure to consumers.

The intensity of the distribution system should also be considered in choosing a retailer. That is, how many outlets should be used, as opposed to which ones? For some products that are mass marketed, an *intensive distribution system* may be appropriate, that is, the product may be available in many, many places. Candy bars are, for example, distributed intensively. They are available in supermarkets, drug stores, convenience stores, and vending machines. At the other extreme is the *exclusive distribution system,* where a product is handled by only one type of outlet. This works well when there is a high degree of customer loyalty, or when the product is purchased infrequently, making it impractical for many stores to maintain an inventory of the product. It is also used for many status items, which by their nature are distributed in exclusive stores. Farm machinery is typically sold through an exclusive distribution system. This arrangement works well because the manufacturers like the kind of control they have in an exclusive distribution system and the use of many distributors would be very costly because of the cost of carrying the expensive inventory.

Up to this point in the chapter we have assumed that the producer has a choice in choosing its retailers. This is not always the case. As we discussed in the previous chapter, it may be a struggle to get any retailer to handle the product. Likewise when a wholesaler is used, the company may have little or no contact with the retailer. Because retailers and wholesalers handle many products—often competing products—the agribusiness must work to see that its products are given the kind of attention they need. This means working with wholesalers and retailers to ensure that the product receives adequate shelf space, that proper inventory levels are maintained, and that it is well promoted.

Internet Wholesalers and Retailers. There is an intense battle taking place between Internet firms and traditional "brick and mortar" wholesalers and retailers. Most routine business-to-business marketing transactions are already handled over the telephone, fax, or the Internet. The Internet will grow in importance in business-to-business and retail marketing transactions because of the increased convenience, efficiencies, and decreased costs that such transactions promise.

Although the Internet still represents a relatively small portion of sales to consumers, it is growing more rapidly than store-based purchasing. This is likely to continue as the technology improves and the pressure on consumers' free time continues to grow. Today's consumers can order almost any product on line, including groceries in larger metropolitan areas. This will create a challenge for the traditional supermarket and other food retailers. Some supermarkets are fighting back by enhancing the shopping experience or providing an Internet shopping service of their own. Supermarkets are growing larger (the largest are over 70,000 square feet) and offer a cornucopia of fruits, vegetables, meats, and dairy products, making it difficult for Internet grocers to keep up with the selection. Don't know how to cook? No problem, they have chefs available for consultation and cooking classes. Don't want to cook? No problem, the deli has a large assortment of food prepared for takeout.

The Internet is also having a profound impact on business-to-business marketing. For example, the tradition-bound produce industry was shaken up in 1999 by the announcement that buyproduce.com would start business as an Internet supplier of fresh fruits and vegetables. Again, the efficiencies offered by Internet commerce will mean that many functions previously handled by wholesalers will eventually be offered electronically. This will reshape the landscape of many wholesale industries, giving rise to new business opportunities and forcing traditional wholesalers to rethink the services they offer and how they are priced.

ILLEGAL MARKETING ACTIVITIES

The laws and regulations governing marketing activities are intended to ensure that firms treat both their customers and competitors fairly. Although the laws are constantly changing, as well as the courts' interpretation of them, common sense and good judgment will keep most food and agribusiness firms within the constraints of the law.

The *Robinson-Patman Act* requires that manufacturers supply all buyers with products at the same price unless price differences can be justified on the basis of cost. Price differences may be acceptable if it can be shown that they do not substantially injure competition or if it is necessary to meet a competitor's price. Price discrimination laws in general do not apply to consumer goods. By ensuring that all customers are treated on an equal basis with respect to price, customers are prevented from using their leverage with a company to secure a lower price, thereby gaining an advantage over their competitors. Other laws have extended this principle to include the equal treatment of customers with respect to the provision of facilities, services, or funds. As an example, promotional allowances must be made available on an equal basis to all customers. A notable exception to the equal treatment rule is the caveat that price differences may be justified on the basis of cost. It is customary for large-volume customers to get a cost-based volume discount.

Manufacturers are not permitted to dictate the retail price of their products. This is referred to as *retail price maintenance.* However, manufacturers may suggest recommended retail prices for their products and they can set a limit as to the maximum price at which their products may be sold at retail. The maximum price ceiling permits the manufacturer to bar retailers from charging customers exorbitantly high prices during periods of acute shortages.

Tying agreements, which require additional purchases of the customer as a condition of the sale, are prohibited. As an example, it would be illegal for a farm machinery manufacturer to require purchasers of its tractors to purchase and use only its implements. It would not, however, be illegal for the manufacturer to build its tractors so that its implements were the only ones that could be used on its equipment.

Manufacturers are not permitted to collude in setting prices, based on the *Sherman and Clayton* anti-trust laws. These laws are based on the traditional belief in the United States that competition is good. Thus any attempt to reduce competition is generally looked upon unfavorably. Anti-trust laws strictly forbid any discussion of price between competitors (farmers marketing through cooperatives are an exception). Firms found guilty of violating price fixing laws are subject to severe civil and criminal penalties. In 1996, Archer-Daniels-Midland (ADM) pleaded guilty to price fixing in the feed additives market. It has paid approximately $200 million in fines and three of its executives were sentenced to jail terms.

Another related practice that has been regarded by the federal government as illegal is the division of a market into territories by firms. Dividing an area into territories, whereby each firm sells exclusively in its area, has been determined to be illegal because it creates, in effect, a geographical monopoly, reducing competition within the area. *Predatory pricing* activities, where a firm sets prices below costs for a prolonged period with the objective of driving a competitor out of business, are also illegal, as are lying and spreading false rumors about another business.

In recent years a good deal of consumer legislation has been passed. The well-worn saying "buyer beware" might be rewritten today to read "seller beware." Consumer protection laws, both state and federal, cover a wide range of topics. Such laws prohibit deception, require that consumers be informed of certain issues, and expressly mandate warranties. For example, the practice referred to as *"bait and switch,"* whereby a firm lures in customers with a low, advertised price and then convinces them to trade up to a more expensive model, ostensibly because the advertised model is out of stock, has been ruled illegal because it is deceptive.

Many companies go one step farther than the law requires by developing a code of ethics that encourages their employees not only to obey the law but act in a manner consistent with the companies' values. ADM is now proud of its ethics program that encourages employees to immediately report any violations of its code of conduct.

SUMMARY

The 4 P framework focuses the marketer's attention on the product and its characteristics, setting the price, choosing the appropriate promotion techniques, and the place (distribution) decision. The 3 C framework, which considers customers, competitors, and the company, is often used in preparation for making these marketing decisions.

The product decision begins with the development of a product for customers in the firm's target market. The product offering is one of the principal ways that a company can differentiate its offerings from those of competitors. Firms that are successful in differentiating their product offerings often face a type of "me too" competition that requires a change in the product offering in order to sustain its competitive advantage.

The product life cycle describes the various stages that most products eventually go through, including development, introduction, growth, maturity, and sales decline. Understanding the stages in a product's life cycle is particularly useful in managing each of the four marketing decision areas.

The prime consideration in establishing a price policy is to meet the company's profit objective. Price differs from the other components of the marketing mix in that it is a direct determinant of the firm's revenue. An important pricing distinction is the difference between the manufacturer's suggested retail price and the realized price. Buyers frequently receive a price discount, price allowance, rebate, coupon, free service, or free gift that reduces the actual price paid.

The pricing decision is typically guided by a pricing policy. The three most common introductory price policies are price skimming, penetration pricing, and introductory price cuts. For mature product offerings, firms may choose from a plethora of pricing policies, including competitive pricing, cost-based pricing, and value-based pricing.

Promotional activities are usually grouped into five categories: advertising, public relations, personal selling, sales promotions, and direct marketing. Advertising media include newspapers, store signs, direct mail, circulars and handbills, Yellow Pages ads, outdoor signs, radio, television, and the Internet. Publicity, often described as free advertising, may be achieved when a firm or its owner, products, or employees are newsworthy. For some products, personal selling is especially appropriate and important. Sales personnel should understand the steps to the selling process, namely planning, prospecting, making the initial contact, presenting the product, handling objections, closing the sale, and following up on the sale. Sales promotion consists of activities that try to make other sales efforts more effective, including in-store displays, coupons, and rebates. Direct marketing is targeted directly at the customer to try to influence the buying decision. Information technology has played an important role in enabling this type of marketing.

The place decision entails the selection of a distribution channel to get the product from the producer to the end consumer. Essentially, businesses have the choice of selling directly to the customer or through intermediaries such as a wholesaler or retailer. Retailers buy goods from manufacturers or wholesalers and sell them to the ultimate consumer. They determine customer needs and satisfy them through their choice of location, goods, promotion, prices, and credit policy. Wholesalers sell principally to retailers and provide a variety of functions including storage, delivery, product servicing, sales promotion, and buyer credit.

CASE QUESTIONS

For each of the following areas determine what strategies you believe are most appropriate for the product and why.

1. Product
   a. Product name
   b. Package size(s)
2. Price
   a. Introductory pricing strategy
   b. Long-term pricing strategy
3. Promotion
   a. Introductory advertising strategy
   b. Introductory publicity strategy
   c. Introductory sales promotion strategy
   d. Introductory personal selling strategy
   e. Introductory direct marketing strategy
4. Place
   a. Where will the product be sold?
   b. How will the product be distributed?

REVIEW QUESTIONS

1. Describe the 4 Ps of the marketing mix.
2. What are the five stages of the product life cycle and their principle characteristics?
3. List and describe the product decisions.
4. How does the package decision differ for business customers versus end consumers?
5. What are the important considerations in the price decision?

6. What are the three commonly used introductory price strategies and the rationale behind them?
7. What are the most common pricing methods and why are they prevalent?
8. Describe the kinds of discounts.
9. Describe the kinds of allowances.
10. What are the principal costs of offering credit?
11. Why is generic advertising appropriate for some agricultural products but not others?
12. What are the major steps in personal selling?
13. Describe the three types of marketing channels.
14. Explain the difference between merchant and agent wholesalers and give examples of each.
15. What are some of the important considerations in the choice of a retailer?
16. How is Internet marketing changing business-to-business wholesaling?
17. List and describe three illegal marketing activities.

CHAPTER 9

OPERATIONS MANAGEMENT

LEARNING OBJECTIVES

In this chapter we will cover the following topics:

- Scope of operations management
- The difference between mass and continuous production
- Important factors to consider in choosing a plant location
- The process of designing a plant layout
- Production planning process
- Methods of estimating production
- Types of production controls
- Distinction between quality control and quality inspection
- Identifying the costs of carrying too much or too little inventory
- Reorder point inventory system
- Material requirements planning, just-in-time, and efficient consumer response systems
- Factors to be considered in the purchasing decision

CASE STUDY

Bob Berk, owner of Flint Hills Manufacturing Company (FHMC), located in Flint Hills, Kansas, is contemplating a complex situation that is confronting his stock trailer manufacturing company. Trailer sales this year have been better than ever, but the company is faced with a working capital shortage and declining profits. Bob is convinced that the most important action he can take to get the company back on track is to reduce the inventory of purchased materials. Purchased material inventory turnover has fallen from 4 to 1 last year to 2 to 1 this year. The company now has $10 million tied up in inventories. Inventory carrying costs are about 33 percent of the value of the inventory on an annual basis, and the latest pro forma income statement shows income before taxes next year will decline to $1 million. Bob figures that if he cuts inventory in half, he can save about half of the current inventory carrying costs. This, in turn, would double profit projections for next year to $2 million.

At first blush, there appears to be no easier and faster way to improve profitability than to decrease purchased material inventory. However, as Bob begins to ponder his decision, he realizes that making a decision will be much more difficult than he had first thought. The company had done an unbelievable job in selling stock trailers in the last year. Not only had sales doubled in the last 12 months, but market share had also risen from 8 percent to 14 percent. And the company accomplished this without increasing finished goods inventory. FHMC was able to support the explosive sales growth by keeping a high inventory of purchased materials coupled with a streamlined procurement system and a highly responsive production system.

Bob felt that the firm's increased purchased materials inventory would be less costly than an increase in the stock trailer inventory because the former did not require a labor investment. From an operations management standpoint, large inventories of purchased materials have two advantages: (1) they allow greater flexibility in planning and scheduling production, and (2) they minimize production disruptions. Also, large purchased materials inventories allow FHMC to buy economically, as the company can take advantage of discounts on bulk purchases. Carload and truckload freight rates are also a fraction of smaller, less-than-truckload rates. Furthermore, administrative costs in purchasing, inspection, receiving, warehousing, and finance tend to be lower on a unit cost basis because high inventory levels allow FHMC to process fewer, relatively large orders for materials and supplies. The same purchase volume condensed into a few large orders reduces the purchase order-contract management workload, and possibly results in better monitoring of the remaining workload, better control, improved timeliness of delivery, and higher quality of material received.

The more Bob thought about the situation, the more confused he became. What is the optimal level of inventory for his company?

INTRODUCTION

The *operations system* is the portion of an organization that is directly engaged in producing a physical good or a service. Firms exist to produce goods and services for consumers and businesses. Farm machinery manufacturers transform metal, plastic, and rubber into tractors; farmers transform seed, fertilizer, and chemicals into grain; paper mills transform pulp into paper; banks provide credit; food processors transform raw agricultural products into processed foods; and farm supply stores provide a wide selection of agricultural inputs. All organizations have operating systems. For food and agribusiness firm manufacturers, as well as farmers, the operations system is easy to locate—just find the plant or the farm and you have identified the bulk of the organization. For nonmanufacturing agribusinesses the operating system may be more difficult to locate, but it is there nonetheless.

Operations management is the management of the processes involved in the production of goods and services. Although most of the terms and techniques used in operations management were developed for manufacturing industries at a time when the majority of all production involved the manufacture of physical goods, they are relevant to almost any type of production. The same principles that are used in the management of physical production may be applied to the routing of paperwork through an office or customers through a store. Therefore, although most of our discussion will be directed toward the physical processes, it is applicable to all types of food and agribusiness firms.

PRODUCTION PROCESSES

Production processes determine the flow of products through the operations system. In both manufacturing and nonmanufacturing operations, products flow in a unit, mass, or continuous process. The plant design, scheduling, and control of production and inventory management are heavily influenced by which type of production is utilized.

The bakery business provides an excellent example of how each of these processes can be implemented. Suppose the baker decides to mix the ingredients and bake each loaf of bread individually according to each customer's specifications. He or she would be implementing a *unit process.* Or the baker could implement a *mass process* by mixing and baking several white loaves of bread and then whole-wheat loaves of bread at a time. Or the baker could implement a *continuous process* by installing a bread-making machine into which ingredients are continuously added at one end while loaves of bread flow out the other end.

The type of product produced dictates the nature of the process. A unit or small batch process is appropriate when the firm needs to individually tailor the good or service to the unique needs and preferences of the customer. Conversely, a continuous or mass process is more suitable if a large number of customers have similar needs and preferences for the product.

Facility and Equipment Layout

With a unit or mass process, the workspace is allocated by the type of work done, with each area being responsible for the performance of a specific function. General-purpose machines with the capability of performing an operation on many different types, sizes, or shapes of products are typically used. Scheduling is particularly important in unit and mass production because each batch of product competes for a limited amount of labor and machine time. Furthermore, a delay in the production of one batch will often cause a delay in the production of other batches, necessitating a complete rescheduling of production. Mass production is usually used for products that are produced in relatively small quantities, where the use of continuous production would result in the plant space, equipment, or labor remaining idle, or used below their capacity much of the time. Many businesses start out using mass production but eventually grow to the point where their volume makes it more efficient to switch to continuous production.

Wineries typically use mass production. They are set up so that the crushers are grouped together, as are the presses, fermenters, storage tanks, and bottling equipment. Careful planning is necessary during the harvest season to ensure that all of the grapes are crushed and pressed on time and that sufficient storage space is available for the wine.

A continuous process is characterized by the uninterrupted production of a single product, or several similar products, having standardized labor requirements, machine setups, and production methods over a long period of time. Because of its unchanging nature, the continuous process allows for the allocation of plant space, on a permanent or semi-permanent basis, to the production of one basic product and the arrangement of machines in a sequence that is most efficient for that product. Highly specialized machines, designed specifically for a particular production process, are often utilized. Continuous production is typically used for products that are produced in high volumes and lends itself well to the use of assembly line techniques.

Commercial egg producers provide a good example of the continuous production method. All of the hens are grouped together in one area of the plant and the eggs are moved along conveyor belts to stations where they are sorted, graded, inspected, and finally packed to await shipment.

PLANT LOCATION AND DESIGN

One of the major tasks in planning for production is choosing a plant location and designing the plant. These tasks are especially important because of the high investment costs that are required and because these decisions will affect production for years to come.

Facilities

The first decision that must be made is to determine whether the business needs just one, a few, or many geographically separate facilities. Having one facility allows the firm to take advantage of *economies of scale* (greater efficiency with larger operations), whereas having several facilities enables the firm to locate nearer to suppliers and customers. An initial indication of the appropriate number of facilities to have can be found by analyzing how costs vary as the number of proposed facilities increases or decreases. As the number of facilities increases, the firm will need to buy more land, buildings, and equipment. On the other hand, with more facilities, transportation costs decrease as the firm locates closer to suppliers and/or customers.

Location

The next decision concerns locating the facilities. New businesses or businesses that are making a major expansion are often faced with choosing a location for their new facilities, a decision that can have a heavy impact on the firm's long-run profitability and its ability to compete. Several factors need to be examined in choosing a location, including: the source of inputs, the location of the firm's customers, the transportation costs of inputs and the finished product, the availability and cost of labor, total production costs, zoning laws, and special incentives offered by specific communities.

For many companies, the most important factor in choosing a plant site is its location relative to that of its major inputs and customers. A firm whose source of inputs is concentrated in one area but whose customers are geographically dispersed will find it advantageous to locate near the source of its major inputs. This is particularly true for products that are highly perishable or that may be damaged in shipping if they have not been properly processed or packaged. For this reason most major beef packers have plants in the Texas panhandle, Kansas, Colorado, and Nebraska. Similarly, a majority of the wineries in the United States are situated in California, and a large number of cheese plants are found in Wisconsin and Minnesota. On the other hand, firms that receive inputs from a wide geographical area, but whose customers are concentrated in one area, will find it advantageous to be situated near their customers. This is especially important for retailers of products where the physical inspection of the product is important for its customers or where close proximity to the customers is needed to ensure product freshness, as is the case with baked goods.

Another important consideration in choosing a plant location is the cost of transporting the inputs and finished products. The cost of transportation is influenced by several factors, including the bulkiness or weight of the product, refrigeration needs, and the mode of transportation, to name a few. Milk is often transported over a long distance by tanker trucks, because milk that has been bottled is much bulkier, that is, it takes up a lot more space relative to its weight than milk that has not yet been bottled. The wasted space makes it much more expensive to ship cartons of milk than the unbottled product. By the same token, reducing the total weight shipped may reduce transportation costs. Wood products would be very expensive if entire trees had to be shipped to cities before they were processed. By removing the waste material, which may comprise a substantial proportion of the raw product, at an early stage, and shipping just the lumber, paper, and other processed products, transportation costs may be greatly reduced. Access to the right kind of transportation facilities can also affect transportation costs. Grain elevators, for example, are almost always located near a major road, and often along a railroad line to take advantage of the lower transportation rates of shipping by the carload.

Three factors need to be considered in evaluating the labor situation in a community: the number of workers available, the skills available, and the labor costs. Food companies, which require workers with a particular skill, should determine the number of workers possessing each required skill in each area under consideration. Locating in an area that cannot provide an adequate labor force can be very expensive. Although workers may be brought in from the outside, this may result in higher initial costs and a higher employee turnover rate. The cost of labor is another important consideration because it varies substantially throughout the country. Both the wage rate and the cost of fringe benefits should be analyzed. A community's union history may also be of interest to some firms. Areas dominated by unions often have higher wage rates, and the presence of a union may result in costly strikes. On the other hand, firms that treat their workers fairly usually find that their labor costs are low relative to the productivity of their workers and have few strikes even when they are unionized.

The characteristics of the community can be an important factor in luring and retaining qualified personnel. The climate, a city's growth rate, and the presence of good schools are a few of the factors affecting the quality of life that may be considered by prospective employees. Most businesses will also want to consider local and state laws as well as the general business climate before selecting a location. The regulations affecting zoning, labor, noise, and the disposition of wastes may not only facilitate the operation of the business but also make a substantial difference in the cost of operations. Many communities offer special incentives for new businesses to locate in their area, such as a low price on land owned by the municipality or special tax treatment.

Once a community has been selected, the specific plant site must be chosen. A list of requirements and desired features will help narrow down the possibilities. Such a list might include the following: water, power, sewer, and transportation facilities, the zoning classification required, and the preference for a rural or urban area, to mention a few. Many cities have industrial parks that can readily provide most of the services required by industrial plants. Industrial parks have the added advantage of being attractive to the city's residents because manufacturing activities are concentrated away from residential areas. If growth is expected, it should be planned for at this stage, especially if it is to be achieved by expanding the planned facilities. Sufficient land should either be acquired or available at a reasonable price adjacent to the plant site.

Plant Design

Finally, the firm has to decide how to design the plant. The following steps should be followed in designing a plant: (1) determine plant size, (2) map out the basic organizational structure, and (3) design the layout of each department or work area. One of the most important decisions the agribusiness will make in designing the plant is determining its size. It represents a long-term commitment and generally requires the investment of large sums of capital. It therefore deserves careful consideration. A firm's production costs may be heavily influenced by the size of its operations. Because of economies of scale, a large plant may result in lower per unit production costs than a small one, as fixed costs are spread out over a greater number of units. However, a plant that is so large that it becomes unwieldy to manage can be more expensive to operate than several smaller ones. Large plants may also benefit from economies of size from quantity discounts that are obtained when supplies are ordered in bulk or by the freight savings obtained by shipping inputs and the finished product in large quantities. Market factors may also play a role in determining plant size. Where the size of the market is the limiting factor, a market analysis is useful to determine a reasonable market share for the firm, which may then influence the plant size. On the other hand, some firms may find that they need a minimum level of production in order to be of interest to buyers of their product.

The seasonal production cycle is particularly important for agribusinesses because of the biological nature of their product, over which they exercise little control. Food handlers and processors are particularly affected by the seasonal problem. In extreme cases a plant may receive its entire production for the year within a period of several weeks. Highly perishable products must be quickly processed to prevent spoilage or deterioration. During the peak season the plant will probably operate around the clock. The cost of building a plant with the capacity to handle the peak production of a product, which may be achieved for only several days per year, must be carefully weighed against the expected revenues. Fruit and vegetable packers have been able to reduce their per-unit production costs by utilizing their facilities for several products with different harvest seasons.

The firm's growth should also be anticipated in the plant design. If growth is expected in the near future, the plant may be built to handle the anticipated capacity, although the extra equipment may not be purchased until the growth is realized. Another consideration related to the anticipated growth is centralization versus decentralization, an important issue as the business expands. A centralized operation is often chosen to take advantage of economies of scale, that is, lower per-unit costs associated with a larger plant size. Firms whose raw materials are limited to a narrow geographic area may have no choice but to centralize their operations. However, there are also advantages to placing facilities in different geographic areas, thereby protecting the agribusiness against localized product shortages, strikes, natural disasters, etc. Firms that must stay in close touch with their customers often find that the best way to do so is by operations that are decentralized, so that they are located near the customers they serve.

The second step in designing the plant is to define its organizational structure. The basic organization will depend on the type of production to be used. When continuous production is used, the plant will be organized along product lines, whereas with unit or mass production the plant will be organized according to the activity performed.

Plants using the continuous production process are organized by determining the placement of each of the product lines and departments that they share in common, such as shipping, receiving, and the warehouse, so as to achieve a smooth flow of materials and final product through the plant. Plants using the mass production process are organized on a functional basis with like activities being grouped together in work areas. The work areas are usually arranged to minimize the movement of products throughout the plant. The departments not related to the day-to-day production activities, such as accounting, personnel, marketing, etc., are usually located away from the areas of heaviest production activity, in order to minimize the disruption of the work in these areas.

The next step is to determine the layout for each product line or work area. There are two basic approaches to facility layout: process layout and product layout.

With a *process layout system,* typically used in unit or mass production systems, the processing components are grouped together in departments on the basis of the kind of function they perform. An advantage of the process layout is that it is flexible in permitting the firm to handle a wide variety of different tasks.

Different tasks can be sent to different departments in different sequences as the processing process dictates. Another advantage is that there is a high degree of independence in operations, which means that if something goes wrong in one department, the other departments can still function. Finally, process layouts have low costs for capital equipment because the grouping of activities by function in an area eliminates the need for duplicate equipment in other parts of the plant.

A disadvantage of the process layout involves the relative complexity and high costs of materials handling. Products must be shuttled back and forth between departments, which is inefficient and expensive. Imbalances in operations can occur as one department becomes overloaded due to its processing requirements as another department sits idle. Finally, process layouts involve relatively high labor costs as the general-use equipment typically found in process layouts requires more labor to operate.

With a *product layout system,* typically used with continuous production systems, the processing components are arranged according to the sequence of tasks that must be performed to produce a specific product. The classic product layout is the assembly line. In the product layout the processing components are located wherever the product's processing requirements dictate, whereas in the process layout the product is moved between the various departments as the process requires.

As you might expect, the advantages and disadvantages of a product layout are the opposite of those for the process layout. Materials handling problems and costs are minimized in product layouts due to the smooth flow of product from each station to the next. Operations along the assembly line can be balanced by assigning the right number of tasks at each work station, thereby preventing work overloads at some workstations while others remain idle. Product layouts rely on specific-use and highly mechanized machinery for a particular good and result in relatively low labor costs.

Product layouts are rigid and inflexible, which typically means that they cannot be used for producing different products or services. They create a high degree of dependency that can cause the entire assembly line to shut down if one workstation runs out of material. Finally, the specific-use and highly mechanized nature of the assembly line results in a relatively high initial cost for equipment.

Several other factors must be considered before the building design is complete. Materials handling, that is raw materials, work in progress, or finished goods, is an important consideration. Emphasis should be placed on minimizing the amount, distance, and complexity of moving materials through the plant. Operations balancing between tasks and work stations and space utilization are important factors in designing an efficient plant. Provisions for both routine maintenance and major breakdowns must be made. It is a costly mistake to overlook the possibility of having to replace a large piece of equipment. It is also important that the plant layout be designed so that growth may be provided for by expanding the plant without the need for a total reorganization and with a minimum disruption of work. Lastly, the overall appearance of the plant must be considered in the design and landscaping of the plant, as well as factors such as security, parking, waste disposal, etc.

PRODUCTION PLANNING AND CONTROL

The objective of production planning and control is the efficient production of a specified product by a given time. To accomplish this goal, production must be carefully planned in order that machine time, labor, and materials may be made available at the right time.

Production Planning

Production planning ensures the efficient operation of the firm over a specified period of time. Customer orders and sales forecasts are the basic input into the *production plan,* which determines the size of the workforce, inventories, and material purchases. The production plan becomes the basis for a master schedule that is used to produce purchase orders and work orders.

The first step in planning production is to forecast future demand. No single forecasting method gives uniformly accurate results. Accordingly, it is desirable to use several methods, with each method acting as a check on the others. All forecasting methods fall into two categories: top-down and buildup. With the *top-down method* the firm makes a series of forecasts.

- First, the firm forecasts the industry's total market potential given overall economic conditions.
- Second, the firm will estimate its market share taking into account competition, pricing, product differentiation, and advertising and promotion.
- Finally, the firm will develop a sales forecast.

With the *buildup method,* the firm uses one or more of the following forecasting techniques: consensus of executive opinion, a sales force composite, users' expectations, and quantitative methods. Consensus of executive opinion involves obtaining opinions from the managers of various departments in the firm or industry experts and developing from these opinions an average forecast. The sales force composite works in a similar fashion except members of the sales division are solicited for their forecasts of future sales in their territories. Salespeople are good sources of information because they are in close contact with the firm's customers. Both the consensus of executive opinion and the sales force composite have the advantage of being easy to develop but have the disadvantage of relying on opinion rather than facts and analysis.

The firm develops a sales forecast from users' expectations by asking their customers how much they expect to purchase in the next operating period. The advantage of using users' expectations is that it is relatively inexpensive; however, the disadvantage to this forecast method is that users' expectations are subject to change. Early order discounts may be given to encourage customers to place their orders early. When used year after year, this may become a reliable guide to estimating total orders.

Many firms rely on quantitative techniques to supplement expert opinions and users' expectations to improve the accuracy of their sales forecasts. Quantitative techniques use historical data on how competition, market demand, economic conditions, and the availability and price of purchased materials influenced past sales to develop a forecast of future sales. Although quantitative techniques are very good at explaining the past, which after all is a fairly easy task, they are often poor predictors of the future. The experienced forecaster will develop several forecasts to yield a range of estimates and, if appropriate, modify the forecast to take into account changes that might cause a deviation in forecasted sales from the past.

Developing the *master schedule* of the work to be performed completes the production planning process. Scheduling is especially important for food and agribusiness firms, many of which are highly seasonal. The problem is compounded because many of the products are highly perishable and thus must be processed quickly. The peak production season must be planned for well in advance. Many agribusinesses schedule maintenance during slack periods so that labor, which would otherwise be idle, may be more efficiently employed. Even so, most agribusinesses whose production is highly seasonal are faced with the decision as to whether or not a trained labor force should be maintained during slack periods. Although laborers who are employed but not productive are maintained at a high cost to the firm, it may still be to the firm's advantage to keep them on because new laborers will require training, be less productive initially, and make more mistakes. Often key supervisors are retained and less important employees are laid off in the hope that some of them will be available when they are needed again.

A variety of factors must be considered in scheduling production, among them: production deadlines, equipment availability and capacity, machine setup and down time, labor availability, the materials required, and the number of defective products produced.

Scheduling for continuous production is the simplest case because increasing or decreasing the amount produced has a minimal effect on other operations in the plant. Operating longer shifts or adding another shift generally increases production. Within limits, it may be possible to increase the speed at which the production process is carried out. However, operating a line at much above the usual rate may actually lower production due to the increased chance of human error, accidents, or machine breakdowns.

Scheduling mass production is more difficult because products are not produced using fixed proportions of machine time, labor, and materials. Furthermore, a product may move in and out of a work area and traverse the plant several times, making it difficult to determine the most efficient schedule for even one product, let alone the production of many products whose schedules are interdependent. For small firms, all that is needed for scheduling is a sharp production manager with the necessary information, a pencil, pad of paper, and a calculator. For large firms the use of a computer and a scheduling program is almost essential.

Production Control

The first step in controlling operations is to gather data concerning how efficiently operations are running. The second step is to transform the data into useful information such as the average amount of time it takes to complete a process. Step three is to compare the data with standards of operation. These standards are normally expressed as a range of acceptable values to see how the data compares with the standards. The fourth step is to identify any discrepancies that may exist between the standards and the actual performance. Step five is to analyze any differences to find out what has gone wrong. The sixth step involves deciding what corrective action should be taken to remedy the discrepancies. The seventh and final step in operations control is to implement the corrective action. The focus at this stage is not only to rectify what went wrong but also, when possible, to redefine the work procedures to prevent the same situation from reoccurring.

There are two types of operation control methods: order control and flow control. Order control is used with mass production, whereas flow control is used with continuous production. With *order control* a batch of product is controlled as it passes through each of the various departments, or work areas, and the different production processes are performed on it. Knowing the location of the batch at any time allows the production manager to determine whether or not the product is being produced on schedule. If it is not, adjustments will be required in the scheduling of machines, labor, and material use to keep production operating efficiently, not only for the batch in question, but for the production of all affected batches.

With *flow control* two things are monitored: the rate of production of the final output and the rate of flow of material to each workstation. The rate of production of the final product is the best indicator of the status of the production line. Work stoppages are quickly reflected there and may then be traced back to their source. The flow of materials to each workstation must also be monitored to ensure that production continues uninterrupted, and to prevent materials from piling up. In a continuous production system, a work stoppage at one point on the line usually results in shutting down the entire line because the work at each station depends on product flowing from the previous station.

Quality Control

The ultimate measure of a firm's production is not the number of products produced, but the number of quality products produced. The goal of producing high-quality products or services should be part of the "business culture" of the company because the cost of producing a defective product is about the same as for one without defects. However, most defective products cannot be sold, or if sold, must be sold at a discount, and those that are sold are often returned. However, the real damage is the injury to the firm's reputation and the loss of future sales.

Quality control is more than the inspection of the final product. It should be part of the production process from the start. Quality standards should be established and care should be taken to ensure quality at each stage of the production process.

Quality inspection is the process of inspecting the finished product to verify that it meets the firm's quality standards. In some instances it may be necessary to inspect every product, particularly where the cost involved in releasing a defective product is high, or the volume of production is low. For most products, however, it is sufficient to inspect samples of the finished product. In mass production, samples may be chosen from each batch, whereas with continuous production samples may be chosen at regular intervals.

Although it is important that quality control be integrated into the production process, the responsibility for inspection should be removed from the person directly responsible for production. The production supervisor, who also has responsibility for inspecting the product, may be tempted at times to overlook products that do not meet quality standards in order to boost output or to meet production quotas. Thus an inspection system that is independent of production will help ensure its integrity.

The *Hazard Analysis and Critical Control Point* system, better known by its acronym, HACCP, is an important innovation in quality control that has been used to dramatically reduce the level of pathogens in food. HACCP works by identifying problem areas and applying controls in critical areas to control, reduce, or eliminate hazards. It is comprised of the following seven steps:

- Conduct a hazard analysis
- Identify the critical control points
- Establish critical limits for each critical control point
- Establish monitoring procedures
- Establish corrective actions
- Establish recordkeeping procedures
- Establish verification procedures

HACCP has been used in food processing plants for years, but recently the federal government has mandated its use in foods that pose the highest risk to consumers. It has been required for meat, poultry, and seafood plants, and may be required for use by processors of unpasteurized juice. The application of HACCP is expected to significantly reduce the number of illnesses and deaths from disease-causing organisms in foods where the HACCP process is utilized.

Inventory Control

Inventories are maintained throughout the operating system. Inputs such as parts and materials, work in progress (or work that has been partially completed), and finished goods are all examples of inventories. Inventory management is important to operations management to keep production running smoothly and to control costs. Too little inventory can result in costly production delays, as machines and employees sit idle, and in lost sales. At the same time too much inventory ties up capital and precious storage space, and may lead to increased insurance rates and property taxes.

The most widely used system for controlling inventory in small food and agribusiness firms is the *reorder point system.* It requires that the rate of product use and length of time required to receive an order be known and that the desired safety margin be established. The following example will serve to illustrate this system.

A canning company uses cans at a rate of 20,000 cans per day. It takes an average of four days for an order to be filled, and it has never taken more than seven days. If it always took exactly four days to receive an order, a minimum inventory of 80,000 cans would be kept, that is, when the number of cans reached 80,000, a new order would be placed. This is the pipeline inventory or inventory required to meet the firm's needs from the time an order is placed until it is received. Then, if everything occurred as planned, four days later, just as the last cans were being used, the new shipment would arrive. Such a system, however, leaves no room for error, which may be completely outside of the control of the firm, as is often the case with shipping delays.

A safety margin is needed to protect against uncertainties in the rate of use, or in the delivery of the product. In determining the safety margin, the costs of running out must be balanced against the carrying costs of the additional inventory. Knowing that it usually takes four days to receive an order, and that it has never taken more than seven days, the firm may determine that it would like to maintain a safety stock equivalent to two days of product use, particularly if a seven day order fulfillment time is extremely rare. Thus the safety stock would be 40,000 cans. On the other hand, if the cost of running out of cans is extremely high, the firm may wish to maintain a safety stock equivalent to three days of product use.

Summing the pipeline inventory and the safety stock arrives at the reorder point. In this case the reorder point is 120,000 cans, corresponding to a six day supply of cans (Figure 9–1).

The reorder point system works best when an item is used at a constant rate, as is the case with continuous production or in nonseasonal wholesale or retail businesses. At the other extreme are materials used in mass production that may be used only infrequently when an order for a product requiring it is placed. In this case a small amount of the material may be kept on hand, possibly enough to get production started while awaiting shipment of the material. A more common occurrence is where a material is used constantly, but the rate of

FIGURE 9–1

Reorder Point Inventory System

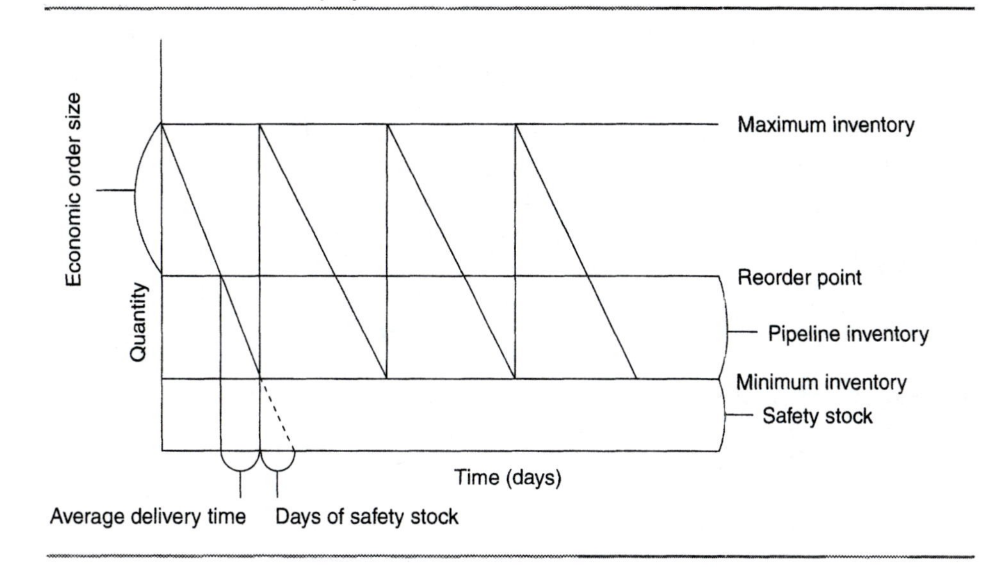

use varies, as is the case with businesses that are highly seasonal. If the rate of production can be predicted, possibly by reviewing past production records or analyzing new orders, the reorder point system may be used with periodic adjustments to the reorder point, depending on the usage rate of the input.

Where sufficient records are available, it is useful to design an inventory control system so that a red flag is sent up whenever the rate of use rises above or falls below a certain level for a given period of time. A very high rate of use may indicate that the pipeline inventory and safety stocks are too low for the firm's needs, whereas a low rate of use may indicate just the opposite. A timely readjustment of the reorder point can prevent costly production delays or result in substantial savings in carrying costs for unnecessary inventory.

Optimal Inventory Level

A high degree of interdependence exists between the *optimal inventory level* and the optimal quantity to purchase at any point in time. Determining the right inventory level and the right quantity to purchase impacts the successful operation and profitability of a firm. As we saw in the Flint Hills Manufacturing case, this determination requires an analysis involving many trade-offs.

Many firms coordinate demand and supply of production materials and supplies required for their operations to such an extent that inventories are unnecessary. These situations will be discussed in more detail later. For many food and agribusiness firms, however, such fine-tuning of operations is not possible. For example, it may be impossible to know future demand with certainty. Furthermore, it may be impossible to guarantee the availability of all purchased materials at a given time.

Inventories serve as buffers between the demand for and the supply of required materials and supplies. They allow the firm to better respond to customer demands for its products and to improve efficiency in operations. For smaller firms, inventories allow for a reduction in the overall cost of purchased materials and supplies through purchasing, transportation, and administrative economies. Inventories also allow the firm to hedge against contingencies such as transportation difficulties, strikes, and natural catastrophes.

Several quantitative approaches exist for determining the optimal inventory level. These are beyond the scope of this book. But the underlying logic of inventory management can be examined. The firm can determine its optimal inventory level for an item by summing the appropriate costs and selecting the appropriate level associated with the minimum total cost. An example of a local farm cooperative selling tires will provide further insight into this concept.

There are four basic costs associated with holding inventories: purchase costs, administrative costs, carrying costs, and stock-out costs. Purchase costs are the actual costs of the tires that go into inventory. The cooperative finds that, on a per-unit basis, as the size of the order is increased, the total cost per tire decreases. This occurs because discounts are given for large orders and per-unit shipping costs usually decline as the size of the order is increased. A large order may also be placed to take advantage of a special price or in advance of an expected price increase. Administrative costs are associated with purchasing, receiving, inspecting, warehousing, and paying the supplier. Consequently, as the size of each purchase order increases, the number of orders that must be placed decreases, and the per-unit administrative costs also decrease. Carrying costs are the costs associated with keeping inventories on hand once they have been acquired. Facility space used for tire storage, interest, insurance, obsolescence, and "shrinkage" (an unexplained decrease in inventory levels) all contribute to the cost of holding the tire inventory. As the size of each order increases, the average number of tires in inventory also increases, and, therefore, total carrying costs increase too. Stock-outs occur when a customer comes into the cooperative to buy a tire and the proper size tire is not in stock. Some customers will wait until the tire can be ordered and received, but others will go to a competitor to buy the tire resulting in a lost sale. By increasing the size of the tire inventory, the cooperative can reduce the incidence of lost sales due to stock-outs. A stock-out situation can also occur when a materials inventory in a manufacturing facility caused the production line to halt causing a decrease in operations productivity.

The cooperative can now develop a table to determine the total inventory costs associated with different order sizes and inventory levels. The cooperative on average sells 10,000 tires per year. Purchase price, transportation rate, and total cost per tire for differing quantities are given in Box 9–1. Administrative costs are estimated to be $50 per purchase. Lost sales costs (based on expecta-

BOX 9–1

Total Inventory Costs for Different Order Sizes and Inventory Levels

	Cost per Tire		
Tires per Order	Per-unit Purchase Price	Per-unit Transporation	Total Per-unit Purchase Cost
0 to 199	$50	$5	$55
200 to 499	$47	$4	$51
500 and over	$46	$4	$50

Total Inventory Cost Calculation Based on Different Order Sizes and Inventory Levels

Order quantity	100	400	500	600
Average inventory	50	200	250	300
Number of orders[1]	100	25	20	16.67
Total purchase costs	$550,000	$510,000	$500,000	$500,000
Total administrative costs[2]	$ 5,000	$ 1,250	$ 1,000	$ 834
Total carrying costs[3]	$ 825	$ 3,060	$ 3,750	$ 4,500
Total out-of-stock costs[4]	$ 2,000	$ 500	$ 400	$ 333
Total inventory costs	$557,825	$514,240	$505,150	$505,667

[1]Number of orders = 10,000 / quantity ordered
[2]Total administrative costs = number of orders × $50
[3]Total carrying costs = 0.30 × average inventory × purchase cost
[4]Total out-of-stock costs = $20 × number of orders

tions) are estimated to be $20 each time a tire shipment is due in because an out-of-stock situation is most likely to occur just prior to the receipt of a new shipment. Tires are not mounted until the tire is on hand so the management does not believe that its inventory policy has any impact on the productivity of its tire operation. Inventory carrying costs are estimated to be 30 percent per year. In this case the optimal inventory level occurs at the minimum total inventory cost of $505,150, which corresponds to an inventory level of 250 tires and an order quantity of 500 tires (the order size). The optimal inventory level often occurs at a point where the order size allows the firm to take advantage of a quantity discount or a lower per-unit shipping rate that occurs when a full truck load or train load is ordered. The cost savings beyond this point may be negligible when compared to the increased carrying costs of holding more inventory.

Inventory Accounting

An accurate accounting of inventory is necessary if inventory control is to be properly exercised and to meet accounting requirements. Two methods are used to keep track of the quantity of inventory in stock: the periodic and perpetual methods.

The *periodic method* requires the periodic counting of inventory, usually quarterly, semiannually, or annually. This system is very simple because the records are only updated periodically. It also has the advantage of being extremely accurate when inventory is taken, although there is much room for error in between periods. It is therefore most useful when an accurate inventory is not needed on a daily basis. It may also involve closing the business if the inventory cannot be conveniently taken when the business is normally closed.

With the *perpetual method* the inventory is adjusted every time a sale is made or an order received. The major advantage of this system is that the records are kept up-to-date, although they will not be as accurate as when the inventory is physically counted, due to inaccurate record keeping, theft, or other losses. For this reason, the use of the perpetual method is usually used in conjunction with the periodic counting of inventory. Computerized accounting systems have made the use of the perpetual method much more common because the principal disadvantage of managing large amounts of records has been overcome.

Industry Profile 9–1 describes 7-Eleven's operational and distribution strategies, two of the keys to its success.

PROCUREMENT SYSTEMS

Procurement is the process of deciding what, when, and how much to purchase, the act of purchasing it, and the process of ensuring that what is required is received on time in the quantity and quality specified. An increasing number of food and agribusiness firms are employing computerized procurement systems to improve the efficiency of the purchasing process and to cut costs.

Material Requirements Planning

Many food and agribusiness firms use a computerized *material requirements planning* program for production scheduling, inventory control, and the scheduling of purchase orders. This program is used in production planning and scheduling to ensure that materials are available when required, to reduce the firm's inventory, and to assist in rescheduling purchasing and manufacturing operations.

Electronic Data Interchange

Materials requirement planning links the manufacturer with its suppliers through a computerized information transfer system called an *electronic data interchange.* The entire system vastly improves purchasing and inventory management by cutting back on paperwork, lead times, data input errors, and inventory accounting. The system works in the following manner.

INDUSTRY PROFILE 9–1

7-ELEVEN

The 7-Eleven chain of convenience stores is the largest such chain in the world. Although it is known among consumers for its customer-friendly format, its operational strategy is just as critical to the company's success.

The lynchpin of 7-Eleven's operational strategy is its distribution system. A successful distribution system must satisfy two competing goals. First, it is important to keep the shelves stocked. Because many customers shop at a convenience store for only one or two items that they really need, an out-of-stock item not only means a lost sale, but also a dissatisfied customer who may not return. The second goal is to minimize the level of inventory. Excessive inventory means increased carrying costs.

In response to the unique needs of the convenience store industry, 7-Eleven designed a distribution system specifically around its requirements. Store stocking lists are maintained for each store and updated monthly. In this way, the product selection can be tailored to each store and slow-moving products are eliminated. Store orders are transmitted to a central computer, which then sends them to the distribution center. At the distribution center, cases are broken down and small order quantities are loaded onto custom-designed trucks with separate compartments for dry, chilled, and frozen products. The trucks make deliveries along routes planned by a computer to minimize travel time. An indicator of the success of this computerized inventory control system is that there is a 99 percent fill rate for orders.

A second aspect of 7-Eleven's operational strategy is location. Location means two things to 7-Eleven, the location of its stores and the location of products within its stores. To be successful, convenience stores typically must be located near a large number of people, on a high-traffic thoroughfare, and with convenient access for entry and exit. 7-Eleven conducts substantial research before opening a new store and it routinely closes stores that fail to meet its sales expectations.

For 7-Eleven, the store itself is essentially its manufacturing plant. The company places a high priority on where items are located. To be convenient to its busy customers, the layout must be intuitive. Most customers want to find the needed product and be on their way. The company continually experiments with different product placements to determine which in-store location generates the highest sales. Impulse goods are carefully chosen and placed to generate additional sales. Because most customers' purchases are very small, increasing the size of the average purchase has a big impact on operational efficiency and profitability.

Another reason for 7-Eleven's success is the way in which it integrates its marketing strategy with its operational strategy. The decision to carry only the top one or two brands in each product category is an example of how marketing and operational needs together determine its product strategy. In this case the need to generate a high rate of sales is balanced by the need to limit the number of products due to space limitations.

For key, large volume materials, the firm negotiates one- to two-year long contracts with a major supplier and a back-up supplier instead of requesting bids every time inventory is purchased. After negotiating the contracts, purchasing is no longer involved in the process unless there are problems or until the next contract is negotiated. Essentially the business and the supplier work as a team to reduce the lead time required to get supplies to the company, thereby reducing inventory needs and administrative costs. Bar coding on supplies automates the documentation process. As supplies are received by the firm, the items are scanned and sent immediately to the production line or to inventory. The computer records the actual inventory level of each item, quantities on order, quantities allocated to the production line, lead times to receive new shipments, and lot sizes. The objective is to establish a continuous-replenishment system where the computer program continuously examines what supplies are being used, what supplies remain in inventory, and what supplies are in transit. The benefits of the materials requirement planning system are:

- reduced inventory investment.
- reduced administrative effort because scheduling, inventory control, and purchasing should be more efficient.
- reduced obsolescence of materials because the computer keeps track of when materials arrive on site and are used in the manufacturing process.

Just-in-Time and Efficient Consumer Response Systems

The materials requirement planning systems, described in the previous section, work by pushing supplies through the manufacturing process based on information generated by the master schedule. In other words, the computer program is designed to schedule production operations and orders from suppliers based on a production plan. By contrast, *just-in-time* computer systems operate by waiting until the materials are needed before a purchase order is placed. At a certain reorder point at each workstation, a production release signal directs the materials from a previous workstation to move to the next workstation in a continuous flow. The system includes materials requisition devices to place orders with suppliers. Just-in-time systems are used in agribusiness manufacturing companies such as John Deere, where tractors are essentially "built to order."

A similar system, called *efficient consumer response*, is used in the consumer products industry. The following example illustrates the way that the efficient consumer response system is utilized in a supermarket. Each individual supermarket is linked to its supply warehouse by an electronic data interchange. As the customer goes through the checkout line, each item purchased is run across a scanner where the bar code is read. This sales information is then fed into a computer system that examines what is being sold at each store, what remains in inventory, and what is in transit. Some of the more advanced computer systems also forecast the future demand for each item. The computer in the supermarket continuously keeps track of all items being sold in the store. As soon as the reorder point for a particular item is triggered, the computer adds that item to the purchase order. The computer automatically forwards the purchase order to a computer in the warehouse, which in turn adds the order to the next shipment headed to the supermarket.

Strategic Supply Alliances

The advent of computerized procurement systems has led many firms to establish alliances with their major suppliers. Through these alliances the firms establish partnership arrangements with their suppliers to ensure that materials are delivered on time and meet the specified quality. To make these arrangements work requires that both the firm and the supplier open their books (usually via the Internet) so that the nature of the procurement costs can be better understood. Once this information is available, operations and purchasing managers in both firms examine the costs of items in an effort to identify costs that can be reduced or eliminated.

Purchasing

Determining the right time at which to make purchases and the right quantity to buy can have a major impact on a firm's success. These decisions affect the firm's responsiveness to its customers, productivity, the cost of purchased materials, and administrative, handling, and storage expenses. In proactive procurement, purchasing personnel must interact with those responsible for forecasting, production planning, and inventory management. The goal of the purchasing department is to ensure that the specified materials are available at the proper time, in the right quantity, of the right quality, and at the best price. The purchasing function entails locating suppliers, gathering information on raw materials and supplies, analyzing this information and possibly testing product samples, placing and following up on orders, and verifying that the delivered goods are what the firm ordered and that they meet the firm's quality specifications.

The purchasing activities must be coordinated with those of inventory control and production management. For example, the purchasing and production departments should jointly determine the specifications for materials to be ordered and the purchasing and inventory people should work together to determine the timing and size of the orders to be placed. At times, conflicting goals may cause a disagreement between departments, as may be the case when the purchasing agent wants to buy a material based on price but the production department's need is for a high-quality good. Interdepartment coordination is necessary to ensure that the decision is made based on what is best for the firm rather than for specific departments. The major considerations in making purchasing decisions are quality, quantity, supplier dependability, and price.

Quality. *Quality* from a firm's standpoint means conformance to required specifications that define specific properties of materials and parts, the efficiency of production, and the performance of the finished product. The desired level of quality begins with the customer and works backward through the firm. Based on customer requirements, the firm defines the requirements for all of the operations necessary to deliver value to the customer. Value in this case means what the customer is willing to pay for a specific level of quality.

When the quality of the input affects the quality of the finished product, the firm has no choice but to purchase materials of sufficient quality to allow it to maintain its product standards, or face lowering the standards. On the other hand, a lower quality input may have no affect on the quality of the output but may mean that the efficiency of the production process is reduced, possibly requiring increased amounts of the input. In this case the cost savings due to the lower price of the product must be weighed against the increased costs due to the reduced production efficiency. The variation in the quality of a material should also be considered because an input of highly variable quality can be very costly in terms of the time lost in sorting out substandard materials or in resetting machines.

Quantity. The quantity decision was discussed in detail under the Inventory Control section relating to the economic order size. Basically, the quantity to be ordered is determined by balancing the lower costs of large shipments, such as quantity discounts and lower shipping rates, against the higher costs of maintaining a larger inventory.

Supplier Dependability. *Supplier dependability* is an important consideration because a product that arrives late or that must be returned to the supplier can be very costly in terms of production delays or by causing the firm to be late in filling its orders. A supplier's reputation and past record as a dependable supplier are good indicators of dependability. Many firms source from two or more independent suppliers, especially when the size of its orders allows it to do so without substantially raising its unit costs.

Price. The price of the product is intimately related to the other factors that influence the purchasing decision, particularly the quality and quantity factors. The price of a material affects the income statement three times: when it is purchased, when it is used in the production process, and when the finished product is sold. All three aspects, its initial cost, how it affects the efficiency of production, and its effect on the quality of the finished product, should be considered in making the purchasing decision.

Purchasing Policies

The purchasing agent, because of his or her ability to influence the buying decision, is a natural target for pressure from those who stand to benefit or lose from his or her decisions. The pressure may take many forms. It may be as simple as a friend, who represents a supplier, suggesting that when two products are basically the same, one friend ought to help another out. At the other extreme some suppliers may attempt to bribe the purchasing agent. The bribe is seldom cash; it usually comes in a more subtle form as a gift delivered to the person's home or tickets to the Super Bowl. The picture is clouded by the difficulty in distinguishing between a bribe and a true gift of appreciation. Should the distinction be based on the intent of the giver or on the item's influence on the receiver, or both? Some firms do not permit employees to accept any gifts from suppliers to remove any question of influence on the buying decision. Although such a policy is not completely enforceable, it nonetheless leaves little doubt as to the company's position on the matter. Many firms assemble a purchasing committee, with rotating membership, to monitor large purchases in an attempt to discourage unethical purchasing behavior.

Purchasing agents may also be subject to another kind of pressure. The firm's salesperson may ask them to purchase from a particular supplier in return for reciprocal business from the supplier. Reciprocity is a common practice among many businesses. However, this practice may be costly to the firm if the supplier's price is high or the quality is low, forcing the firm to evaluate whether the extra cost is worth the risk of losing the firm a customer.

SUMMARY

Operations management includes the processes involved in the production of goods and services. Production processes determine the way in which the product flows through the operations system. In turn, the type of product that is being produced dictates the nature of the process. With unit and mass processes, the workspace is designed based on the type of work being done, with each area being responsible for the performance of a specific function. By contrast, continuous processes are characterized by the uninterrupted production of a single product having standardized labor and machine requirements.

There are three basic concerns that must be addressed in deciding where operations facilities should be located: the area or region, the community or locality, and the specific site or sites. For most food and agribusiness firms, the most important factor in choosing a plant site is its location relative to that of its major inputs and its customers. Another important factor in choosing the plant location is the cost of transporting inputs and finished products. The availability of a qualified labor force and the cost of labor are the primary locality considerations, because many food and agribusiness firms require workers with a particular skill. The principal site location considerations include the availability and cost of water, power, sewer, and transportation facilities, as well as zoning requirements.

Once the site has been established, the firm has to decide how to design the plant. Fundamental concepts such as materials handling, operations balancing, space utilization, and attractiveness are affected by the layout decision. Two common layout approaches include the process layout and the product layout. In a process layout, processing components are grouped together in departments by the function they perform, whereas they are arranged according to the sequence of tasks to be performed in a product layout.

The purchasing decision actually begins with production planning. Likewise, production planning begins with a forecast of future demand. A production plan is then developed that includes the size of the workforce, inventory levels, and the size and timing of materials purchases. Once the production plan has been developed, the operations manager develops a master schedule, which, in conjunction with inventory information, is the control mechanism for releasing work orders and purchase orders.

An increasing number of food and agribusiness firms are employing computerized material requirements planning programs to facilitate operations management. Such programs are used for production scheduling, inventory control, and the scheduling of purchase orders. They ensure that materials are available when needed, thereby reducing the firm's investment in inventory.

Many food and agribusiness firms have also adopted computerized just-in-time systems or efficient consumer response systems to continuously replenish materials and inventory. Just-in-time systems work by scanning the bar codes of materials moving through the production line. Similarly, efficient consumer response systems operate by scanning the bar codes of products purchased. With both systems the computer continuously keeps track of product sales, the amount of product remaining in inventory, and the amount of product in transit. When a reorder point is reached the firm's computer signals the supplier's computer of the need to replenish inventory. The item is then added to the next shipment. The advent of computerized ordering systems has forced many food and agribusiness firms to negotiate long-term contracts and alliances with their suppliers.

Operations managers are also concerned with determining optimal inventory levels. In doing so they must balance many factors, including the cost of lost sales due to stock-outs, administrative costs, lost production productivity due to a shortage of materials and supplies, the cost of carrying inventory, and the cost of transportation associated with different order sizes.

Purchasing agents must ensure that materials and supplies are delivered on time, in the right quantity, and at the right quality and price.

CASE QUESTIONS

1. Discuss how Bob Berk would determine the optimal inventory level for Flint Hills Manufacturing Company.
2. Specifically, what factors should be considered in determining the optimal inventory level?

REVIEW QUESTIONS

1. Describe mass and continuous production processes and give an example of each.
2. Explain the importance of each of the factors affecting plant location.
3. Describe the steps in designing a plant.
4. Why is scheduling more difficult for mass production than for continuous production?
5. Describe order and flow control.
6. What is the difference between quality control and quality inspection?
7. List the costs of maintaining an inventory that is too high or too low.
8. Explain the reorder point inventory system.
9. Describe the periodic and perpetual methods of accounting for inventory.
10. Describe how materials requirement planning, just-in-time, and efficient consumer response systems work. How does the materials requirement planning system differ from the efficient consumer response system?
11. Describe the key considerations for each of the principal purchasing decision factors.

CHAPTER 10

MANAGING ORGANIZATIONS

LEARNING OBJECTIVES

In this chapter we will cover the following topics:

- The six primary functions of management and their basic characteristics
- Characteristics of a mission statement, long-term goals, and short-term goals
- The control function
- Terms and concepts used in organizational theory

- The difference between a line position and a staff position and the authority and responsibilities associated with each
- Principles of delegating responsibility and authority
- The basic concepts of an organizational structure
- The need for a formal organizational structure and the role of the informal structure
- Appropriate use of activity-based accounting including cost centers and profit centers
- Major laws and regulations that affect the way companies manage their human resources
- Implications of Equal Employment Opportunity regulations in human resources decisions

CASE STUDY

John Brandon was recently hired as an assistant to the general manager of a large, vertically integrated agribusiness that grows fresh market vegetables, fruits, nuts, and field crops. The firm also operates its own fresh fruit and vegetable packing shed, runs a nut-processing plant, and has its own product sales force. The firm was recently formed when a major nonagricultural company purchased three farming companies, located primarily in southern Arizona and southern California, that were experiencing cash flow problems. The general manager was sent by the parent company to integrate the three recently purchased companies into an efficient, single organization. Although the general manager is experienced and has a degree in agriculture and an MBA, he has not worked directly in agribusiness management. John graduated with a degree in agribusiness management 10 years ago and had experience as an operations analyst and assistant manager for a large farming company based in California's Central Valley. The general manager instructed John to provide him with a suggested organizational structure for the newly formed company. He said he wanted the company organized into efficient management units with the number of administrative levels kept to a minimum. He wanted authority delegated as broadly as possible. Part of John's assignment was to prepare an organizational chart.

After a careful study of the information in this chapter, you should be in a position to assist John in preparing his report to the general manager.

INTRODUCTION

The essence of management is simple: We all manage every day of our lives, and we soon know it when we don't manage properly! Successful management means that the business's objectives are accomplished with the available resources. It is a popular misconception that managers do little work. What people may mean is that they don't do much physical work. Managers are charged with the most important responsibility of all—achieving the firm's objectives. It is the manager's job to assemble and combine the physical, financial, and human resources to accomplish the firm's objectives.

There have been many attempts to describe what managers do, including characterizations of the roles the manager fills and definitions of management by area of responsibility, such as marketing, finance, production, and human resources. The task is further complicated because the responsibilities differ for operating management, middle management, and senior management. In this text, the *functional approach* is used to describe management. The functional approach, although not totally inclusive, has the advantage of being descriptive and of using the terms most business people use.

Managers perform six basic functions: planning, controlling, organizing, coordinating, staffing, and directing. Over the years these functions have remained the same, but the way managers have carried them out has continually evolved. Historically, the manager simply carried out the orders of his or her boss. The thinking, decision making, and responsibility resided at the top of the organization and flowed downward. Some organizations still use this type of management system, including the military and the typical franchise restaurant. Procedural manuals prescribe the actions to be taken and the decisions to be made for virtually every situation. Today, most managers are employed in organizations where empowered workers are organized as formal and informal work teams. What managers do hasn't changed, but how they do it is dramatically different. Managers spend more time planning and organizing the work to be done and less time directly supervising workers.

PLANNING

One of the primary functions of management is planning. Indeed, the conduct of the other management functions is based on the premise that the business has a plan. A plan may be likened to a road map that specifies where the firm is going and how it intends to get there. It is comprised of multiple components that define the firm's purpose, objectives (both long term and short term), strategy, action plans, and policies and procedures.

Mission Statement

Planning starts with the specification of the firm's mission. The *mission statement* should reflect the company's mission or fundamental purpose. Unlike the firm's strategy or objectives, which may change from time to time, the mission statement should be enduring. Rather than requiring a change as new products are developed or customers change, a well-defined mission should guide the company through turbulent times. The mission statement should accomplish three things. It should define what business the company is in, specify the direction for the company, and effectively communicate the firm's mission to its employees, customers, and the public.

A good definition of the company's business may be thought of as answering the *what, who,* and *how* questions. The *what* question is typically answered by defining the products or services offered by the company. Alternatively, the *what* question may be answered by defining the customer needs that are met by the company. The latter approach is more durable; products change and evolve much more quickly than customer needs. The *who* question is answered by defining who the target customers or groups are. In other words, this part of the definition specifies whose needs the company will meet. The target customer group may be defined as a group or groups of customers with specific needs, or it may specify a geographic area, such as a region, country, or group of countries. Lastly, the *how* question is answered by describing what technologies the business uses or what functions the business performs. The latter approach is often preferred because technology changes so rapidly in many industries, whereas functions performed by businesses change relatively slowly. The functions should include those aspects of production and distribution that the company will perform. A winery might define its business as the production of premium quality wines for the American consumer, using grapes produced in its vineyards and by contract farmers, and sold through its tasting room and distributors.

The second aspect of the mission statement is to specify the direction in which the business is headed. Whereas the business definition is a statement of where the company is currently, the direction of the business should chart the future course for the company. This must be coupled with effective communication so that the company's mission will inspire, challenge, and motivate managers and employees. Statements that a firm will maximize profits or increase sales are inadequate because they are too general and fail to engage the firm's stakeholders.

Let's take a look at a leading food company's mission statement and see how it stacks up against our criteria. McDonald's mission statement, which it calls its vision, states: "McDonald's vision is to be the world's best quick service restaurant experience. Being the best means consistently satisfying customers better than anyone else through outstanding quality, service, cleanliness and value." McDonald's mission is clearly customer focused. It indicates that its customer base is global and that it targets those customers who have a need for quick service. It is least specific in the technological area; it specifies only that it is a quick service restaurant. McDonald's mission also indicates that it strives to satisfy customers in four areas: quality, service, cleanliness, and value. This statement gets high marks for simplicity and clarity. Lastly, McDonald's mission states that it will be the "world's best quick service restaurant experience." Although this statement is short on specifics, such as financial or market share targets, setting a goal of being the "world's best" leaves little doubt as to the company's aspirations.

Objectives

Objectives are used to translate the firm's direction, as laid out in the mission statement, into specific performance targets. Well-written objectives have several characteristics. They should be clear, quantifiable, and include a specific time frame. The business's objectives, taken as a whole, should allow the company to achieve its purpose, through the achievement of key results. As such, the objectives should include both long-term and short-term targets. They should cover the entire organization and include corporate, company or division, and department or functional area objectives. Some of the areas that a business's objectives typically cover include:

- Financial: profitability, stock performance, and sales growth
- Marketing: new products and services, market share, customer satisfaction, and external relations
- Operations: inventory management, efficiency measures, and technology and equipment
- Human resources: productivity, hiring, and training and development

Strategy

Another important element of the planning process is the firm's strategy. Defined simply, the firm's *strategy* is a comprehensive plan for the long-term success of the business. It should describe how the business will achieve a competitive advantage over its competitors and sustain that advantage over time. It is developed after an analysis of the external environment, industry environment, and competitors, as well as the company.

Comprehensive Planning

Collectively, the mission statement, objectives, and strategy are the core of the long-term planning process. However, to be effective they must be successfully implemented. This requires that action plans be drawn up and that the appropriate policies, programs, procedures, and budgets be established to ensure that the organization works as a cohesive whole. The planning process should thereby ensure that resources are allocated and that managers' and all employees' efforts are focused toward achieving the purpose set out in the company's mission statement.

See Industry Profile 10–1 for an example of how having a clear strategy helped the Coca Cola company increase its profitability.

INDUSTRY PROFILE 10-1

THE COCA COLA COMPANY AND ROBERTO GOIZUETA

The Coca Cola Company, with roughly 50 percent of the world's soft drink market, owns the world's best known brand. Although it is known mostly as a marketing icon, its success may also be attributed to its solid management and the leadership of the man who led Coca Cola for 16 years, Roberto Goizueta.

Goizueta was born in Havana, Cuba in 1931. After receiving a degree in chemical engineering from Yale, he returned to Cuba in 1953 where he went to work for his family's sugar business. A year later he responded to a help wanted ad and was hired by Coca Cola. Then, when Fidel Castro came to power, he fled Cuba with his young family, $40, and 100 shares of Coca Cola stock.

In 1981, when Goizueta took over the reigns at Coca Cola as CEO and Chairman of the Board, the company was widely diversified with a market value of $4.3 billion. However, the company lacked direction and had posted lackluster earnings for years. In addition to its soft drink business, Coca Cola was involved in such diverse interests as shrimp farming in South Africa, wine, plastics, and industrial boilers.

The hardworking Goizueta is credited with turning the company around by giving it a strategic focus. Early in his tenure as CEO of Coca Cola he unveiled his vision for the company. He wrote a mission statement, outlined a set of ambitious objectives, and defined a set of specific strategic steps to accomplish his goals. The central mission was to create value for Coca Cola's shareholders by emphasizing the company's most valuable asset, the Coca Cola brand. The essence of his strategy was to refocus the firm on its core business, soft drinks, and to grow the business overseas.

By the time Goizueta died, in 1997, he had successfully transformed The Coca Cola Company. It was heavily focused on soft drinks and related products. Eighty percent of its profits came from overseas and it had a market value of $152 billion. Perhaps most significantly, Goizueta had increased the return on equity from 20 percent in 1981 to 60 percent in 1997.

CONTROLLING

The control function helps ensure that plans are met and that results are achieved. *Controlling* consists of monitoring performance, comparing actual results to expected results, and taking corrective action when necessary. Controls are designed to measure progress toward established goals through carefully chosen performance measures. These performance measures are then compared to previously established goals or interim targets. When a deviation is observed, the cause can be determined and corrective action taken. Although controls are used at the termination of the planning period to determine if the firm's objectives were met, it is critical that they be employed throughout the planning period so that progress toward individual goals may be monitored and evaluated, and plans and strategies may be reformulated if necessary.

ORGANIZING

Organizing is the process of assigning people and allocating resources to accomplish the firm's objectives. Managers decide where the firm is going through the planning process, and they deliver on these plans through the organizational process.

When only one or two people are involved in a business, organizational theory is not very important. If you operate your own business, you are the boss and sole employee. You do the work of both the chief executive officer and the janitor as you see fit. There is no need for a formal relationship between the employer and employee. However, most food and agribusiness firms are much larger in size, requiring the efforts of many individuals working cooperatively to accomplish all of the activities necessary to make a profit. An organizational structure becomes necessary if the business requires the services of more than one person. The larger the company, the greater the need for a formal organizational structure. Organizational theory encompasses a body of knowledge directed toward building an organization that can efficiently accomplish the objectives of the business.

For a business to operate efficiently, all individuals involved must know what is expected of them, what responsibility and authority they have, and who their boss (supervisor) is. Organizational theory provides a basis for assigning tasks and understanding the resulting authority-responsibility relationships in a manner consistent with accomplishing the company's long-term objectives.

Because organizational theory deals with the human factors of production, it is infinitely more complex than most physical production processes that involve only machines. Machines generally perform in a consistent, predictable manner. They do not have individual personalities or "good days" and "bad days" that impact productivity, although some mechanics might argue this point. It is the human element, the interpersonal relationships, in and outside the workplace, that make organizing the human elements of the production process interesting, challenging, and often frustrating. If an agribusiness does not have a harmonious, well-organized work force, it probably won't achieve its profit potential.

The *formal organization* has a hierarchy of authority, specific rules and regulations for work, standardized training, and division of work. Determining this organizational structure requires activities to be grouped together and the formal relationships between departments and personnel to be specified in order to facilitate the operation of the business. A well-thought-out organizational structure contributes to the success of a food or agribusiness company by providing for its efficient operation.

The company's formal organizational structure defines the responsibilities of each department, the lines of supervision, the formal relationship among personnel and resources, and the flow of work through the company. Presumably the company is organized in an optimal way that facilitates getting the

BOX 10–1

General Guidelines for Organizing Businesses

1. Clear lines of authority should run from the top to the bottom of the organization.
2. No one in the organization should report to more than one supervisor. All individuals in the organization should know to whom they report, and who reports to them.
3. The responsibility and authority of each supervisor should be clearly defined.
4. Responsibility should always be coupled with the corresponding authority.
5. The accountability of higher authorities for the acts of subordinates is absolute.
6. Responsibility should be delegated to the lowest practical level in the organization.
7. The number of levels of authority should be kept to a minimum.
8. The work of every person in the organization should be confined as much as possible to the performance of a single leading function.
9. Whenever possible, line functions should be separated from staff functions, and important staff activities should be adequately emphasized.
10. The number of positions a supervisor can coordinate is limited.
11. The organizational structure should be flexible so that it can be adapted to changing conditions.
12. The organizational structure should be kept as simple as possible.

work done. If not, the company's president may decide to reorganize. Companies are regularly reorganized after periodic evaluations of the company's performance. This is an indication that there is no one best way to organize a company and that the organizational structure must evolve to meet the continually changing business environment.

In this section on organizing, we address some of the most important principles used in designing business organizations, beginning with the formal organizational structure and ending with the informal organization. Twelve general guidelines, shown in Box 10–1, are provided to guide managers in designing and managing an effective business organization. These guidelines are much more appropriate than fixed laws, principles, and rules because of their flexibility and adaptability to diverse situations. These are by no means the only guidelines that might be used; in fact, some guidelines used in one situation may conflict with guidelines used in others. Hence, the application of the guidelines must be tempered by the specifics of a particular situation, and judgement must be used to balance tradeoffs.

Line and Staff Personnel

In a very general sense, most positions in a business can be categorized as being either a line position or a staff position. A person in a line position has di-

FIGURE 10–1

Example of a Line Relationship for a Fertilizer Manufacturer

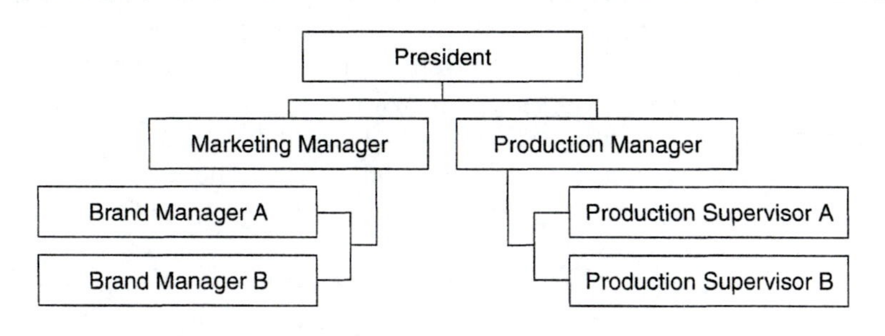

rect authority and responsibility relationships pertaining to the principle product or service of the business. By contrast, staff personnel perform a supporting role and their work is outside of the direct chain of command responsible for the principle product or service of the company.

Line Personnel. Line personnel accomplish the primary purposes of the organization and are directly involved in generating income. For example, in an agribusiness that manufactures and sells fertilizer, the president, production manager, and marketing manager are *line personnel.* All three positions are involved daily in accomplishing the goals of the business, that is, the manufacture and sale of fertilizer. The formal relationships among these individuals are illustrated in Figure 10–1.

It is the production manager's responsibility to see that the fertilizer is manufactured according to specifications and in the most cost-efficient manner possible. The production manager has the authority to organize the necessary resources and command action on the part of all individuals involved in the production process. It is this authority to command action and the associated responsibility that identifies a person as being in a line position or line relationship. In Figure 10–1, the production manager is shown as having direct responsibility for two production supervisors.

The marketing manager has similar line responsibilities, but of course they are limited to the business's marketing activities. The marketing manager makes decisions on prices, types of marketing channels, dealers, credit policies, and many other elements of marketing. The authority to make the marketing decisions places the marketing manager in a line position. Figure 10–1 indicates that the marketing manager has two brand managers reporting to him or her. The president of the fertilizer company is also in a line position because the president has the final authority and responsibility for all business decisions.

The interrelationships among the president, marketing manager, production manager, as well as the managers' subordinates, as shown in Figure 10–1, represent an "*organizational chart.*" The marketing and production managers are listed below the president and are connected to the president by a solid line. This *solid line* represents an authority relationship. The president directs the activities of both the marketing and production managers. Thus, a solid line runs from the president to each manager. This is an important convention used in organizational charts. Solid lines represent authority-responsibility relationships. One can look at a properly drawn organizational chart and quickly identify who reports to whom. For example, there is no solid line directly connecting the marketing manager with the production manager. Hence, the marketing manager does not report to or supervise the activities of the production manager.

Being in a line relationship is sometimes referred to as being in the *chain of command,* a term that is very familiar to anyone who has served in the armed services. For example, privates in the U.S. Army learn very quickly in basic training that they are at the bottom of the chain of command. They do what their corporal directs them to do. The corporal in turn follows the orders of the sergeant, the sergeant follows the commands of the lieutenant, and so on, up to the commanding general. This same principal applies in business organizations as well. However, in most business organizations the chain of command is generally not followed as closely or enforced as strictly as it is in the military. Nonetheless, it is important to understand the chain-of-command concept and the importance its plays in defining the relationship between positions.

Staff Personnel. A *staff position* differs from a line position in that the staff personnel are not within an organization's direct chain of command. Staff positions are more specific in focus and tend to be advisory in nature. Most staff positions provide technical or specific advice or services. Staff personnel support line managers and employees at every level of the organization. The successful organization develops staff departments only to the size required to adequately service the needs of its line departments. If a company is not large enough to warrant the full-time service of a staff position, these services are usually obtained by retaining the needed services from an outside firm.

Examples of staff personnel include: (1) advisory positions, such as lawyers and economists, (2) service positions, such as strategic planning, information technology, research, human resource, or maintenance personnel, and (3) control positions, such as accounting, quality control, or health and safety personnel. Individuals in staff positions make recommendations to the line managers, who have the responsibility for making the final decisions. For example, lawyers might provide legal advice on such issues as fair trade, packaging laws, or constraint of trade, but they do not make decisions for the marketing department. They also cannot demand that their advice be followed.

Similarly, company economists might forecast future economic conditions that will affect sales or possibly the cost of acquiring capital or production inputs, but they do not make marketing or purchasing decisions. They provide information for the marketing and purchasing managers to consider.

FIGURE 10–2

Example of a Staff Relationship for a Fertilizer Manufacturer

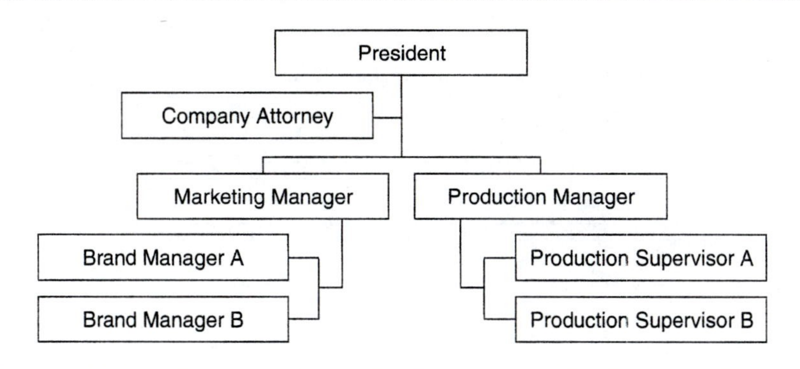

The fact that individuals in staff positions are not in the chain of command—that is, they are not directly responsible for major business decisions—might lead one to conclude that staff positions are unimportant. This is definitely not the case. The use of staff positions allows the organization to access highly technical information and expert advice for use in decision making without having to provide all the line managers with highly specialized skills, such as legal interpretations or economic forecasting. In most situations the time limitations and the need to acquire many other skills simply do not allow line managers the luxury of having in-depth knowledge in such diverse fields as law and information technology.

It is equally important that technical advisors be given an opportunity to provide input into the decision-making process but not be charged with the responsibility of making decisions. Not unexpectedly, individuals trained in highly technical areas might tend to focus too narrowly and might not consider the broad range of important factors in the full decision-making process. Although legal and economic information may be highly important for some marketing decisions, many other factors must also be considered.

A staff position, as represented on an organizational chart for our hypothetical fertilizer company, is illustrated in Figure 10–2. The company attorney is connected to the president by a solid line. This means that the attorney reports directly to the president and that the president is his or her immediate supervisor. Note, however, that the company attorney is outside of the chain of command because he or she provides advisory services.

In many organizations, the company attorney would be asked to provide advice to the functional or division managers. Because it would not be practical for all communications to pass through the president's office, the attorney would most likely work directly with these managers. This is known as a functional relationship, or as it is more commonly termed, a *dotted-line relationship.* The

FIGURE 10–3

Example of a Functional Relationship for a Fertilizer Manufacturer

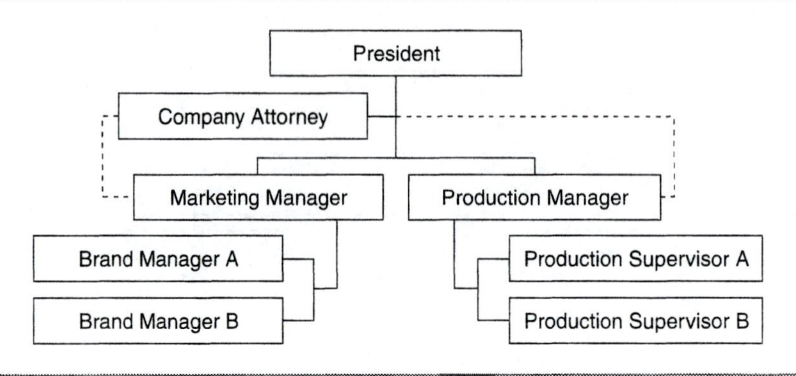

name derives from the way the relationship is depicted on the organizational chart, by means of a dotted line, as shown in Figure 10–3. In this case, the company attorney still reports directly to the president, but he or she is also able to provide advice directly to the production and marketing managers. Direct access to information goes both ways; managers could request information or advice from the company attorney, and the attorney could offer advice directly to the managers on his or her own initiative without prior approval from the president.

In modern organizations the use of dotted-line relationships has become common because of its practicality. A principal advantage of the dotted-line relationship is that it bypasses the chain-of-command process. Strict adherence to the chain of command would require that all requests and information be channeled through the appropriate supervisor for direct approval. The teamwork structure used in many firms and the need for rapid decision making has encouraged the widespread use of dotted-line relationships. Another advantage of the dotted-line relationship is that it encourages the making of decisions by those managers who are most directly involved in the decision process. Although this is usually an advantage, it can also be a disadvantage, as is the case when there is disagreement. In most cases, the decision making reverts back to the chain of command and the manager with the most direct responsibility for the decision makes the call. Likewise, highly critical decisions for the firm are not typically made by people in dotted-line relationships, but passed up the chain of command.

In our hypothetical example, shown in Figure 10–3, the company attorney has a functional relationship with the marketing and production managers to provide an efficient means for information transfer. Because the attorney has no responsibility for marketing, he or she does not make the marketing decisions even if he or she provides all of the important information used in making a particular marketing decision. The decision is the responsibility of the marketing manager. However, when there is disagreement between the mar-

BOX 10–2

Characteristics of Personnel in Line and Staff Positions

Line Personnel	Staff Personnel
Have authority to direct the actions of others	Provide advice
Responsible for carrying out entire activities, from beginning to end	Perform studies and make recommendations but do not carry out the entire activity that leads to a decision
Follow the chain of command	May advise across departmental lines
Generally identified with the activity performed	Generally focus on company issues from the viewpoint of a specific discipline

keting manager and company attorney on legal matters, the president would most likely be responsible for making the final decision.

The efficient operation of most agribusinesses, particularly those having numerous employees dealing with a diverse set of tasks, requires the effective use of both line and staff positions. Both are important and contribute in different ways to the accomplishment of the company's objectives. The distinct nature of line and staff positions is summarized in Box 10–2.

Departmentalization

The work tasks of most food and agribusiness firms are sufficiently large and complex to require the time, energy, and expertise of many individuals. It is neither advisable nor practical to have everyone doing the same thing. The basis or method of separating and assigning the responsibility for accomplishing the many tasks into separate operating units is called *departmentalization.* Departmentalization involves identifying the necessary tasks, separating work units into efficient sizes, and assigning authority-responsibility relationships.

Companies departmentalize their activities in many different ways. However, most of these arrangements can be grouped under the following four general categories: function (activity), product (for example, crop or livestock), customer, and geographic area. The particular method or methods a company chooses depends largely on the type of company, the product or service it sells, the general philosophy of the top managers, and the company's history.

In many food and agribusiness firms the type of departmentalization used is the result of historical events. Inappropriate organizational structures often result from rapid growth, either internally or through acquisitions or mergers, and the subsequent failure by top management to reorganize the firm's activities based on the needs of the new organization. Most successful food and agribusiness firms have a clear plan or philosophy regarding departmentalization.

Functional Departmentalization. An agribusiness organized along functional lines might have departments with titles such as production, marketing, harvesting, irrigation, processing, quality control, sales, and research. The activities assigned to these departments are associated with accomplishing a specific function. For example, a marketing department might be assigned the tasks of organizing sales, preparing advertising campaigns, designing product packaging, and supervising the sales force. All of the activities associated with the marketing function are assigned to the marketing department.

A very large farming company growing perishable specialty crops might have a department charged with the responsibility of harvesting all of the crops. Similarly, a large farming company in an arid area may assign the responsibility for all crop irrigation to a single irrigation department. The major consideration in a functional structure is that all activities associated with performing a certain function are grouped together and placed under the authority of one manager.

An organizational chart for a large cattle feedlot that is departmentalized functionally is shown in Figure 10–4. The activities of the feedlot are divided among five departments: cattle, yard, feed mill, feed procurement, and office. These departments are assigned all of the tasks relating to their specific functions, including cattle transactions—buying and selling cattle; yard—feeding and moving cattle, maintaining the health of the cattle, and taking care of the yard facilities; feed mill—processing feed and feeding cattle; feed procurement—purchasing feed; and office—managing office activities such as accounting and maintaining production and sales records.

The rationale behind the functional approach to assigning activities to departments is the expected efficiency gains from specialization. The philosophy is that by focusing employees' training and experience on specific activities, such as buying and selling cattle or feed, or taking care of sick cattle, employees will become efficient at what they do. They might also become more efficient by learning to take advantage of the latest technology. Specialization can yield substantial gains in efficiency. For this reason functional departmentalization is in widespread use among food and agribusiness companies.

The major disadvantage of the functional approach is that department personnel may become so focused on achieving efficiency within their particular functional area that it becomes detrimental to the company as a whole. For example, purchasing the cheapest cattle or achieving the lowest labor cost for feeding cattle may not lead to the highest profits for the company.

Product Departmentalization. Another common means of organizing used by food and agribusiness firms is to departmentalize based on the type of product or service the firm sells. A company that uses product departmentalization might have departments with names such as fertilizer, seed, chemical, tractor,

FIGURE 10–4

Example Organizational Chart for a Large Cattle Feedlot Departmentalized by Function

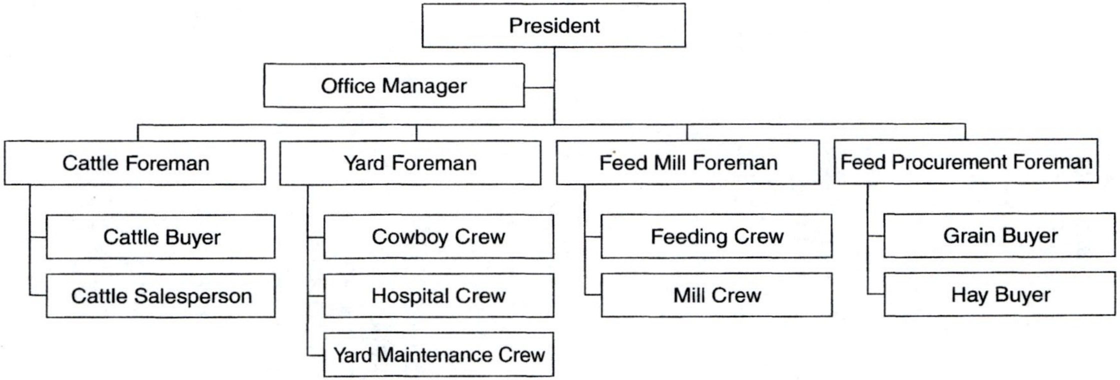

- President
  - Office Manager
  - Cattle Foreman
    - Cattle Buyer
    - Cattle Salesperson
  - Yard Foreman
    - Cowboy Crew
    - Hospital Crew
    - Yard Maintenance Crew
  - Feed Mill Foreman
    - Feeding Crew
    - Mill Crew
  - Feed Procurement Foreman
    - Grain Buyer
    - Hay Buyer

FIGURE 10–5

Example Organizational Chart for a Supermarket Departmentalized by Product Line

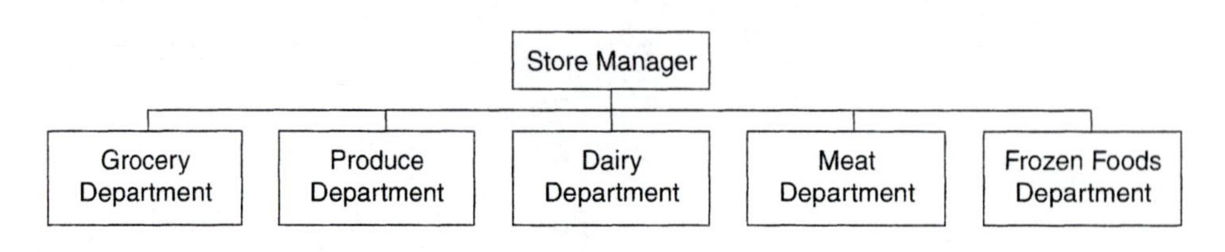

harvester, and tillage. A food company might organize by product or type of product such as refrigerated and frozen products. A large farming company growing several crops may assign departmental responsibilities based on the type of crop, for example, tree crop, row crop, or vineyard. This crop-based departmentalization is equivalent to product departmentalization. Thus, the focus for the company's business activities is on the product or service the firm provides.

Product-based departmentalization involves grouping diverse functional activities, such as manufacturing, selling, and shipping, under one administrative unit. The rationale for this method is that the cumulative effect of coordinating all particular product functions under one administrative unit outweighs the efficiency that might be gained through functional departmentalization. Product-based departmentalization is particularly appropriate for firms that have differentiated products because it allows each product group to focus attention on the target customers' needs.

Figure 10–5 shows an organizational chart for a supermarket using product departmentalization. Responsibility for the store's activities is assigned to the managers of five departments: grocery, produce, dairy, meat, and frozen foods. Generally, grocery stores and other retail stores that sell many products or product groups use product departmentalization. Very large stores may have 15 or more separate departments.

Customer Departmentalization. Customer departmentalization is also commonly used to separate a firm's activities into manageable units. A business using this method might have departments entitled wholesale division, retail division, and direct sales. The wholesale division would handle sales made to wholesalers, the retail division would handle sales made to retailers, and the direct sales division would handle sales made directly to customers. Businesses that manufacture consumer products often use customer departmentalization. Likewise, large advertising agencies often use customer departmentalization when they place account executives in charge of all activities associated with specific clients. Large law, accounting, and consulting firms also rely heavily on customer departmentalization. This works well for these companies because the needs of the customer groups differ widely. Because the amount of business done with each customer or customer group is usually large and im-

FIGURE 10–6

Example Organizational Chart for a Farm Management Company Departmentalized by Customer

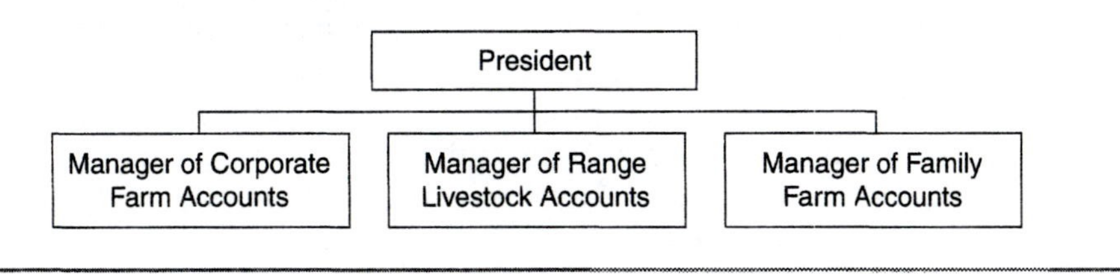

FIGURE 10–7

Example Organizational Chart for an Agricultural Chemical Company Departmentalized by Geographic Region

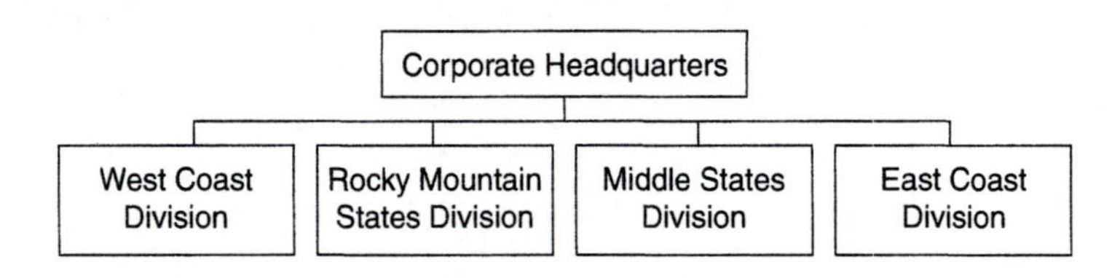

portant to the profitability of the firm, it is important that each client's needs are cared for and a close personal relationship is maintained. Figure 10–6 shows an organizational chart for a farm management company that uses customer departmentalization.

Geographic Departmentalization. Some food and agribusiness companies organize their activities based on the concept of geographic departmentalization. Such firms might have West Coast, Rocky Mountain, Middle States, and East Coast divisions, as illustrated in Figure 10–7. The idea behind geographic departmentalization is that customer needs may vary by geographic area and it is more efficient and cost effective to service customer needs within, rather than across, regions. However, communication problems caused by distance and the high level of travel expenses are important factors that should be considered before deciding to departmentalize geographically.

Choosing a Method of Departmentalization

There is no one best system of dividing up the work tasks and assigning responsibility to departments that fits all situations. Each method has its advantages and disadvantages, so a decision on which method is most appropriate should be based on the company's needs and the products or services it provides. Geographic departmentalization may be best for one company, whereas functional departmentalization may work best for another company, even though both companies are in the same industry.

It is important to note that the choice of one method of departmentalization does not preclude using other methods. Hybrid systems, which employ more than one method of departmentalization, allow management to design an organization customized to meet the firm's individual needs. For example, a company may departmentalize by product at the company level but departmentalize by function within the product divisions, or by geographic region within the sales division. In fact, large conglomerates may employ all four methods of departmentalization.

Service and Staff Departments

Accounting, personnel, public relations, and sometimes purchasing activities are frequently placed in *service departments.* They are called service departments as opposed to activity departments because their central focus is to assist the business or other departments in performing specific necessary functions not directly related to the firm's primary revenue-producing activities. Similarly, *staff departments* are sometimes designated within a firm. These departments differ from service departments in that they provide information such as legal advice rather than services such as accounting or payroll.

The proper organizational placement of staff or service departments can be highly important to the efficient operation of a company. For example, a purchasing department for an agricultural equipment manufacturer that buys production inputs for both the tractor manufacturing and harvester assembly departments must be situated so that both line departments have equal access to its services. Hence, the manager of the purchasing department should not report directly to either the manager of the tractor manufacturing department or to the manager of the harvester assembly department. Suppose, for example, that the manager of the purchasing department reported directly to the manager of the harvester assembly department but was additionally charged with servicing the needs of the tractor manufacturing department. In this case, the purchasing department manager would most likely favor the needs of the harvester assembly department because the harvester assembly manager controls the purchasing manager's salary and ultimately his or her employment at the firm.

The proper organizational placement of a service department, such as purchasing, that serves the needs of more than one department is illustrated in Figure 10–8. The purchasing manager reports directly to the vice president for production but has a functional responsibility, as indicated by the dotted lines, to service the needs of the tractor manufacturing and harvester assembly divisions.

Span of Control

The *span of control* is an important organizational principle stating that there is a limit to the number of employees that any one manager can effectively su-

FIGURE 10–8

Example Organizational Chart for an Agricultural Equipment Company with a Service Department

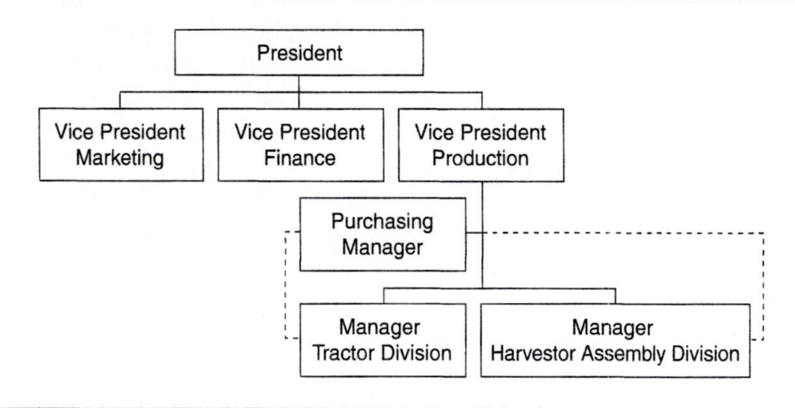

pervise. This principle acknowledges that the manager must spend some time with each employee. If a supervisor has a very *wide span of control,* then he or she supervises the activities of many people. A manager with a *narrow span of control* supervises only a few subordinates.

Although there is a limit to the number of activities or employees one supervisor can effectively manage, the concept is often difficult to apply. The number of tasks or employees any one supervisor can effectively control varies greatly from one person and situation to the next. It depends on the type of task being performed, the level of training and education of the supervisor and subordinates, and the ability of the supervisor. Some individuals can handle many diverse tasks simultaneously whereas others can juggle only one or two jobs at a time. Nonetheless, several general guidelines concerning the span of control will prove useful.

A wide span of control is possible and advisable in most assembly line situations, where each employee must perform only a few repetitive tasks. Likewise, a wide span of control is generally advisable in situations where the subordinates are highly educated. Such individuals require a great amount of freedom to do their job, so a narrow span of control often unnecessarily constrains them and stifles their motivation and creativity. On the other hand, in situations when the cost of making an error or wrong decision is very high, such as when a large account may be lost because the customer's needs are not met, a narrow span of control might be advisable.

Division of Labor

The concept of division of labor or labor specialization is useful in establishing individual positions and the tasks assigned to those positions. By subdividing all activities into their most basic tasks, managers can determine the most efficient grouping of these tasks into jobs. Based on this analysis they can write specific job descriptions, establish expected results, and measure job performance. The fundamental tradeoff facing most managers is between the increased efficiencies gained when employees are allowed to specialize and the monotony associated with the performance of repetitive tasks. The manager of a feedlot, for example, applies the concept of division of labor by assigning employees to the office, feedmill, hospital, and pens. Assigning employees to a specific department allows them to utilize their skills and become familiar with their job assignment. Many companies use job rotation or otherwise limit the degree of specialization to reduce the boredom involved in repeatedly performing the same task.

Informal Organization

A complex and dynamic network of interpersonal relationships between the organization's members characterizes the *informal organization* within a company. The informal organization is heavily influenced by the company culture and company politics. These relationships cannot be diagrammed like the relationships in a formal organization, but they are just as important. They are different for every company and society and continuously change with the company's personnel. Ideally, the formal and informal organizations are mutually supportive and reinforce each other. When they are in conflict, the effectiveness of the organization to pursue the company's objectives suffers.

The informal organization is a byproduct of social interactions among people. These interactions occur because human beings are social creatures and because the relationships people form cannot be dictated by an organizational chart. Informal interactions between coworkers also help the firm's employees learn the company's unwritten rules and overcome obstacles to success. Finally, and most importantly, the informal organization is a source of information for the company's employees. Information is power in any organization. Information released through the formal communication systems, such as interoffice memos, e-mail, newsletters, and bulletin boards, is useful; but it does not have the influence and impact of information exchanged through informal communication channels.

The informal organization is based on a broader base of personal power in addition to the power associated with a formal position of authority. *Power* is the ability of one person to influence the behavior of another. Managers in a company have a position of formal authority called *positional power.* Because of their position in the company they have the power to reward other individuals, for example, by salary increases or promotions. They also have the power to coerce other individuals by punishing them or removing rewards. Informal leaders reward coworkers by accepting them, recognizing their contributions to the company, providing assistance, and sharing information. They influence coworkers by avoiding or excluding them from group activities or discussions. Usually, informal leaders derive power either because group members like them or want to be like them, or because they hold some special knowledge or skill that contributes to the group's collective success. The most effective managers recognize the need to manage through both formal and informal networks.

COORDINATING

The purpose of the *coordinating function* is to ensure that the firm's departments and their managers function together as a team, so that they work efficiently to achieve the overall objectives of the organization. Conflict among departments that works to the detriment of a business is commonplace in modern-day organizations. Managers must see to it that individual employees as well as departments are motivated so that their goals are in line with the firm's goals. They must further ensure that the appropriate policies are in place and that communication is structured so that managers have the information they need to manage effectively.

Real-world business organizational structures are highly complex and must be carefully designed to optimize company profits. Recognizing the complexities of the human element in management is very important. The challenge is to design organizational arrangements that are as simple as possible yet sophisticated enough to ensure that all managers are given the proper incentives as well as the authority and latitude necessary to perform their jobs to the best of their capabilities. The ultimate goal is to coordinate departments so that they attain optimum efficiencies and maximize profits.

In a small company you can generally judge efficiency (the effective use of the company's resources) by looking at the bottom line of the income statement—net income. However, in a large organization it is possible to have several units or divisions making money and several operating poorly or inefficiently. How does one identify the divisions or departments that are not performing up to standards? One method is to designate each unit, department, or division as either a profit center or a cost center. Then, in addition to keeping accounting and management information records for the company as a whole, records are kept for each division or management unit. This can be visualized as the formation of a set of individual smaller companies, each of which is of an efficient management size, which together make up the company. This concept is often referred to as *activity-based accounting.*

Cost Centers

A *cost center* is an accounting concept whereby a particular unit or division of a company keeps track of its costs by appropriate categories, for example, labor, equipment, materials, or supplies. The division then allocates its costs to the profit centers that use its services. The allocation may be made on a per-acre, per-ton, per-unit, or per-hour basis, or on any other meaningful unit-of-cost measurement. Generally, a budget is prepared before the accounting period begins, accurate records are kept, and then actual costs are compared with budgeted costs. If actual costs are significantly out of line with budgeted costs, the source of the deviation is investigated.

An equipment division of a large company serves as an example of how cost centers operate. As a cost center, an equipment division would set up depreciation schedules for all items as well as schedules for regular maintenance (for example, oil changes and tune-ups), major repairs, and overhauls. It would also keep records for costs such as labor, parts, supplies, and on all repairs made on each piece of equipment. When a service is performed, a work order would be written up in the same way that an auto repair shop writes up a work order when servicing a car. At the end of the year, all of the costs associated with each piece of equipment would be charged to the relevant profit center, for example, the farming division, based on usage.

In our hypothetical farming company the cost allocation might work something like this. Assume that the equipment division estimates its expected costs for the year in December. The division estimates that it costs $20 per acre to operate a large tractor, $8 per acre to use a disc, and $25 per acre to operate a combine. During the year, the profit centers are charged these rates for the use of equipment. At the end of the year, the equipment division determines its actual costs and then adjusts the costs charged to each of the centers using its services (profit centers). The company also compares actual costs with budgeted costs. If the actual cost of operating the combine was $30 per acre, management would try to determine the source of the overrun. Was it due to higher fuel charges that were beyond the division's control or was it due to inefficient equipment use or wasted labor and supplies? The investigation process serves as a technique to hold managers responsible for unnecessary cost overruns and allows rewarding for effective cost control.

Profit Centers

Profit centers are designed to track both the costs and revenues generated by a unit or division of a company. Actual revenues are used for transactions where money changes hands, as is the case when a department makes a sale to an external customer. For internal customers, profit centers charge "market rates" for products produced or services provided. Like cost centers, profit centers keep track of their costs to determine their profit level. If revenues exceed costs, the profit center makes a profit. If costs exceed revenues, the unit suffers a loss. In the previous example, if the equipment division were a profit center, it would bill its internal customers at the going market rates for the use of its equipment. This is known as a *transfer price* because it would represent an internal transfer of funds.

One of the major issues regarding the administration of profit centers is the determination of costs and revenues. Because managers are typically rewarded based on the performance of their profit center, they are inclined to argue for a reduction in the allocation of costs to their unit. Likewise, profit center managers stand to benefit by inflating their revenues.

Some of the major questions faced with operating profit centers are: How are "market prices" determined? Is it possible to obtain comparable rates to use in fair pricing? Do cost or profit centers better serve the company's overall objectives? It is also important to remember that price and cost are not the only considerations. The quality and timing of the service are also important. In some cases the timing of a service is more important than saving a few dollars.

Cost versus Profit Centers

Historically, units that incurred costs but that did not directly generate revenues were managed as cost centers. Units that incurred costs and directly generated revenues were run as profit centers. It is an open question as to whether service divisions in large companies, that is, units that incur costs and indirectly generate revenues, should be profit centers or cost centers. In recent years many businesses have favored the use of profit centers because they were believed to lead to the most efficient use of a firm's resources. Thus, service units that perform a purely supporting role, such as human resources or accounting, are typically treated as overall service centers. The expenses related to overall service units, such as accounting or human resources, are typically included in the general and administrative expenses and are not allocated to cost or profit centers. Units that are more directly linked to the generation of revenue may be classified as either cost or profit centers, although those units most closely related to the major business activity of the firm stand to benefit the business the most from being operated as a profit center.

To be successful, profit centers must be properly designed and administered. A profit center that is not charged the full cost of doing business may look extremely profitable and grow based solely on its inflated profits. Depending on the true level of costs, this may be to the detriment of other departments or the firm as a whole. Similarly, inaccurate transfer prices may lead to the overutilization or underutilization of a department.

Figure 10–9 illustrates the use of cost and profit centers in a large farming operation. In this example, the feedlot and farming divisions should clearly be operated as profit centers. The office, on the other hand, should be operated as a service center, because it serves the firm in a supporting role. The chemical and equipment divisions could be operated as either cost or profit centers. We have chosen to designate the chemical division as a profit center because it

FIGURE 10–9

Example Organizational Chart for a Large Farming Company Illustrating the Use of Cost, Service, and Profit Centers

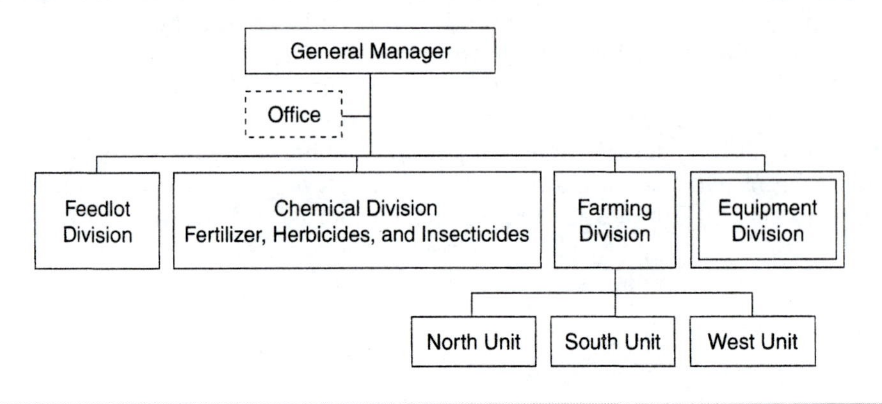

Note: Solid-line borders denote profit centers, double-line borders denote cost centers, and dotted-line borders denote service centers.

sells chemicals to other farms in the area in addition to the "home farm." Because the equipment division only provides equipment to the "home farm" and because of the difficulty in allocating its expenses to other units, it has been designated as a cost center.

Accountability and Departmentalization

Another important consideration in departmentalizing activities is ensuring accountability. Wherever possible, the company's activities should be departmentalized so that accountability may be achieved. *Accountability* entails being able to trace the results of good or poor performance to the actions of specific individuals so that good performance may be rewarded and poor performance corrected. In an agribusiness or farming context, measures of performance might include yields, profits, or cost control.

The principle of accountability is well illustrated in the following agribusiness example. Suppose the farming division of the large farming company shown in Figure 10–9 is organized by geographic area where the North Unit, South Unit, and West Unit are each under the control of an area farm manager. Each farm unit has similar soils and water quality, and grows the same crops. The managers have the authority to control all of the activities on their farm units. As is shown in the organizational chart, each farm unit is treated as a profit center. Cost and revenue records are maintained for each farm unit so that profits can be measured. For example, assume that the North Unit and

West Unit are profitable in a particular year while the South Unit loses money. An effort should be made to determine the cause of the South Unit's loss. The manager of the South Unit must be able to explain the loss and what measures have been taken to ensure that the unit will regain profitability. In other words, the manager of the farming division must hold the manager of the South Unit accountable for the unit's results.

Accountability is easier to maintain under some organizational structures than others. It is particularly difficult to maintain when work is organized by function. In the example shown in Figure 10–9, if functional departmentalization were used so that the North Unit, South Unit, and West Unit managers were replaced by an equipment manager, an irrigation manager, and a farming manager, it would be very difficult to determine accountability for results. In a bad year, the farming manager may say that it was the fault of the irrigation department manager, explaining that irrigation was done on the day the harvest was scheduled, thereby delaying harvest. The irrigation department manager may place the blame on the equipment department manager, saying that the equipment manager was a week late in making harvest equipment available. The equipment manager may respond that he or she was not notified that harvest equipment was needed.

As this example illustrates, a principal challenge of functional departmentalization is the high level of coordination that is required to ensure that the various functional departments operate to serve the best interest of the company. In this case it would require the establishment of policies governing scheduling and the priority of various types of work, good communication between the functional managers, and effective supervision by the farming division manager.

STAFFING

The staffing function entails managing the human resource needs of the company. It includes hiring, training, developing, evaluating the performance of, and rewarding the firm's employees. Because of the importance of this function to food and agribusiness companies, we have devoted an entire chapter to human resources management, Chapter 11. In this section, we address the most significant laws and regulations relating to staffing, of which everyone in a management position should be aware.

The organizational structure determines the positions in an organization and their relationship to one another. It is through staffing that these positions are filled. This is one of the most important functions a manager performs because it is often the employees—the human element of a business—that distinguish one business from another. A manager can have no greater impact on the business and its future than through the selection of employees.

The basis for employment in the United States is the *employment at-will doctrine.* This doctrine states that if there is no agreement or contract to the contrary, employment is considered to be at will. That is, either the employer or the employee may terminate the relationship at will as long as state and federal employment laws do not prohibit the basis for termination. In addition, the manager must ensure that all job candidates are treated fairly, as mandated by state and federal laws.

An *employer* is someone who contracts with another person who agrees to perform work for the employer. An *employee* is someone who works for a salary or hourly wage and whose actions while employed are controlled by the employer. An employee can work full-time, part-time, or as a temporary employee. This distinction is important because full-time employees are covered by numerous labor laws and generally receive benefits, whereas part-time and temporary employees typically do not. An *independent contractor* is a person who contracts with the employer to perform a task according to the contractor's own methods, and who is not under the employer's control regarding the physical details of the work. Independent contractors are especially useful when the employer needs a specific skill or technical knowledge for a special project lasting only a short time. Common examples of independent contractors are painters, electricians, plumbers, and business consultants. The employer outlines the job to the contractor, who then determines how to accomplish it on a given schedule. The distinction between employees and independent contractors is important because independent contractors are not subject to payroll taxes and most employment laws from the hiring firm's perspective.

Title VII of the Civil Rights Act

Federal employment laws govern most employment situations, starting with the hiring process and continuing through the termination of the employment relationship. A large body of employment law was established with *Title VII of the Civil Rights Act of 1964.* This act, and the amendments that followed, made it unlawful for an employer to discriminate against anyone who is a member of a protected group in recruitment, selection, compensation, performance appraisal, training, promotion, transfer, demotion, firing, or layoff. (Box 10–3 summarizes the protected groups covered by several federal employment laws.) Members of protected groups are defined by several characteristics, including race or color, national origin, religion, gender, marital status, parental or potential parental status, age, physical or emotional handicap or disability, whistleblowers, and Vietnam War veterans. Freedom from discrimination on the basis of sexual orientation is not protected by federal legislation but is protected by the laws of some states and localities. Title VII applies to the federal government and private employers, state and local governments, and educational institutions employing 15 or more people. Native American Indian reservations and religious institutions and associations are exempt from the Act.

BOX 10–3

Federal Employment Laws

Federal Law	Purpose	Employers Covered
Title VII of the Civil Rights Act	Prevents discrimination of employees on the basis of race, color, religion, gender, or national origin	15 or more employees
Americans with Disabilities Act	Disabled employees may not be discriminated against and employers must provide reasonable accommodation to disabled employees	15 or more employees
Affirmative Action	Intentional inclusion of women and minorities in jobs where they have been underrepresented	As a remedy to a discrimination lawsuit and employers conducting $10,000 or more of business with the federal government, voluntary otherwise
Immigration Reform Act	Prevents discrimination against employees on the basis of national origin or citizenship status	Four or more employees
Fair Labor Standards Act	Requires minimum wage and overtime pay and sets minimum employment age at 16 for most jobs	All employers

The *Equal Employment Opportunity Commission* (EEOC) is the lead agency for monitoring job discrimination and handles most matters of employment discrimination arising under federal laws. Employees who feel they have been the victims of employment discrimination may file a charge or claim with the EEOC. Nonfederal government employee claims must be filed within 180 days of the discriminatory event. After the complaint is filed with the EEOC, the EEOC serves notice of the charge with the employer. The EEOC then investigates the complaint by talking with the employer and employee and any other witnesses. The EEOC will work with the employer and employee to find an administrative remedy. Employees who are not satisfied with the results of the EEOC investigation are then free to bring a lawsuit against the employer, asking for compensatory and punitive damages.

In alleging discrimination, an employee may file a lawsuit based on disparate treatment or disparate impact. *Disparate treatment* is used in cases of individual discrimination. The employee (or job applicant) bringing suit alleges that the employer treated the employee differently than the employer treated other employees because of the employee's race, religion, gender, color, national origin, or other characteristic. To avoid problems, employers should consistently treat all employees the same. Employers should also think carefully before singling out an employee for punitive action, making sure that the action is justified and that the reason for the action is clearly related to job performance.

Disparate impact is used in cases of group discrimination. An employee alleges that the employer's policy has a disparate or adverse impact on a protected group. Disparate impact cases pose the greatest challenge to the employer. Any policy, procedure, or screening device used to decide who receives an employment benefit can serve as the basis for a disparate impact claim. This includes decisions on hiring, termination or layoffs, promotions, training, raises, and employee benefit packages.

Several federal agencies have adopted a set of uniform guidelines to provide standards for ruling on the legality of employers' selection procedures. The Uniform Guidelines on Employee Selection Procedures states that disparate impact occurs when the selection rates for protected groups is less than 80 percent of the highest-scoring majority group. The guideline is called the *four-fifths rule.* As an example, let's assume 100 women and 100 men apply for jobs at a new factory, and that all the men and half the women were hired. The hiring procedures used would be determined to have a disparate impact on the women because the women did not receive jobs at a level of at least 80 percent of the male group. The employer would now be required to show that the hiring procedures and decisions were based on factors related to the applicants' qualifications for doing the job.

An employer can defend against the charge of unfair employment practices that exclude members of a protected group by proving one of the following:

- *The selection method was valid; that is, it accurately predicted an applicant's success on the job.* For example, let's assume that in the hiring for a new fertilizer factory, each applicant was required to demonstrate that he or she could successfully lift a 100-pound sack of fertilizer. Because this criterion is job related and is predictive of success on the job, the job requirement would be permitted even though it may have a disparate impact on women.
- *The selection method was based on a compelling business necessity.* Assume, for instance, that a prospective employee challenged a check-cashing business for hiring proportionally more white non-Hispanics than Hispanics. The employer had an employment policy of requesting credit histories of all job applicants because only those applicants with good, stable credit histories were qualified to handle large sums of money. Because this would be deemed to be a business necessity, this policy would be permitted, even though it may cause a disproportional negative impact on some protected groups.

- *The selection method was a bona fide occupational qualification (BFOQ).* BFOQ constitutes legal discrimination. The employer in a BFOQ case must be able to prove that the basis for preferring one group over another goes to the essence of what the employer is in business to do. The employer must also show that it is impractical to determine if each member of the group who is discriminated against could qualify for the position. Some professionals (for example, police officers and airline pilots) face mandatory retirement when they reach a given age because the skills demanded in the profession deteriorate with age. In these cases, retaining the employee could jeopardize public safety.

Americans with Disabilities Act

The *Americans with Disabilities Act of 1990* (ADA) made it illegal for organizations to discriminate against a qualified individual with a disability in regard to all employment decisions. Like Title VII of the Civil Rights Act, the ADA applies to the federal government and private employers, state and local governments, and educational institutions employing 15 or more people. Under the ADA, employers must make *reasonable accommodations* to their workplaces to make them amenable to disabled workers as long as the changes do not create an undue burden for the employer. In other words, employers are required to restructure their workplaces and job descriptions to allow disabled individuals access to employment. For example, an employer may have to adapt the workspace to make it accessible to wheelchairs or redesign a job to eliminate tasks not essential in the performance of the job. For instance, if an employee is unable to stand and can't reach the top of the supply cabinet, the employer is required to accommodate the employee by relocating the supplies to a lower level that is accessible to the employee. The employer is not required to reassign essential job functions or to make changes where the cost of accommodation would result in undue hardship.

The ADA defines *disability* as "a physical or mental impairment that substantially limits one or more of the major life activities of an individual, a record of such an impairment, or being regarded as having such an impairment." There is no definitive list of impairments considered to be disabilities; thus, some cases must be handled individually. However, ADA generally covers loss of senses, HIV infection and AIDS, arthritis, epilepsy, cancer, heart disease, loss of limb, paralysis, loss of organ systems, mental retardation, and mental illness. The ADA does not cover treatable diseases such as behavior disorders, obesity, compulsive gambling, and alcoholism or drug addiction. However, individuals who have successfully completed treatment for treatable diseases are protected. In other words, you cannot deny employment benefits to an individual with a previous history of alcoholism or drug addiction, but you can hold an alcoholic or drug addicted person to the same standards of performance and behavior required of other employees.

Affirmative Action

Affirmative action is the intentional inclusion of women and minorities in jobs where they have been historically underrepresented. It requires employers who contract to furnish the federal government with goods and services of $10,000 or more not to discriminate in employment practices on the basis of race, color, religion, gender, or national origin. Companies may become involved in affirmative action voluntarily, as a remedy to a discrimination lawsuit, or as part of the employer's responsibility as a contractor to the federal government. Affirmative action is an executive order and as such does not stem from the labor laws passed by Congress.

In addition to not discriminating in employment, small contractors of the federal government are required to post notices that they are equal opportunity employers. Larger federal contractors (those doing $10,000 or more of business with the federal government) are required to develop a written affirmative action plan. An *affirmative action plan* includes goals for the inclusion of women and minorities in the workplace based on a workplace assessment. The *workplace assessment* compares the percentage of women and minority employees in all types and levels of employment in the company with the percentage of such employees in the general workforce. Under the affirmative action plan, the company intentionally hires more women and minorities if it has fewer women and minorities in a particular job than would be expected based on their availability. The affirmative action plan establishes goals regarding the number of employees needed to remedy the situation and establishes a timetable for achieving the goals.

There are no specific requirements as to how affirmative action is to be accomplished. As a result, companies differ on how they address affirmative action. This has led to confusion over the best way to handle affirmative action, and charges of reverse discrimination and related legal action have occurred. The courts have made it clear that affirmative action plans should not be used to replace nonminority employees or to hire unqualified people. As a result many companies have instituted workplace diversity programs that sensitize employees to differences among people in the workplace. As the workplace becomes more diverse, it is important that employees learn to accept each other and value each other's contributions.

Diversity

Diversity has become an important management issue. As discussed in the Title VII and Affirmative Action sections, diversity not only affects a firm's staffing efforts, but also the management of the organization. Put simply, *diversity* means differences. Diversity becomes a management problem when employees do not accept or tolerate differences in the workplace.

Traditionally, accepting a job meant accepting an organization's culture. If employees wanted to be successful, they had to assimilate, or fit in. Because they wanted to keep their jobs, employees were motivated to avoid conflict and to conform, even if it meant tolerating inappropriate behavior directed at them. Assimilation reduced the need for managers to solve diversity-based problems.

In the 1990s, many employees were empowered by antidiscrimination laws and the strong economy that kept unemployment low. The makeup of the workforce rapidly changed as more women and minorities entered the workforce. The demographic makeup of the United States indicates that far from abating, this trend will accelerate. In the future the percentage of jobs held by white, middle-class males will shrink. There will be an increasing number of single-parent, dual-career, and blended families due to divorce and remarriage. There will be an increasing number of temporary, part-time, and home-based employees. There will be more displaced employees due to corporate mergers and restructuring.

Diversity is a fact of life in modern companies in the same way that corporate culture or company politics is. It exists as part of the informal organization of a company, and there is little a manager can do to control it. Rather than ignoring or resisting diversity, successful managers are finding ways to use diversity as a benefit in the same way that they use corporate culture to get more out of their employees. We now live in a global economy where diversity can be a source of competitive advantage and a positive influence for the company. Fostering a company culture that embraces diversity will lead to a more satisfied and productive workforce.

Immigration Reform and Control Act

The *Immigration Reform and Control Act of 1986* (IRCA) places the burden of checking the status of aliens on the employer. Employers must ask every employee (not just those who appear to be of foreign origin) for documentation of the employee's identity and eligibility to work in the United States. The documents should not be checked until after the offer of employment is made and accepted. There are numerous documents employees can use to prove identity and/or eligibility for employment. Most employers request a driver's license to prove identity and a social security card to prove eligibility.

The employer does not have to verify the authenticity of the documents. In fact, IRCA specifically states that employers may be wrongfully discriminating against an applicant if they refuse to honor documents that appear to be genuine. IRCA prohibits employers of four or more employees from discriminating against employees on the basis of their citizenship, and from hiring those not legally authorized for employment in the United States. IRCA does not specifically preclude employers from exercising a preference for U.S. citizens over non-U.S. citizens, although this is inadvisable because it may violate Title VII of the Civil Rights Act. If the applicant is of foreign origin, the employer must fill out an Employee Eligibility Verification Form, otherwise known as an I-9 form, which reports information to the Immigration and Naturalization Service (INS).

Fair Labor Standards Act

The *Fair Labor Standards Act* (FLSA) governs wages, hours, and the employment of children in the workplace. It also prohibits pay differentials based solely on gender. Generally, its coverage applies to all businesses. Employers are required to pay employees at least the federal minimum wage. Children ages 14–16 are allowed to work at jobs such as delivering newspapers or certain agricultural work when it does not interfere with their health, education, or well-being. Otherwise, children younger than age 16 cannot work, with 18 being the minimum working age for hazardous jobs. FLSA also requires employers to keep records on wages and hours; however, no forms are required.

DIRECTING

The *directing* function entails the effective supervision of employees, delegation of authority and responsibility, and organizational leadership. It is through the directing function that managers exercise active leadership to accomplish the work of the organization.

Supervising Employees

Managers execute plans through supervising the actions of their subordinates. This entails determining what tasks need to be accomplished and who will be assigned them. An important element of supervising is the willingness to delegate the responsibility for completing a task and then granting the authority necessary to complete an assignment. It also includes setting expectations, following up on delegated responsibilities to ensure that work is properly completed, evaluating performance, and ensuring that good performance is well rewarded. It is the manager's responsibility to ensure that employees understand the organization's objectives and to motivate them to work to the best of their abilities.

Delegation of Responsibility and Authority

Delegation is one of the key elements in directing employees because it is through delegation that managers accomplish objectives through the efforts of others. It is delegation that allows a manager to assemble a team of employees with specific expertise to focus on a task. No single topic of organizational theory has received more attention in the management literature than the delegation of authority and responsibility. In spite of the volumes of literature available, many managers, including those in the food and agribusiness industry, have had little or no formal education or training in this important aspect of management science.

Responsibility is an obligation to perform work activities. The concept of responsibility gives employees their focus. When responsibilities are unclear, the organization suffers because the work does not get done. If a task is not specifically mentioned in a person's job description, or more importantly, not specifically assigned, the supervisor is taking a risk that the task will not get done. *Authority* includes the right to decide, to direct others to take action, or to perform certain duties in the process of achieving organizational goals. The process of assigning responsibility and granting the related authority is called *delegation.*

It is essential that responsibility and authority be delegated jointly and that they be commensurate with each other. When a subordinate is given the responsibility of accomplishing a task, it is his or her duty to see that the task is accomplished. However, a subordinate who is given the responsibility for a task without the necessary authority cannot be held accountable for the results. For example, marketing managers who do not control the sales force will likely achieve poor results because they do not command authority over one of the most essential resources necessary for successfully marketing the company's products. The failure to grant the necessary authority leads to frustrated employees because they are powerless to accomplish the tasks for which they are responsible.

Delegating Responsibility to the Lowest Organizational Level

Many managers are reluctant to delegate responsibility. This often results from being "burned" by the actions of subordinates. Mistakes by subordinates reflect poorly on managers because managers are ultimately accountable for the actions of their subordinates, even when they have delegated responsibility to those subordinates. It is a natural response to be hesitant to delegate following a serious mistake by a subordinate. Many managers also believe, sometimes correctly, that they can perform a task better and more efficiently than their subordinates. As a result, it is common for managers to delegate too little.

Not delegating responsibility may result in better performance for a manager in the short run, but it is detrimental in the long run. The result is an overworked manager and underutilized subordinates. The best results are achieved when all employees are utilized to their fullest. This does not happen by chance. The manager must monitor the performance of a task to make sure that the employee who is assigned the task actually performs it well and in a timely fashion. Carefully choosing the right employee for the task and investing in the employee's training is required to ensure that the employee has the competency to perform the task. Furthermore, setting appropriate expectations and providing the right level of intermediate oversight are important to achieving good results.

Responsibilities should be delegated to the lowest organizational level at which they can be competently performed. This not only frees managers to do other tasks but also prepares subordinates for even greater responsibilities. Moreover, whenever possible, each employee should be accountable to only one supervisor. This concept is referred to as *unity of command.* Unity of command simplifies communication and accountability throughout the organization.

Leadership and Empowerment

Up to this point we have emphasized the *management* aspects of the directing function. However, managers may also achieve results by exercising leadership and empowering employees. Supervising is accomplished through the formal mechanisms of an organization. *Leadership,* on the other hand, is the power to influence people by developing a shared vision, setting an example, and empowering others to act. It does not depend on the formal organizational structure, but on the individual's ability to inspire others. The most successful managers are also good leaders.

Empowering employees is a powerful leadership tool. *Empowerment* results from passing down responsibility and authority, or power, to subordinates. Empowerment gives people more freedom in how they do their work. Empowered employees have the opportunity to work with customers without the need to consult or communicate with their supervisors. As companies have downsized and reduced the ranks of middle managers, the trend has been to broaden the span of control in management. Broad spans of control are workable when the organization has a culture that empowers its employees. The evidence shows that if people are given the opportunity to perform a wide range of interesting activities, they will rise to the occasion and grow into the job. People want more opportunity to use their skills, learn more, and do more. If they are constrained by too much control, they will not expand their abilities, at least not to the same extent as they would if given a greater opportunity to exercise independent judgement.

SUMMARY

The primary functions of management are: planning, controlling, organizing, coordinating, staffing, and directing.

Planning begins with defining the firm's mission or purpose, and includes defining specific long-range and short-range objectives for the company. It also entails formulating a sound strategy for the firm to give it a sustainable competitive advantage.

The purpose of the control function is to ensure that the company's plans are met. Controls compare actual performance to expected performance. It is particularly important that controls be designed to measure the progress toward meeting goals so that mid-term corrections may be made quickly if the firm gets off course.

Organizing is the process of assigning people and allocating resources to accomplish the firm's objectives. The formal organization reflects the defined relationships between supervisors and subordinates. Most organizations are formally organized based on functions, products, customers, or geography. The informal organization is a complex and dynamic network of interpersonal relationships among an organization's members, and it envelops the formal organization. Ideally, the formal and informal organizations are mutually supportive and reinforce each other. When they are in conflict, the company suffers.

It is the manager's job to coordinate the actions of the departments and work groups within the organization so they function as a team. Conflict among departments and work groups is an inevitable byproduct of competition within the company. It is important that the proper incentives are in place to ensure that departments and individuals all work toward furthering the company's goals. An effective means of coordinating disparate units is to designate them as either profit, cost, or service centers so that the contribution of each department can be measured separately.

Staffing entails managing the human resource needs of the company and includes ensuring that all laws and regulations pertaining to staffing are followed. Every manager is impacted by the Equal Employment Opportunity guidelines. It is therefore crucial that managers understand the implications of major employment laws and regulations. The most far-reaching law affecting hiring practices is Title VII of the Civil Rights Act, which forbids discrimination on the basis of race, religion, color, gender, or national origin. The Americans with Disabilities Act requires employers to make a concerted effort to reasonably accommodate the needs of disabled employees. Affirmative action encourages managers to consider women and minorities for jobs where they have been historically underrepresented. Diversity in the workplace is now a fact of life. Whereas managing in a diverse workplace presents a challenge, it also offers an opportunity for managers to embrace diversity as a means of being competitive in a global economy.

The directing function includes supervising employees, delegating authority and responsibility, and exercising organizational leadership. A key management principle is that effective managers must assign the responsibility as well as the necessary authority for completing tasks to their subordinates. Responsibility and authority should be delegated to the lowest level at which the tasks can be competently performed. This frees the manager for more important tasks and prepares employees to take on more important responsibilities. Directing also entails exercising leadership and empowering employees. Employees who feel empowered are most easily motivated to be productive and creative.

CASE QUESTIONS

1. Prepare an organizational chart that John Brandon can present to his boss. The chart should suggest how the three farming companies could be integrated into one efficient unit. Use the following information in preparing your chart.
   a. The farm consists of a total of 24,000 acres. The 8,000-acre unit, which is planted to almonds, is located 38 miles south of the other 16,000 acres. The 16,000-acre unit is one contiguous block.
   b. The nut-processing plant is located on the 8,000-acre almond unit. The processing plant manager also purchases nuts from other growers in the area, in addition to processing nuts on a custom basis for other growers.
   c. A fresh fruit and vegetable-packing shed is located on the 16,000-acre tract.
   d. The main office is located on the 16,000-acre tract.
   e. The crops being grown on the 16,000-acre tract include 2,000 acres of table grapes, 2,000 acres of carrots, 1,000 acres of asparagus, 2,000 acres of melons, 4,000 acres of stone fruits (peaches, plums, and nectarines), and barley for agronomic rotation purposes.
   f. At present there is a full-service shop on the 8,000-acre almond farm and two full-service shops on the 16,000-acre tract. The two shops on the 16,000-acre unit are a carryover from the merger of the two previously separate farming companies.
   g. Each of the three previously separate farm units has its own set of farm machinery and equipment, much of it in need of replacement.
   h. At the present time the operation has two separate accounting, marketing, and personnel departments, one at the headquarters unit and one at the almond unit.
2. Designate each of your management units, including geographic areas, functional activities, and crops, as profit centers or cost centers.
3. What overall service departments would you recommend?
4. How many equipment departments and/or farm shops do you recommend? Why?

5. **What are the arguments for and against maintaining more than one human resources, marketing, and accounting department?**
6. **Explain how you would set up the financial accounting and management information system to achieve accountability.**

REVIEW QUESTIONS

1. What are the principle functions of a manager and how do they relate to each other?
2. What are the key components of a mission statement?
3. Why should controls be used throughout the planning period?
4. Explain the differences between staff and line positions in a business organization.
5. Define what is meant by the chain of command in a business organization. Give an example.
6. Why is departmentalization of work activities necessary for most companies?
7. Describe the four major means of departmentalizing work in organizations.
8. Would you expect the span of control to be greater for a supervisor in a vegetable packinghouse or a partner in a farm management consulting firm? Why?
9. What is meant by the informal organization? How can it work to the benefit and detriment of a company?
10. In a retail food store, would you expect individual departments, such as meat, produce, dairy, and frozen foods, to be profit centers or cost centers? Why?
11. Why are some departments designated as overall service departments whereas others are referred to as line departments?
12. What does it mean to say that a company's employment policies cause disparate treatment or disparate impact? What are the three ways a company can defend itself against a disparate impact claim?
13. How are companies expected to respond to disabled applicants and employees when they make "special" requests?
14. Explain why it is necessary for individuals to be given the authority to command the required resources if they are to be given the responsibility for completing a task.

CHAPTER 11

HUMAN RESOURCES MANAGEMENT

LEARNING OBJECTIVES

In this chapter we will cover the following topics:

- Importance and scope of human resources management
- Functions of job design and analysis
- Essential elements for recruiting and hiring good employees
- Components of a good orientation program
- Characteristics of a successful training and education program
- Identifying and correcting workplace harassment
- Evaluating, compensating, and promoting employees
- Handling grievances
- Procedures involved in transferring and terminating employees
- Challenges faced by food and agribusiness firms

CASE STUDY

Diane Harrison is the owner and manager of Diane's Delicacies, located in St. Petersburg, Florida. The company specializes in the production of gourmet food products such as jams, condiments, and baked goods. Diane founded her company after working for a large food processing firm for 15 years, where she had risen to the position of production manager at one of their major plants.

Diane's Delicacies has been successful during its first three years. Until now, Diane has been responsible for all management functions of the firm. However, increased production demands have meant that she has had to spend more and more time on the production floor just to keep up with her orders. As a result, many other aspects of the business have suffered from her lack of attention.

Her biggest problem has been the high turnover in employees. Although she has only 24 full-time employees, 35 people have quit in the past year. Most of her employees feel that they are overworked and underpaid. Because the company has frequently been behind schedule, the employees have often been asked to work overtime without advance notice. The majority of those that have quit left for better paying jobs with other firms. Recently, Diane has been forced to hire the first person that walked in the door just to keep all of her positions filled. As a result, most of the employees, including the production supervisors, have little experience and no training. Several costly mistakes have occurred and twice during the last six months an entire batch of product has had to be thrown out. This chapter should help you identify Diane's major problems and set up a human resources management program for her firm.

INTRODUCTION

Human resources management is the management of one of the most important and often overlooked resources among food and agribusiness firms. Small businesses often confuse not needing a full-time human resources manager for not needing to manage its human resources. Even if all of the basic human resources functions, such as recruiting, hiring, compensation, and promotion are performed, employees may be performing below their potential. Human resources programs should contribute to the selection of the best-qualified people and the highest levels of accomplishment. Furthermore, human resources policies should be designed in accordance with the firm's strategy. Human resources management must support the firm's pursuit of its long-term objectives, it's quality and cost goals, and ultimately it's sustainable competitive advantage.

For the most part, managing human resources in a food or agribusiness firm is similar to that of other firms. The management of human resources is largely governed by the same laws, regulations, and management principles, regardless of the type of firm. The most significant differences in human resources management are associated with the size of the firm, since the applicable laws and regulations often depend on the number of people employed by the firm. Where food and agribusinesses companies face unique problems we point them out. We have also included a section at the end of the chapter that addresses special human resources issues affecting the food and agribusiness industry.

HUMAN RESOURCES PLANNING

Human resources planning is the process of organizing the company's personnel needs and developing systems for meeting those needs. It can be thought of as the process of anticipating changes in the company and preparing for these changes by hiring new employees with the appropriate qualifications, retraining existing employees, and eliminating positions. Thus, human resources planning is about creating the framework for meeting the company's employee needs—today and in the future—as the company changes.

A principal objective of human resources planning is to hire the "right" person who will be a good fit in terms of the company's culture, who has aspirations that will mesh with the company's, and who has the skills needed to do the job. To accomplish this goal, it is necessary to have a clear understanding of the company as it is now as well as where it is going. Thus, whether the company is large or small, meeting human resources needs is a company-wide function. In large companies with a human resources department it is the responsibility of this department to work with other departments throughout the company to assist them with their personnel needs. The human resources department helps other departments develop a human resources plan, manage human resources issues, and monitor employee progress, in addition to providing support for their routine human resources work. In smaller companies without a human resources department the company's top managers or owners typically handle human resources planning as well as many of the other human resources functions (Box 11–1).

JOB ANALYSIS

Job analysis provides the foundation for many other human resources functions, such as recruiting, hiring, training, evaluation, compensation, and promotion. It is the foundation of all the human resources functions. Like human resources management in general, job analysis is an often overlooked management function, because it is not necessary to keep people employed. However,

BOX 11-1

Functions of Human Resources Management

Writing job descriptions
Recruiting employees
Selecting and hiring new employees
Orienting new employees
Training and education
Evaluating employees
Determining compensation
Promoting employees
Handling grievances
Handling harassment cases
Making transfers
Handling terminations
Developing and enforcing human resources policies and procedures

it is critical to ensuring that a company's human resources are managed efficiently and effectively. Job analysis is used to:

- Comply with labor laws
- Develop recruitment strategies, selection procedures, and interview questions
- Develop training programs and job restructuring
- Conduct performance appraisals and determine compensation
- Ensure worker health and safety

Job analysis involves determining exactly what a job involves, such that the job can be distinguished from other jobs. Information obtained in a job analysis, either from the employee or the supervisor, includes work activities, equipment and environment, formal and informal relationships with other employees, expected performance, and the supervision given or received.

Job analysis is also important from a legal standpoint. Equal Employment Opportunity guidelines require that a thorough job analysis be performed in order to validate selection criteria for the job. For example, if a firm uses some level of formal education as a selection criterion, then the firm should be able to show, by means of a job analysis, how the duties and responsibilities of the job can only be performed by someone who has attained the required level of formal education. The results of a job analysis are a job design, job description, and job specification.

Job Design

Job design is the process of either identifying the components of a job before it is occupied or redesigning a currently filled job. An important consideration in job design is the job environment, including the conditions, location, physical environment, social environment, and supervision needs under which the job is performed. Attention to job design is important because it affects motivation, performance, satisfaction, and the physical and mental health of the employee.

Historically, job design as a management function finds its origins in the late nineteenth century "scientific management" of Frederick Taylor. Taylor felt that jobs should be broken down into small elements and that a scientific way to perform each job should be determined. Time and motion studies were used in factories to discover which work methods wasted the least amount of time and motion. Jobs were standardized to reduce training time and costs, and specialization became commonplace as a means of ensuring that workers became very good at their tasks. Scientific management transformed the craft system into the industrial factory system.

Modern job design concepts include job enlargement, job enrichment, quality circles, autonomous work groups, and alternative work schedules. *Job enlargement* involves adding more tasks to a job, which decreases specialization. Job enlargement can be thought of as a horizontal expansion of a job. Job rotation is a form of job enlargement. In a packing facility, instead of being assigned to one task such as unloading produce, a worker may spend one day grading onions, another day driving a forklift, and another day assembling packing boxes. Although job enlargement alone may not be enough to reduce tedium or boredom, it can contribute to developing employees with multiple skills.

Job enrichment vertically expands the scope of the job, creating more job depth. An enriched job will involve more responsibility for decision making, planning, and control. For example, employees may be involved in the decision as to how their job might best be accomplished or evaluated. Sales personnel might decide how to approach their clients, follow up on sales calls, and evaluate different sales techniques. Job enrichment makes the job more interesting and more growth orientated for the employee. From the company's perspective, the employee becomes more valuable to the company by adding higher-level tasks and responsibilities to the job.

Quality circles are groups of employees who meet on a regular basis to share ideas with management for improving productivity and cutting costs. Bonuses are often paid for usable ideas.

Autonomous work groups are groups of employees that take a project from beginning to end. In contrast to the assembly line approach, where a factory worker may perform a single task repeatedly, like installing a piece on a machine, a group of workers might build the entire machine.

Alternative work schedules include flexible working hours, job sharing, compressed workweeks, part-time employment, and working out of the home. *Flexible work hours* permit employees the opportunity to adjust their work schedule to accommodate individual needs. Generally, a two-hour period at the beginning and end of the traditional eight-hour workday is designed as flexible time. Employees are free to decide when, during flexible time, they want to arrive and leave as long as they work the required number of hours. *Job sharing* is a system in which two people share one full-time job. Both flextime and job sharing have become increasingly popular with families who must arrange childcare or in urban areas where travel during rush hour can be very time consuming.

Compressed workweeks allow employees to work more hours per day and fewer days per week. This is an attractive option for people who have long commutes. Common alternatives include the four-day week in which employees work four, ten-hour days, or the nine-hour day where employees get every other Friday off.

The advent of computer and information technology has permitted many employees to work out of their home without having to make the daily commute to the office. Many companies support this option, known as *telecommuting,* by providing employees with a laptop computer.

Lastly, permanent part-time employment has been on the rise in recent years. This is popular with retired professionals or people considering retiring who would rather not give up working completely. Many people with young families also find this option attractive.

Job Description

The *job description* states the tasks, duties, and responsibilities of the job. The job description should include the following items:

- The exact job title
- A brief job description
- A list of tasks and duties to be performed
- Requirements for the job
- Hours of work
- Method of evaluation
- Salary range
- Line of supervision
- Prospects for promotion

Job descriptions help prospective employees envision what they would be doing in the job. New employees use their job description to understand the requirements and opportunities associated with their job. For employers the job description is useful in advertising the job and in the hiring process. Job descriptions are also used by employers to determine compensation levels by comparing the relative importance of, skills needed for, and relative difficulty of different jobs. Lastly, the job description fosters communication between the employee and the employer and minimizes misunderstanding between the two.

Job Specification

Whereas the job description describes the demands of the job, the *job specification* describes the characteristics and qualifications of the person expected to fill the job. These characteristics include skills, knowledge, and abilities necessary to perform the job, certifications earned, and previous experience that would enable the employee to perform the job. Job specification is a critical part of job analysis because selecting successful employees depends on being able to identify what makes an employee successful. A job specification for an employee of a fresh produce packing facility might include manual dexterity, good hand-eye coordination, forklift operation experience, and the ability to stand for long hours.

RECRUITING

When a position becomes open, it may be filled either by recruiting people from outside the firm or by transferring or promoting someone from within the company. The decision as to which of the two methods should be used depends largely on the position to be filled, the qualifications of potential applicants within the company, and the cost of recruiting.

Because recruiting can be expensive, present employees are usually preferred over outsiders when an adequate supply of well-qualified applicants are available from within the firm. A major advantage of hiring an insider is that he or she is a known quantity. His or her performance, both strengths and weaknesses, should be well documented if thorough, regular evaluations have been conducted. Promoting or transferring workers who have excelled at their jobs is also good for employee morale in addition to being a means of rewarding productive workers. This is especially true for lower-level positions for which no special skill or education is required. Employees in such positions usually advance through hard work, and promotions are typically based in part on seniority. Internal recruitment is especially appropriate when knowledge of the organization, its people, and procedures is important. Of course, a disadvantage of hiring internally is that filling one open position simply creates another open position.

The two most common ways of hiring an employee within the company are through job posting and searching through employee skills and succession plans. *Job posting* consists of a written notice to all current employees announcing vacancies within the company. Job posting is done through internal memos, posting on electronic bulletin boards, and publication in the company newsletter. Searching the employee skills inventories and succession plans enables managers to identify all qualified employees for the purpose of bringing the job opening to their attention. Many times managers will informally identify an employee they think is qualified and approach the employee as well as his or her supervisor to see if a transfer is possible. This policy should be discouraged because it shows favoritism and does not promote an open process, which can lead to employee resentment.

At times it is desirable to hire someone from outside the firm. When no qualified applicants are found within the company, or when it is thought that recruiting outside of the firm may yield better-qualified applicants, the extra recruiting cost will be justified. Managerial positions, in particular, often benefit from bringing in "new blood."

Some of the most common sources of new workers are the recommendations of employees and friends, state and private employment agencies, colleges or trade schools, classified advertisements, trade associations, and competitors. The choice of a method (or methods) of recruiting will depend on the difficulty in finding qualified personnel, the level of training or education required, the level of experience desired, and most importantly, the type of job. Recruiting is much like advertising a product in that to be successful you must know your audience. You must know both where to reach them and know how to interest them in working for the food or agribusiness company.

Classified Advertisements

One of the most common means of attracting new employees is classified advertisements, or classified ads as they are more commonly called. The help wanted section of the classified ads in the local newspaper is usually scanned daily by the unemployed. It is especially appropriate when no special skill is required or when the skill is readily available in the publication's area of circulation. Advertisements in trade magazines or journals are most effective when a highly trained or experienced individual is needed. It has recently become common for companies to post employment opportunities on the Internet as well as their company's web site.

State and Private Employment Agencies

Employment agencies can be an efficient means of locating qualified personnel. These agencies maintain files on people actively looking for work, including their qualifications, experience, and other relevant factors. Historically, state employment agencies have been a good source of skilled or semiskilled employees. Because people who are receiving unemployment benefits must register with these agencies to maintain their unemployment benefits, state agencies have a constant stream of newly unemployed workers. Public agencies offer their services to job applicants and employers for free.

Private agencies (or head hunters) are usually a better source of potential employees for high-level positions or when a particular skill is needed. Some employment agencies even specialize in specific types of employees, such as managers or accountants. Executive search firms are private employment agencies that specialize in placing high-level executives. All private agencies charge a fee for their services when an applicant is hired. Some employment agencies charge the job seeker a fee; however, with most employment agencies the fee is paid by the employer. Employment agencies provide a service to both employers and job seekers. They typically interview applicants, determine their employment interests and qualifications, administer job skills and knowledge tests, and run background checks. When an employer has a need for a new employee, they contact the private employment agency and specify the job they need to fill, the job's requirements, and the qualifications potential employees must possess. The agency then checks its records and sends only those applicants that best fit the job description for an interview.

Trade Associations and Shows

Trade associations may be a good source of locating experienced personnel. Trade association representatives are usually in close contact with many of their member firms and may know of people who are out of work or who are looking to change jobs. They often publish newsletters or magazines in which available positions are advertised and host trade shows where a company can find good employees in addition to marketing its goods and services. Companies usually print recruitment brochures that can be distributed to prospective applicants.

School Recruiting

When special education or training is required, colleges, universities, or trade schools may be the best source of new employees. Many large employers send recruiters to these schools before graduation to interview for any open positions. Small firms that frequently have only one position to fill often contact school counselors to get their recommendations and then contact the individual personally. School recruiting can also serve as a form of advance screening because only the best candidates are brought in for an interview. Schools are a good source of employees for positions where a minimum of experience is needed, because most graduates have limited work experience, although this is changing as older students with experience are returning to continue their education.

Referrals

Referrals of employees or friends may also be a good, although somewhat limited source of new employees. Because it is difficult to assess a person's personality in a short interview, recommendations of trusted people are often an effective means of finding people who will be a good fit with the organization's culture. Someone who knows the company and the potential employee can often create a better fit than a random search.

Competitors

Competing firms are a good source of qualified and experienced workers. However, it must be used judiciously if the agribusiness does not want to acquire the reputation of raiding its competitors. Most companies do not object too strongly to a competitor hiring one of its employees when the move is beneficial for the employee. This is particularly true when the move involves a promotion or substantial raise that would not be possible if the employee continued in his or her current job. It should, however, be remembered that it is both illegal and unethical to hire a competitor's employee with the intention of stealing company secrets. Employment contracts sometimes contain what is known as a "*noncompete clause.*" Such clauses forbid employees from working for a competitor in the same industry within a specified period of time following their termination with a firm.

Internships

Internships are often used by students to obtain experience, job references, and contacts in their chosen field. Internships are also beneficial for the employer as it allows the company to bring in and evaluate potential employees on a short-term, cost-effective basis. Internships must be handled well so that they serve as a learning experience for the student as well as an effective recruitment tool for the company.

SELECTING NEW EMPLOYEES

The importance of an effective selection process cannot be overstated. The cost to the firm of recruiting, selecting, and training new employees is high. Therefore the firm already has a substantial investment in each employee before he or she has done anything productive for the company. Furthermore, it is often difficult to dismiss employees who have unsatisfactory performance. Once a list of candidates has been assembled, the difficult task of selecting which one(s) will be hired begins.

Job Application

Job applicants should be asked to fill out a job application. The purpose of the *application form* is to obtain the information necessary for a preliminary screening of the applicants and to provide the interviewer with some basic information about the applicant. The application should allow for the comparison of applicants on the basis of their knowledge, skills, and abilities. It should not discriminate unfairly by obtaining information that would identify members of protected groups. Thus designing an application requires the employer to consider federal as well as state and local employment laws.

Application forms should include only those questions that will elicit job-related information. Questions regarding age, marital or parental status, birthplace, residence, race, sex, military status, height and weight, whether you have friends or relatives who work for the company, arrest and conviction records, citizenship, physical, mental, or emotional handicaps, questions related to social, religious, or political affiliations, or previous salary should not be asked. Exceptions to the rule would be in cases where the company has a business necessity or a bona fide occupational qualification. For example, if you are hiring a bartender, it is permissible to ask applicants if they are over 21 because serving alcohol is a requirement of the job and to legally do so you must be at least 21 years of age. You may ask applicants if they were convicted of a crime (but not arrested for committing a crime) that relates to the job. For example, you could ask applicants applying for employment in a bank whether they were convicted of forgery, embezzlement, or writing bad checks. Box 11–2 contains a list of questions that should not be asked on a job application form.

BOX 11–2

Questions Not to Ask on a Job Application Form

Marital status	
Age	You may ask if the applicant is 18 or over, or 21 or over if the job requires a minimum age.
Parental status	
Birthplace	You cannot ask because of possible national origin or immigration issues.
Residence	You may ask for address but not whether the applicant owns or rents.
Arrest record	
Felony record	Felony inquiries must be job related, however you can indicate that you will do a criminal records investigation.
Military status	You cannot ask the type of military discharge received, however you can ask if the applicant has military experience.
Social, religious, political	You cannot ask any questions about personal affiliations, however you can ask about professional and job-related affiliations.
Disability, health	Avoid all medical questions until after an offer of employment has been made, unless there are physical requirements related to the job.
Workers' compensation	Avoid workers' compensation history inquiries until after an offer of employment has been made.
Citizenship	You may only ask if an applicant can provide proof of their right to work in the United States.

BOX 11–3

Job Applicant Information Required to Meet Equal Employment Opportunity Laws

Employees of this company are treated equally without regard to their race, color, religion, sex, ethnic background, national origin, age, marital status, military status, or non-job-related handicap or mental health. We are gathering the following information to comply with government regulations as an equal opportunity employer. Completion of this information is not required, but it is important in evaluating and monitoring our hiring practices. This form will be permanently separated from the remainder of your job application before the application is seen by anyone involved in hiring decisions. We appreciate your assistance.

Date:

Position applied for:

How did you become aware of the position?

Please circle the categories that apply to you:

1. Male Female
2. White African American Hispanic Asian Native American
3. Veteran Disabled Veteran Disabled Person

Companies with 15 or more employees must keep records for all applicants for at least one year to ensure that they are complying with employment laws by not discriminating against protected groups. Information on protected groups should be requested on a form separate from the job application. To ensure that this information is used only for compiling affirmative action statistics, the information should be kept separate from the employee's personnel file and anyone involved in the hiring decision should not have access to the information. (Box 11–3).

Questions that may be asked of job applicants include name, address, telephone number, educational background, licensing and professional affiliations, previous employment history and dates of employment, and professional references. It is also important to include some specific job-related questions that will help determine which employees should be brought in for an interview.

Many employers add statements to protect themselves in case an applicant is disgruntled with the hiring process. Requesting that an applicant sign an *information release form* to give the employer permission to check references can be useful in obtaining full cooperation from previous employers. This should be done in such a manner that the applicant does not feel pressured into signing the release. The employer should also inform applicants if a criminal

BOX 11–4

Application Form Statements

Applicants should be asked to sign and date each statement as shown in statement 1.

1. I promise that the information contained in this application and in my resume is true and complete. Furthermore, I am aware that if the information is false, I can be eliminated from consideration for this job or if I am hired, my employment can be terminated.

   Signature: ______________________________

   Date: ______________________________

2. I understand that all statements in my application and resume may be investigated and I authorize the company or its agent to contact anyone who might be able to comment on my ability to perform the job. I understand that a consumer credit check might be performed and that I can receive a copy of that report upon request. I understand that a criminal record check might be performed.
3. I authorize any person, educational institution, and employer to provide you with job-related information that may be useful in making a hiring decision. I release any reference or organization from legal liability in providing job-related information to the company or its agent.
4. I give permission to the company to conduct preemployment drug screening, and a preemployment medical examination with the understanding that the results of the medical examination will not eliminate me from consideration unless the examination indicates that I am unable to perform the essential activities of the job.
5. I understand that the company has a "zero tolerance" policy regarding job-related drug and alcohol abuse and harassing or threatening fellow employees and customers. Violations of this company policy may result in immediate dismissal.
6. By signing, I acknowledge that if I am hired, my employment will be at will and subject to termination by me or the company for any reason as long as the reason is not prohibited by law.

check will be conducted. If a credit check will be run, a special notice must be provided. Employment-at-will, false information, and work rule statements are also appropriate. Examples of these statements are given in Box 11–4.

Interviewing Job Applicants

Screening the applications may save the employer a great deal of time, especially when there are a large number of applicants, because it is likely that some or most of the candidates will be unqualified for the position. Those who appear to be best qualified may then be called for an interview. If there are a large number of applicants who appear to be qualified, even after an initial screening of the applications, two interviews may be necessary. The first interview (a prescreening interview) is used to narrow down the number of qualified applicants based on several criteria that are determined to be important. Conducted by telephone, the prescreening interview is an inexpensive way to discuss the job with the applicant and to make a quick determination regarding the applicant's suitability. The goals of the prescreening interview are to provide both the applicant and interviewer with additional information. The interviewer should gather information to evaluate whether the candidate is qualified for the job. Applicants should be given additional information about the job to help them decide whether they are still interested in pursuing the job opportunity. The final interview is then easier because the number of interviewees is smaller (normally three to six applicants are chosen), permitting a more thorough interview and facilitating comparisons between applicants.

The purpose of the interview is twofold: for the employer to obtain as much information about the applicant as possible, and for the applicant to learn about the job and the employer. A good working relationship will depend as much on the employee's satisfaction with the employer as it will on the employer's satisfaction with the employee. The interview should be used to complement the job application. Any questions that were left blank or partially answered should be addressed in the interview. Often responses on the application give rise to other questions, such as when a problem area, like frequent job changes, is identified. Finally, information that cannot be gleaned from the application form should be obtained, particularly behavioral characteristics, such as personality, motivation, and interpersonal skills.

Because the ability of the interviewer to accurately assess a candidate's potential for success in a job requires a great deal of skill, many firms hire trained interviewers. However, this is not feasible for all food and agribusiness companies. The remainder of this section is devoted to some general guidelines that will prove useful to the untrained interviewer.

Although there are many methods and styles of interviewing, the direct approach is probably best for the inexperienced interviewer. The questions should be written down in advance to ensure that important points are covered and that each applicant is asked the same basic set of questions. With time an interviewer will develop a style that is best suited to him or her. Because most people are very nervous and tense when going into a job interview, the interviewer should generally attempt to put the applicant at ease. This will create a more relaxed atmosphere, and the applicant will feel freer to be his or her usual self. An exception to this rule might be when it is important to assess how well an applicant behaves under stress. Areas of common interest, the applicant's hobbies or sports activities, or how the local basketball team is doing are examples of the type of light conversation topics that may be used to break the ice.

Once rapport has been established between the interviewer and interviewee, the interviewer may proceed to cover the prepared questions. There are a couple of points that should be remembered when preparing questions. First, all questions should be directed at determining the applicant's ability to perform the job. Specifically, prepared questions should relate to the job description and job specifications. Second, there should be a way of evaluating the applicant's answers to each question. In other words, the answers to questions should be predictive of job performance. For this reason, it is a good idea to determine ahead of time how low, average, and high job performers might respond to each question.

The interviewer should also give the applicant a chance to learn about the company and ask any questions. It is not uncommon for interviewees to draw a complete blank when asked if they have any questions. This does not necessarily mean that they have no questions, but more likely that they can't think of them at the time. In such a situation the interviewer may want to tell applicants about the company, answer frequently asked questions about the job, or bring up subjects about which the interviewees are likely to have questions. Sometimes applicants are introduced to their potential supervisor. The supervisor's input is especially important when the job calls for a close working relationship between the supervisor and employee.

Testing

Testing is sometimes used to screen a large number of applicants. Whether used to screen applicants for hiring, promotion, or other personnel decisions, testing should be used with caution. Tests must be shown to be valid (that is they measure what they are supposed to measure) and reliable (they are consistent). Companies have employed ability, personality, drug-testing, and performance tests to screen applicants. Because of the difficulty in proving that many of these tests are valid predictors of job performance, and the fact that their use has resulted in discrimination lawsuits, many companies use professional testing services.

Ability and Performance Testing. Ability testing is the most commonly used form of testing. Such tests are easy and inexpensive to administer and interpret and, when properly designed, can be highly predictive of job performance. Mental ability and conceptual skills can be measured using intelligence, achievement, and aptitude tests. Common examples of these tests include basic mathematical ability tests used by agribusinesses hiring farm management consultants and writing tests used by agricultural communication firms. Sensory ability such as vision and hearing tests and physical ability such as strength, flexibility, coordination, and stamina tests are also used for jobs requiring these physical abilities. Performance tests are designed to measure the same attributes as ability tests by asking the applicant to perform the key specific tasks that the job demands. An agricultural communications company may give applicants a performance test by asking them to write an article that would normally be assigned to an employee for its magazine. The performance test in this example would provide direct evidence of an applicant's ability to do the job.

Personality Tests. Personality and other forms of psychological tests are more controversial. These tests are used by companies to determine if the applicant is a good fit for a particular job as well as with the company's culture. These tests are complicated to develop and interpret and should be administered only by professional testing organizations. Personality tests are used to measure various interpersonal skills. Among these are communication skills, listening skills, skills that determine the applicant's ability to sense other's needs, knowledge of the applicant's motives, preferences, and values, ability to handle difficult people and situations, adaptability and ability to change, and ability to criticize constructively. Personality tests are also used to measure judgment, energy levels, decisiveness, ability to handle stress, leadership, work style, creativity, motivation, and the applicant's ability to influence others.

Drug Testing and Medical Exams. Preemployment drug testing and medical exams are legal but should only be conducted after the applicant is given a conditional job offer. Drug and alcohol preemployment testing is popular because it is easy not to hire applicants who test positive for drug and alcohol abuse. Drug and alcohol preemployment testing is reliable, relatively inexpensive, and sends a message to potential employees that the company is serious about rules against drug and alcohol abuse. However, the Americans with Disabilities Act, which makes testing and firing employees with disabilities more difficult, may protect employees who become drug and alcohol abusers on the job. In such cases it is advisable for the company to consult with an attorney specializing in labor law before taking action. Preemployment medical examinations are also permitted; however, employment decisions must be based strictly on job-related outcomes because the Americans with Disabilities Act prohibits discrimination against ill or disabled individuals. An employer can not renege on a hiring decision based on the results of a medical exam unless the results indicate that the applicant will not be able to perform the essential features of a job.

References

References can be useful in selecting an employee and are an important part of the hiring process. Previous employers can provide useful information that cannot be obtained from other sources, because they have had the opportunity to closely observe the applicant in a work setting. It should be kept in mind that the applicant typically lists those references that are likely to give him or her a favorable recommendation. Thus, previous employers not listed as references should also be contacted to verify information given by the listed references.

Checking references is also important in verifying information that the applicant has provided. There are employee screening companies that, for a modest fee, help companies check on a prospective employee's educational and work history, social security number, driving, credit, and criminal records. Because an employer is legally responsible for employee negligence and illegal activities that can be attributed to employment with the company, record checks are doubly important.

It can be difficult to obtain useful information from an applicant's current or former employers because most state laws require only that employers verify dates of employment and the position last held. A notable exception is that employers are allowed to provide information concerning former employees who have caused harm to individuals or the company, provided it was related to the employment. A former employee does not have the basis for a lawsuit against a former employer if the information provided is true, offered without malice, and is job related. Practically, to get the cooperation of former employers references must often be assured that the information he or she gives will remain confidential.

The courts have ruled that if the reference check is done in-house, there is no requirement to provide the applicant with information from references. However, the company must respect the privacy of the applicant. This means that the information must be solicited in good faith, must be given to an individual with a bona fide business need to know, and must be used only for the purpose for which it is intended. The information must be limited to the questions asked and must be job related.

When possible, the references should be contacted by telephone as opposed to a letter. Written references are only useful to verify employment history. The best information will often be obtained over the telephone and the caller may judge not only what is said, but also the manner in which it is said. The company representative should explain to the reference that the purpose of the call is to verify employment information obtained from the applicant, that the reference interview will be confidential, and that the applicant has signed a release form granting the firm permission to contact references. Lastly, the company representative needs to document and save the reference checks in a file kept separate from the employee files. Employees have access to their employee files and the company must respect the confidentiality of the information contained in the reference checks.

ORIENTING NEW EMPLOYEES

Once the hiring decision has been made, new employees should receive an orientation. Each new employee represents an investment for the food or agribusiness company. The orientation process is the first step in making a new employee a valuable member of the work force. It includes acquainting new employees with the company, introducing them to their supervisor and fellow employees, getting them started in the new job, and evaluating their initial progress. In small companies the person in charge of human resources may handle the orientation on an informal basis. In larger companies it is usually handled more formally. The most common method is to schedule group orientations at regular intervals. New employees are then asked to report for the first time on the day of the next orientation. This method is efficient for large companies because the same material does not have to be repeated for every new worker.

As with hiring, the presentation of the company and its policies should be conducted by one person to ensure a thorough, consistent coverage of all important information concerning the food or agribusiness firm. The information about the company should include: company history, products and services offered, company regulations, work hours, salary policy, overtime policy, benefits, vacation and holidays, sick leave, and promotion policy.

New employees should be given a tour of the company's facilities. This will not only assist them in finding their way around the physical plant but also give them an idea of how they fit in with the overall operation of the firm. This is a good opportunity to introduce new employees to the other employees of the company. Depending on the size of the firm, this may mean introducing them to everyone that works there or simply to their supervisor and coworkers. Every attempt should be made to make new employees feel at home. The initial relationships that are established can be very important in helping new employees adapt to their new job.

The supervisor is usually in charge of introducing the new employee to the job. For simple jobs the supervisor may explain the job or perform a demonstration before allowing the new employee to perform the task. Either the supervisor or a coworker familiar with the new employee's job should be available to assist him or her during the orientation period.

Industry Profile 11–1 describes some of McDonald's human resources policies that have led to its success.

TRAINING AND DEVELOPMENT

The objectives of the training and development program are to prepare employees to perform their jobs and to encourage their future development. *Training* is conducted to teach a specific task. New or transferring employees must be trained to develop the skills necessary to perform an unfamiliar job. Training programs may also be used to reinforce previous training, to increase skill levels, or to keep employees up to date with the latest developments in the field. Training should not stop when the employee can perform the job satisfactorily, but rather be a continuing process whereby all employees are encouraged to develop their abilities to their fullest potential. By encouraging the development of their personnel, companies ensure that they will constantly have at their disposal employees who are qualified for promotions and ready to accept greater responsibilities.

Food and agribusiness firms that invest in their people will be rewarded by the improved performance of their employees due to their increased skills and knowledge. Employees that are challenged to do their best work, and provided the opportunities and rewards for doing so, can be expected to make the maximum contribution to the firm as well as exhibit increased job satisfaction.

INDUSTRY PROFILE 11–1

MCDONALD'S CORPORATION

McDonald's Corporation is renowned for its high standards in customer service and its consistent food quality. Its success in maintaining these standards over the several decades it has been in business is due largely to its sound operating principles and its commitment to managing its human resources.

Customers who have eaten at McDonald's throughout the world are often surprised to find that the same attention to detail they find in their local McDonald's may be found anywhere. The food tastes the same, hamburgers and fries are served with a smile, and the restrooms are always clean. McDonald's commitment to these attributes, known as QSC&V, which stands for quality, service, cleanliness, and value, has become legendary in corporate circles.

McDonald's starts with an extensive training program for all of its employees. Its training manual is more than 600 pages long. Successful applicants for new McDonald's franchises must complete a 12-month training program. Most of its managers have graduated from its Hamburger University, located at its headquarters in Oak Brook, Illinois.

Another key to McDonald's success is the extensive job analysis it has conducted. By thoroughly understanding each job and its components, McDonald's designed an assembly line-style system that maximizes efficiency and consistency. Where necessary it has automated functions, such as the filling of drinks and automatic fryers.

McDonald's continually monitors all of its stores through monthly visits. At each visit the store is given a score for each of the QSC&V categories. The company then uses this information to improve the performance of individual stores.

McDonald's has also been a leader in promoting a diverse workforce. It has reached out to minority workers, retirees, and people with disabilities. In addition to providing the company with a good source of workers, its employment practices have generated much positive publicity and good will for the firm.

New Job Training

Many employees will need to be trained for a new job, whether they are new to the company or not. In some cases a simple explanation followed by a period where the trainee asks questions and practices performing the new task may be sufficient. When more in-depth training is required, two methods are commonly used: the apprenticeship and the company school.

Apprenticeship. Under the apprentice system, an experienced worker is assigned to a new worker to teach him or her the necessary skills. This centuries old system was once used in professions such as blacksmithing or printing to pass on skills acquired over many years by the master to the apprentice, often his son. Today, it is used mainly to teach complex skills that are best learned by doing. The training period will usually extend from a very short period of time, such as a month for simpler skills, to a year or even years for more difficult skills. It is important to consider the personalities involved when assigning an apprentice to an experienced worker, as well as the ability and patience of the employee who will do the training.

Company School. Many large companies, which often have several new workers in training at the same time, have found it economical to operate a company school. Trainees are taught a skill in a structured program usually lasting for a specified period of time, by professional instructors or experienced workers. Because the training is done outside of the main production area, the trainees do not interfere with normal production, and thus their mistakes are less costly. However, care must be taken to simulate job conditions as closely as possible. Talking about a job is no substitute for doing it! A major disadvantage of this method is that some trainees may not get the personalized attention they require, because of a program's rigid structure and large number of trainees.

Ongoing Training and Development

Ongoing training and development is usually provided by utilizing outside resources, because only the largest food and agribusiness companies have the resources and employee talent to efficiently provide a diverse training and development program. Furthermore, the use of outside resources greatly increases the options available to the firm. Outside resources may generally be classified into two types: learning materials and organized courses.

Learning Materials. *Learning materials* include videotapes, training manuals, and more recently the Internet, to name a few of the more commonly used resources. They cover a wide variety of topics on almost any imaginable subject. The major advantages of using resource materials are that they are relatively inexpensive and provide a great deal of flexibility because the actual presence of the expert is not required. Although sellers of these services often seek out potential users, trade associations, other businesses in the same industry, and extension personnel may be helpful in locating these services.

Organized Courses. *Organized courses* range from seminars or workshops that last from one day to a week, to short courses that may last several weeks, to courses offered by trade schools, colleges, or universities that may last several months. The major difference between using an organized course and learning materials is the presence of an instructor. Difficult or complex material may be more easily learned when taught by an instructor, because the presentation can be directed to the level and needs of the students. Food and agribusiness firms that can afford to sponsor such programs may increase employee participation by offering the course at a convenient time at the company facilities or sometimes over the Internet. Many companies reimburse all or part of the cost of attending outside courses, particularly those that are job related, as an incentive to their employees.

Many other meetings, conventions, seminars, conferences, and congresses are available in many different areas of the food industry. Although they differ greatly in terms of their focus, they share a common purpose in that they provide a forum for the interchange of ideas and interaction with colleagues.

MANAGING EMPLOYEES

A key factor in managing employees is keeping morale high. High morale occurs when an employee has a positive attitude towards fellow employees and wants to be a responsible, productive member of the team. Building employee morale involves sharing the company's vision and mission, empowering employees, and building employees' self-esteem by encouraging them to take risks and further their job training. Morale building also involves establishing a safe and healthy workplace by establishing rules that govern employee behavior.

For the most part, employee morale is the responsibility of the employee's direct supervisor and outside the scope of this book. However, all employers are responsible for ensuring that their employees are treated fairly in the workplace and that all federal, state, and local statutes are observed. The focus of this chapter is on the policies and procedures of which managers involved in human resources management should be aware. The specific laws and regulations governing fair employment practices are covered in detail under the staffing function in Chapter 10.

Antiharassment Policy

In addition to employers being liable for employment discrimination under Title VII of the Civil Rights Act, the employer may also be liable for harassment that occurs in the workplace. An employer is considered responsible for harassment if he or she permits the harassment of an employee by supervisors, other employees, customers, or suppliers. *Harassment* is considered intimidation, insult, or ridicule based on race, color, religion, gender, or national origin. Although Title VII applies only to companies with 15 or more employees, some states and localities also have rules applying to smaller employers as well as prohibiting discrimination based on sexual preference.

Title VII holds an employer liable for harassment when the workplace is permeated with discriminatory intimidation, ridicule, and insult that is sufficiently severe and pervasive to alter the conditions of the victim's employment and create an abusive working environment. Most courts use a *reasonable person standard* in deciding whether conduct is harassment, that is, would a reasonable person in the same situation as the employee find the conduct hostile or intimidating. The best approach is to prevent harassment from occurring by maintaining a positive workplace environment in which discriminatory behavior is not permitted. Many companies have gone further and initiated training and sensitivity programs for employees in management positions. When harassment is reported or a complaint is filed, the report should be immediately investigated and corrective action should be taken, if appropriate.

Sexual Harassment. Unwelcome sexual advances, requests for sexual favors, and other verbal or physical conduct of a sexual nature constitute *sexual harassment* when the following three criteria are met:

- Submission to such conduct is necessary in order to get or keep a job.
- Submission to or rejection of such conduct is used as a basis for employment decisions such as raises, promotions, and demotions.
- The conduct interferes with an individual's work performance or creates an intimidating, hostile, or offensive working environment.

The courts have recognized two basic types of sexual harassment. *Quid pro quo sexual harassment* occurs when the employee is required to engage in sexual activity in exchange for workplace entitlements such as promotions, raises, working hours, or other benefits. This type of sexual harassment is easy to recognize and is linked to supervisors and managers who have the authority to hire and fire, and to impact the terms and conditions of employment. Employers are responsible for a supervisor's quid pro quo sexual harassment when the employer knows or should have known about the harassment. The supervisor is the employer's agent and representative and the acts of the supervisor are considered to be the acts of the employer. Consequently, employers should carefully choose supervisors and limit the supervisor's ability to give raises or promotions by having in place a performance appraisal system with adequate justifications, monitors, and checks.

Hostile environment sexual harassment occurs where the sexually oriented conduct creates an offensive and unpleasant working environment. To determine whether an environment is hostile, it is generally required that five conditions be met:

- *The harassment is unwelcome by the employee.* Voluntary sexual relationships or activities that are desired or welcomed by the employee are not considered sexual harassment.
- *The harassment is based on gender.* Harassing conduct need not be motivated by sexual desire and can occur by someone of the same or a different gender. The critical issue is whether members of one gender are exposed to disadvantageous terms or conditions of employment to which members of the other gender are not exposed.
- *The activity is considered to be severe or pervasive.* It typically must be more than an isolated occurrence. The more frequent the occurrences and the more offensive the behavior, the more likely that the severe or pervasive requirement will be met.

- *The harassment affected a term, condition, or privilege of employment.*
- *The employer had knowledge of or should have been aware of the sexually hostile working environment and took no prompt action.*

In order to prevent sexual harassment, as well as limit the company's liability stemming from sexual harassment claims, all businesses should have in place policies, programs, and procedures designed to discourage sexual harassment, encourage the reporting of sexual harassment by anyone who experiences or observes it, and promptly investigate any reports of sexual harassment and take corrective action. Whereas all employees should understand that sexual harassment in the workplace will not be tolerated, it is especially important that employees in management positions understand and be supportive of the firm's position. Not only do they have the most power, and therefore the possibility to abuse that power, but they are in the best position to influence the workplace environment.

Other Employment Guidelines

In addition to antiharassment policies, many businesses establish a variety of workplace policies including safety rules, absence and tardiness policy, prohibition against the use of alcohol or illicit drugs, dress codes and personal appearance rules, and rules about keeping the employer's and customers' sensitive information confidential.

Employers should also have procedures for dealing with employees who are performing below expectations. This may involve putting them on a performance improvement plan designed to bring their performance up to standard. It should also include the consequences of not meeting the plan's goals, including, ultimately, dismissal.

To ensure that all employees are made aware of the firm's policies and procedures, all employment and workplace guidelines should be compiled into a manual. Many firms distribute these manuals to new employees at their orientation. Copies are typically maintained in the human resources department or office and are available to all employees. New or modified policies or procedures may be communicated by a notice to employees or posted on company bulletin boards.

HANDLING GRIEVANCES

Employee complaints or grievances range from a minor dissatisfaction, such as the employee who feels that his or her office is too small, to a major problem, such as the case when an employee feels that he or she was the victim of discrimination. A formal policy should outline the procedures to be followed and guarantee that the employee's grievances are given just consideration. Every attempt should be made to settle the matter fairly within the company. The loss of a worker is not only costly but is often bad for employee morale. Moreover, legal action taken by an employee can be very expensive for the firm and result in unwanted publicity, regardless of the outcome for the firm. The fact that the company attempts to settle the dispute is in itself a sign of the firm's good will and commitment to its employees, even if an agreement cannot be reached. Every effort should be made to resolve the dispute in a timely manner. Not only will this be appreciated by the affected employee, but it is best for employee morale to put the matter in the past and get on with business as usual.

Employees should be encouraged to report grievances either to their supervisor or to the human resources department or manager. The option to report grievances to the human resources management is particularly important because grievances often involve an employee's supervisor. Every effort should be made to protect the employee's privacy and thereby avoid unnecessary embarrassment for or retribution against the employee. Minor issues or misunderstandings can often be handled immediately by discussing the issue with the employee and explaining the company's position. If the employee is not satisfied with the explanation, then a formal grievance should be filed. The firm's policy should clearly define the process, including who will hear the complaint and make a decision, the appeals process, and any provision for arbitration. An outside arbitrator, mutually agreeable to both sides, can be useful in resolving disputes that cannot be settled within the firm. Furthermore, the use of an arbitrator may avert a costly lawsuit that can be damaging to the firm's image.

Issues of discrimination and sexual harassment require a more formal process because of a potential violation of the employee's civil rights. Serious issues should be handled by someone who is trained to handle such matters to ensure that all applicable laws and regulations are followed. The process starts by asking the employee, in private, to describe in detail exactly what happened. Next, all witnesses are identified and interviewed and all relevant files and documents are reviewed. The interviewer should ask open-ended questions that don't require witnesses to confirm or deny allegations. The investigation should be conducted in private and all statements should be kept confidential. The entire process should be conducted on a need-to-know basis, that is, only information relevant and necessary to conducting the investigation should be asked. The interviewer must remain objective throughout the investigation and keep detailed notes for documentation purposes. After a conclusion has been reached, the interviewer should follow up with the complainant. When disciplinary action is taken against an employee, it should not be divulged, even to the complainant, in order to protect the confidentiality of the disciplined employee.

Progressive Discipline Systems

Progressive discipline is a system where the severity of the penalty increases each time an employee breaks the rules. For example, in the initial instance of an employee breaking the rules, the employee's supervisor might talk with the employee about the improper behavior and issue a verbal warning including the consequences of future occurrences. If the unacceptable behavior is repeated, the next step would be to issue a written warning. A third infraction might result in suspension or termination. The employer should document each infraction and have the employee sign and date it. The employee should be allowed to record his or her disagreement with the procedure on the form. Usually, after a specific time period passes without another infraction, the employee gets a fresh start. A later infraction would start the process again with a verbal warning. Some misconduct is so severe that employers may move directly to suspension or termination. Examples would include fighting or assaults, threatening coworkers, destruction of property, and theft. However, it is generally not a good policy to terminate an employee on the spot, because the situation is often different than it first appears. Employers should make sure that they have a clear picture of what happened and who was responsible before initiating disciplinary action.

PERFORMANCE APPRAISALS

Periodic evaluation is important for all employees; with new employees it is essential. Frequent evaluation of new employees allows the supervisor to recognize and correct unsatisfactory performance before work habits become ingrained. It also demonstrates to the employee that the company is interested in his or her work and therefore that it is important. A probationary period is an effective means of guaranteeing that new employees are properly evaluated and that they receive the proper feedback. An evaluation should be conducted after a fairly short period of time, such as a week or a month, at which time the employee's performance, including both positive and negative aspects, should be evaluated and discussed in person with the new worker. Many employers have found that initial problems are often the result of poor communication and can be easily corrected. Future evaluations should be conducted at less frequent intervals until the end of the probationary period, at which time a final decision on retaining the employee is made. If the employee is not performing satisfactorily after the probationary period, a transfer or dismissal should be considered. Because the company has a sizeable investment in the employee by this time, problem employees are often given a second chance when there is a reasonable hope that they may be successful elsewhere in the firm. Successful employees are often rewarded by a raise that coincides with the change in their status from probationary to permanent.

The performance appraisal process is one of the most important human resources functions because it serves as the basis for many other human resources decisions. Equitable decisions concerning the employee's compensation, promotions, and even transfers and terminations depend on accurate and objective evaluations. It is particularly important to the employee because the evaluation directly affects promotion decisions and salary increases. For the firm, the evaluation is a crucial step in rewarding employees' performance and in motivating them to do their best work. Performance appraisals are also an important source of feedback to the employee. The performance appraisal system's primary functions are: (1) to establish performance criteria, (2) measure past job performance, (3) justify salary, bonuses, and other rewards, and (4) identify the development needs for the employee to improve job performance and prepare for future responsibilities.

The criteria to be used in evaluating an employee should be specified at the beginning of the evaluation period. The criteria may be standard for all employees, determined by the supervisor, or by a joint agreement between the supervisor and the employee. Regardless of which method is used, it is important that the employee be made aware of the criteria that will be used to evaluate him or her and that the criteria accurately measure expected performance.

All employees should be evaluated on a regular basis. Frequent evaluations conducted quarterly or biannually may be most effective because they are often the only source of feedback to the employee. In practice, most food and agribusiness firms evaluate their employees once a year. New employees should be evaluated more frequently than experienced workers because they need more guidance. Because their work habits are being formed, they are more receptive to change.

Each employee should be evaluated using the criteria specified at the beginning of the period and the evaluation should be put in writing. When a common set of criteria is used for all or a group of employees, a standardized form should be used to ensure that employees are evaluated consistently, and that all important points are covered. The form should be flexible enough so that information not on the form, but relevant to the employee's performance, may be reported. The manager should conduct the evaluation when the manager can devote his or her full attention to it. It is important to remember that although managers often regard evaluations as a nuisance, it is very important to the employee, because it affects his or her livelihood. Furthermore, the evaluation and reward process, when properly implemented, can be a powerful motivating tool. Historical performance evaluations also serve as an invaluable source of information to managers selecting employees for promotion, transfers, and special training.

The last step in the process is a frank discussion of the evaluation with the employee. The supervisor should write a memo to the employee's file describing the conversation and have the employee sign and date it. Good performance should be commended. Shortcomings should be discussed in an effort to find their cause and possible solutions. Most employees want to be good employees and they should be given every opportunity to succeed. The employee should be given the opportunity to respond to the evaluation and participate in the discussion of the criteria that will be used for the next performance appraisal period. Allowing employees to participate in the process encourages them to work constructively with their manager rather than viewing him or her as an adversary. Furthermore, the employee is often in the best position to know how his or her performance may be improved.

DETERMINING COMPENSATION

Employee compensation consists of the financial payments and benefits paid to the employee, or on their behalf, in return for the services they provide to the firm. Having a compensation policy that specifies the criteria used in determining employee compensation will ensure that compensation is fairly and consistently administered.

A good place to start in determining the level of employee compensation is to research the pay scale in the local community. The wages paid by similar food and agribusiness firms in the area, or to employees performing comparable work, will provide a general idea of what it will take to attract qualified personnel to work for the company. When highly trained or experienced personnel are not available in the local community, it is necessary to pay a salary competitive with what qualified workers earn elsewhere in the region or nation. Other factors that may influence the wages paid to employees are the desired level of employee turnover, the company's public image, whether or not employees are unionized, and the area's cost of living.

Salary adjustments are typically made each year after the annual performance evaluation and typically include an adjustment based on cost of living and the employee's performance evaluation. The cost-of-living adjustment is awarded to most employees to keep the purchasing power of their salaries in line with inflation. Determining an employee's raise based on their subjective performance evaluation is much more difficult. Many supervisors feel more comfortable giving all employees equal raises. However, most firms find that rewarding superior and mediocre performance equally results in a mediocre work force. Outstanding individuals will be lost to other firms that better reward their workers' efforts. Proper employee performance evaluation and compensation based on the results of the performance evaluation ensures that good performance is adequately rewarded and encouraged.

In addition to the regular salary, additional compensation is sometimes used to reward employees when the company has a good year. Some companies make it a practice to give bonuses as a gift, with the Christmas bonus being the most common example of this.

The *profit-sharing plan,* which ties bonuses to the company's profit level, is a common bonus plan. It encourages employees to work together as a team, because the reward is based not on individual performance, but on the overall performance of the firm. On the other hand, some companies believe that they get the most out of their employees by basing bonuses on individual performance. Many companies now use a hybrid system whereby the bonus pool is determined by the overall performance of the company, but individual performance is used to determine each employee's reward.

Some companies have had success by rewarding employees with stock and stock options in the company. The theory is that employees will work harder and be less wasteful when they have an ownership stake in the company. *Employee stock options* give the employee the option to purchase shares of stock in the company at a specific price for a designated future period of time. The stock options are typically priced at or slightly below the current market price at the time they are given. The intent is to motivate employees to work to improve the firm's profitability and stock price.

Benefits include any compensation offered by the agribusiness other than wages, salary, and bonuses. Most benefits are optional, however, federal and state laws require employers to comply with workers' compensation requirements, pay state and federal unemployment taxes, and pay the employer's share and withhold the employee's share of FICA taxes. Workers' compensation coverage provides for the payment of income and medical expenses when employees have work-related injuries, accidents, illnesses, or diseases. Unemployment taxes provide benefits to unemployed workers, and FICA taxes provide social security and disability benefits to employees.

Employers must allow employees time off to vote, serve on jury duty, and perform military service. In addition, most companies have leave policies that include paid or unpaid holidays, vacation, sick leave, personal leave, funeral leave, and maternity/paternity leave. The *Family and Medical Leave Act* requires employers with 50 or more employees to permit employees to take up to 12 weeks of unpaid leave each year for a birth or adoption or for the care of sick children, spouses, or parents.

Benefits such as retirement plans and life, disability, health, vision, dental, and prescription medication insurance are not required by law but are standard benefits in most large firms. Although the word fringe is often used in conjunction with these and other benefits, most companies have found that a good benefits package is necessary if the firm is to acquire and retain top-notch personnel. Other benefits may include childcare, subsidized lunches, recreational facilities, legal assistance, financial planning, long-term health care insurance, and post-graduate education. Many benefits are tax deductible to the company and tax exempt to the employees.

PROMOTING EMPLOYEES

Another way in which employees are rewarded for good performance is through promotions. One of the most difficult decisions the firm has to make when filling vacant positions is whether to promote one of their own employees or to hire someone from outside the organization.

The *promotion policy* should specify the criteria to be used in promoting employees. The job description is useful in assessing the candidates' qualifications based on the job requirements and the specific responsibilities to be performed. An objective promotion system will help ensure that the best possible candidate is identified and that all candidates are treated equitably. Employees who are passed over for a promotion are understandably displeased. However, a fair, equitably administered system helps to minimize complaints. Such a system also serves to protect the firm should it ever be sued over unfair treatment of or discrimination against an employee.

Many factors must be considered in promoting employees. The employee's past performance is an important consideration, especially when the responsibilities of past jobs and the new job are similar. However, it is a common mistake to assume that a person who has been successful at one job will therefore be successful when promoted to a more important job. This widely held belief is the basis for the Peter Principle that states that a person is eventually promoted to a level at which he or she is no longer competent. It is essential that each candidate be assessed in light of what is expected of him or her in the new job. A promotion will usually mean increased responsibilities. Some employees will be placed in the position of supervising others for the first time. Therefore, such factors as leadership capabilities and the ability to get along with others should be considered. When two candidates appear to be equally well qualified for a job, seniority may become an important consideration. Many people change jobs frequently and those employees that exhibit loyalty to the food or agribusiness company should be rewarded.

MAKING TRANSFERS

Unlike a promotion, which is a move from a job at one level to one at a higher level, a *transfer* is a lateral move between jobs at the same level. Either the employee or the employer may initiate a transfer. Employees request transfers for many reasons, such as the preference of one job over another, to move to another location, or due to a personality conflict with a fellow employee. It makes good sense to grant the employee's request, if it is reasonable and does not represent a large cost or inconvenience to the firm, because an unsatisfied employee is likely to be less productive than one who is satisfied.

Employers initiate transfers to place employees where they will be more productive or because their skill or experience is needed elsewhere. Employees are often asked to transfer when a new plant opens and the experience of employees from other plants is needed. In the case of a problem worker, the firm may salvage its investment in the employee by finding a place where he or she can make a positive contribution to the firm.

HANDLING TERMINATIONS

Terminations take at least three forms:

- Voluntary termination, which occurs when employees leave due to retirement, death, for another job, or other voluntary causes
- Layoff, which may be temporary or permanent
- Dismissal

Some terminations are inevitable, particularly voluntary terminations that are often the result of factors beyond the firm's control. On the other hand, good management of a firm's human resources can help to reduce the number of involuntary terminations.

Layoffs

Due to the seasonal nature of agriculture, food and agribusiness firms are particularly vulnerable to wide fluctuations in sales and production. The number of workers required varies greatly during the year and often between years. Many food and agribusiness companies have a permanent staff that they employ year round and hire temporary help during the peak season. Thus, when production falls below a certain level for a prolonged period of time, it is necessary to lay off some employees to reduce expenses. Needless to say, layoffs should be avoided whenever possible. They are very difficult on employees as well as their families, and the company will find it hard to recruit high-quality employees if it cannot offer them continuous employment. In many areas of the food sector, planning can help to smooth out the peaks and valleys of production during the year, thereby reducing the need for seasonal help and the layoff of permanent employees.

If the company decides to lay off some but not all of its employees, it must be sure that the selection process does not discriminate on the basis of age, sex, or race. When choosing between many equally qualified employees for layoffs, the employer should be prepared to show that the employees remaining in the downsized business reflect the workforce pool of the community. Many agribusinesses choose to lay off employees on the basis of seniority. However, most employers would rather keep their best employees and lay off those who are less productive regardless of seniority. By conducting regular performance reviews, an employer can eliminate the positions of employees whose performance has been documented to be less than satisfactory.

Dismissal

Employees are usually dismissed as a last resort. Every worker should be given a reasonable chance to become a productive member of the work force. Through the evaluation process the employee should have been informed that his or her performance was unsatisfactory and of the consequences of substandard performance. Before an employee is dismissed, the employer should thoroughly investigate the reason for dismissal and have documented the situation using fair rules and procedures. The employer must have a valid, nondiscriminatory business reason for the action coupled with enough documentation to prove it.

Termination Policy

A *termination policy* will help the firm ensure that all employees are treated equitably and in a uniform manner when they leave the firm. A termination policy that is fairly administered will help minimize the negative impact on employee morale. Termination policies typically address issues such as severance pay, continuation of employee benefits, assistance offered by the firm in finding a new job, and other assistance the firm may offer.

Businesses frequently negotiate a *severance agreement* with employees who have been dismissed or permanently laid off. In addition to being a standard industry practice, a negotiated package may prevent future lawsuits and other forms of retaliation. As part of the agreement, the employee should be asked to sign a severance agreement stating that he or she will not hold the company liable for any damages. Many companies permit their employees, especially higher level ones, to resign. To be effective, the severance agreement must be a voluntary waiver, in writing, and signed by the employee.

The termination policy should address the items that may be included in a severance package, including the method for determining the amount of severance pay, the status of the employee's benefits, employment assistance provided by the employer, and the employer's agreement to not contest payment of unemployment benefits and to provide a job reference for the employee. Additionally, the termination policy should indicate how much notice is expected of employees who leave voluntarily.

Severance pay is a lump-sum payment to the dismissed employee that usually covers at least two weeks' pay. A practice followed by many firms is to offer terminated employees one week's pay for every year of service they have with the company. Senior employees are often given more generous severance pay, either by negotiated agreement or based on their contract with the firm.

The employer should pay the dismissed employee for all accumulated vacation at the time of his or her dismissal. Many employers also pay for any unused sick leave. Some companies offer to continue medical and life insurance benefits for a specified period of time.

In recent years it has become common for employers to provide career assistance to employees who have lost their jobs. This may include career counseling, resume preparation, and other assistance to help them find employment. This assistance is normally contracted through private placement services for which the firm agrees to pay.

Exit Interview

Exit interviews should be conducted with all employees that leave the company. The format of the exit interview will depend on why the employee is leaving the firm. Although the discussion in this section is directed toward an exit interview for someone who has been dismissed, because it is the most difficult situation, many of the same principles apply to all exit interviews.

There are five objectives that the interviewer should seek to accomplish in the exit interview. These are:

- To the extent possible, ensure that the employee leaves the firm with a positive attitude.
- Determine the cause of the problem leading to the employee's dismissal.
- Allow the employee to express his or her opinions about the firm.
- Inform the employee of any benefits to which he or she has a right.
- Clear up any loose ends including the collection of keys, credit cards, uniforms, and other company property, and settle expense accounts.

It is important that the exit interview be conducted by an objective third party, especially if the employee is not leaving the firm voluntarily. Most frequently, the person in charge of human resources, or a member of that department, conducts the exit interview.

When the employee has been dismissed, the exit interview will most likely be an unpleasant situation. The employee may be charged with emotion and reluctant to cooperate. It is the interviewer's function to make the most out of a bad situation. Although it might not be possible to ensure that the employee leaves the firm with a positive attitude, at the very minimum the interviewer should try to convince the dismissed worker that he or she was treated fairly. The exit interview, as well as the dismissal, should be done in private and in person. This shows respect for the employee and provides an opportunity for discussion to clear up any questions.

One of the key elements of the exit interview is determining the cause of the problem, to prevent future problems. The interviewer should maintain a neutral role, and be objective both in the asking of questions and recording of answers, if useful information is to be obtained. Open-ended questions such as, "How do you feel your supervisor treated you?" rather than "Did your supervisor treat you fairly?" will be the most productive. The interviewer should question the employee concerning his or her attitudes toward the training, orientation, compensation, and promotions he or she received, as well as his or her perceptions of the firm's policies concerning these items. The interviewer should ensure that the employee understands the reason for the dismissal and the employee should be asked for his or her explanation and interpretation of the events leading up to the dismissal.

The interviewer should explain the severance package including severance pay and any other benefits the company is offering. The severance agreement should be presented to the employee along with a final paycheck. The interviewer should explain the company's job reference policy and collect all items belonging to the company that are in the employee's possession. Company items may include keys, beepers, company car, company credit cards, and work in progress.

The information obtained in the exit interview should be combined with that acquired through other sources, such as supervisors, coworkers, and former employees, and used to improve the management of firm's human resources.

COMMON MISTAKES

One of the most common mistakes, found mostly among small food and agribusiness firms, is the lack of a formal human resources policy that addresses each of the human resources functions. Putting together a good human resources policy requires the manager's commitment and the allocation of sufficient time to properly complete the task. Many managers find themselves too busy handling "emergencies" to develop a comprehensive human resources policy. This usually results in doing the minimum necessary for the organization to function, but does not allow the food or agribusiness firm to get the most out of its people. It also results in a lack of consistency in treating employees, which may lead to complaints of favoritism or discrimination and a generally low morale and motivation on the part of employees. One solution to this problem is to hire an outside consultant specializing in setting up human resources management systems for small businesses. Although this may seem like an expensive alternative in the short run, the potential rewards are increased productivity, improved employee morale, and lower employee turnover.

Another common problem is the failure to give adequate orientation and training to new employees. It is understandable that managers are anxious for new employees to become productive members of the work force as soon as possible. The employee that leaves or has been replaced suddenly leaves a gap that needs to be filled and this places an additional burden on the supervisor and remaining workers. The tendency is to provide only the basic information necessary for the performance of the job, as opposed to a complete orientation and training program, which the employee needs to be productive in the long run. This can lead to costly mistakes for the firm and an employee who feels frustrated and overwhelmed by his or her new job. This problem can be avoided by offering a standard orientation and training program for all new employees and making sure that they have access to a supervisor or coworkers who can offer the necessary guidance.

A related problem is the inclination to provide similar performance appraisals and compensation to employees in similar positions. This often arises out of a desire to avoid conflict or the appearance of showing favoritism, especially when the supervisor must apply subjective standards in the evaluation procedure. However, evaluating and rewarding all employees equally will eventually lower the motivation of all employees, because superior performance is not rewarded. It should also be remembered that such treatment is not fair to employees who have excelled. This problem may be avoided by having clear performance appraisal and compensation standards and procedures. The firm's human resources specialist may render valuable assistance by providing oversight in the evaluation process and determination of compensation.

CHALLENGES FACED BY FOOD AND AGRIBUSINESS FIRMS

Food and agribusiness companies face several human resources challenges due in large part to the biological nature of the products that the industry produces. These challenges include the need for a seasonal labor force, the diversity of the labor force, and the need to ensure the safety of food products.

Seasonal Production

The seasonal nature of agricultural production presents a difficult staffing issue for firms whose production revolves around seasonally produced products. This effect is felt most severely by processors of seasonal products; however, input suppliers, such as equipment or packaging material suppliers, are also affected to some extent.

To address the problem of seasonality in production, the firm should first analyze its production processes to understand what may be done to mitigate the impact of seasonality. Sometimes it is possible to store the raw product for later processing. Dried fruit producers often dry the product to a low enough moisture level so that it can be stored, thereby reducing the need for immediate processing. The judicious planting of varieties that ripen over an extended period of time may also extend the processing season.

Another alternative is to automate processes to reduce the need for manual labor. The major considerations involved in the decision to automate are the relative costs, quality of the finished product, and the difficulty in finding seasonal labor. The higher initial cost of labor-saving equipment is often offset by increased efficiency, lower defect rates, and the certainty that processing won't be delayed because of labor shortages.

When all alternatives to reducing the impact of seasonality have been exhausted, the firm must focus its efforts on effectively managing its seasonal labor force. Many agribusinesses find that they employ the same seasonal workers year after year. Whereas some workers prefer seasonal work, many employees choose seasonal work because they lack other options. For this reason, it is typically most difficult to find seasonal workers when the unemployment rate is low. Agribusinesses that treat their seasonal workers well and establish a good rapport with these workers find it easiest to meet their seasonal staffing needs.

Applicant Pool

Because many of the jobs in the food and agribusiness industry require few if any skills, they are typically at the low end of the wage scale, often paying minimum wage. Extreme competitive pressures, due to the commodity nature of many products, also serve to keep wages low in the industry. These low-paying jobs attract workers that have few employment opportunities, including those workers who are unskilled, speak little or no English, and may lack the proper papers to work in the United States.

Managing such a diverse workforce presents difficult human resources challenges. It is imperative that employers verify each employee's right to work in the United States. Hiring an undocumented worker or failing to keep proper records can subject an employer to criminal prosecution and heavy fines. The guidelines for hiring workers established under the Immigration Reform and Control Act of 1986 were covered in Chapter 10.

Another challenge often faced by food and agribusiness companies, particularly those in high immigration areas, is that many employees may speak little or no English. Furthermore, the workforce may be comprised of people from multiple cultures, whose customs differ greatly from their employers. Many firms have successfully met these challenges by hiring bi- or multi-lingual supervisors and translating key materials into a second or even a third language. Employers who develop an understanding and respect for the different cultures of their employees will typically improve their relationship with their workers and develop a more loyal employee base.

Food Safety

Ensuring the safety of the food products produced by the food and agribusiness industry is of critical importance. Failure to do so can result in injury or death to the consumers of unsafe products and grave consequences for the producing firm. Producing a safe food product requires that employees involved in the production process be well trained and that they follow established safety guidelines.

This challenge is exacerbated when employees lack English language proficiency or when differing customs make implementation of food safety guidelines difficult. Many firms in the food industry have found that training programs conducted in their employees' native language and constant reinforcement of food safety guidelines are the most effective approach to ensuring their products meet food safety standards.

SUMMARY

The function of human resources management is to effectively staff and administer the employees of the food or agribusiness firm. Although not all businesses need a human resources department, it is important for every business to manage its personnel in a consistent and fair manner. Because human resources management is heavily impacted by Equal Employment Opportunity guidelines, it is crucial for human resources managers to understand the implications of the Equal Employment Opportunity regulations and court decisions for hiring, compensation, promotion, and other human resources functions.

The foundation for all other human resources functions is job design and analysis. Job design involves putting together the components of a job in such a way as to maximize employee productivity. Job design no longer stresses maximum efficiency based on time and motion studies and scientific management principles, but increasingly focuses on ways to maximize employee participation and satisfaction as a way to achieve higher levels of performance.

Job analysis involves examining components of a job in order to produce job descriptions, specifications, and the basis for evaluating performance. Job descriptions are written documents detailing job components. Job specifications describe the characteristics required of individuals who perform a job.

Recruiting employees either from inside or outside the organization is necessary to staff positions. Job candidates are asked to fill out job applications, which are screened to ensure that the firm interviews the best-qualified candidates. Those candidates who are hired should be properly oriented and trained. Many agribusinesses have ongoing training and development programs, through which their employees are kept up to date and prepared to accept greater responsibilities.

One of the most important issues currently facing human resources managers is discrimination and harassment in the workplace. Human resources managers should be well versed in this issue to ensure that all employees are treated fairly and with respect. It is important for the firm to establish a "zero tolerance" policy against employee harassment to protect the company from potential lawsuits and maintain a focused, productive workforce.

Employees should be evaluated at least once a year, based upon previously established performance criteria. Compensation increases are typically based on both cost-of-living adjustments and performance evaluations. Employee evaluations are also used to evaluate employees for promotions, although the most important consideration should be the employee's potential performance in the new job.

Grievances should be handled quickly and fairly. Whenever possible the matter should be resolved within the company. Transfers, which are lateral moves within a company, may be requested by the employee or by the employer when it is felt that the employee may be better utilized in a different position or location. Terminations may be both voluntary and involuntary as in the case of layoffs and dismissals. The termination policy should spell out both the employee's and employer's rights and responsibilities. Upon leaving the firm, all employees should be given an exit interview.

Food and agribusiness firms face several unique human resources challenges. These include dealing with a seasonal labor force due to the seasonal nature of production, a diverse workforce often characterized by low skill levels and lack of English language proficiency, and the difficult task of ensuring that their workers follow established food safety guidelines.

CASE QUESTIONS

After rereading the case at the front of the chapter, it should be apparent that Diane has no human resources management program. Although the information in the case is limited, you should nonetheless be able to recognize several problems.

1. Based on the information you are given, what are some of the possible mistakes that Diane has made?
2. Design a basic personnel management program for Diane for each of the human resources management functions relevant to her company. Your discussion should include the names of the functional areas and the specific steps that Diane should take in each area.

REVIEW QUESTIONS

1. Why is it important that one person be responsible for the human resources management in the firm?
2. Select a job where you are currently employed or have worked in the past, and redesign it using modern job design concepts discussed in this chapter.
3. Prepare a job description for a position in a local food or agribusiness firm or for a position you have held in the past. Be sure to address all nine factors discussed in the chapter.
4. Choose a position in a food or agribusiness firm. What sources would you use for recruiting employees for that position? Why?
5. Describe the two basic approaches to training new employees. In what situation is each approach most appropriate?
6. Why is it important to evaluate candidates for promotion based on criteria for the new job as opposed to performance on the current job?
7. Design an interview guide for an exit interview.
8. What constitutes sexual harassment?
9. How should a human resources manager conduct an investigation into an allegation of sexual harassment?
10. Which employee benefits are required by law and which are optional? Why is it important for a food or agribusiness firm to provide optional benefits?

APPENDIX

TABLE A–1

Future Value of $1

Future Value = $(1 + i)^n$

Number of Periods	1%	2%	3%	4%	5%	6%	7%	8%	9%	10%
1	1.0100	1.0200	1.0300	1.0400	1.0500	1.0600	1.0700	1.0800	1.0900	1.1000
2	1.0201	1.0404	1.0609	1.0816	1.1025	1.1236	1.1449	1.1664	1.1881	1.2100
3	1.0303	1.0612	1.0927	1.1249	1.1576	1.1910	1.2250	1.2597	1.2950	1.3310
4	1.0406	1.0824	1.1255	1.1699	1.2155	1.2625	1.3108	1.3605	1.4116	1.4641
5	1.0510	1.1041	1.1593	1.2167	1.2763	1.3382	1.4026	1.4693	1.5386	1.6105
6	1.0615	1.1262	1.1941	1.2653	1.3401	1.4185	1.5007	1.5869	1.6771	1.7716
7	1.0721	1.1487	1.2299	1.3159	1.4071	1.5036	1.6058	1.7138	1.8280	1.9487
8	1.0829	1.1717	1.2668	1.3686	1.4775	1.5938	1.7182	1.8509	1.9926	2.1436
9	1.0937	1.1951	1.3048	1.4233	1.5513	1.6895	1.8385	1.9990	2.1719	2.3579
10	1.1046	1.2190	1.3439	1.4802	1.6289	1.7908	1.9672	2.1589	2.3674	2.5937
11	1.1157	1.2434	1.3842	1.5395	1.7103	1.8983	2.1049	2.3316	2.5804	2.8531
12	1.1268	1.2682	1.4258	1.6010	1.7959	2.0122	2.2522	2.5182	2.8127	3.1384
13	1.1381	1.2936	1.4685	1.6651	1.8856	2.1329	2.4098	2.7196	3.0658	3.4523
14	1.1495	1.3195	1.5126	1.7317	1.9799	2.2609	2.5785	2.9372	3.3417	3.7975
15	1.1610	1.3459	1.5580	1.8009	2.0789	2.3966	2.7590	3.1722	3.6425	4.1772
16	1.1726	1.3728	1.6047	1.8730	2.1829	2.5404	2.9522	3.4259	3.9703	4.5950
17	1.1843	1.4002	1.6528	1.9479	2.2920	2.6928	3.1588	3.7000	4.3276	5.0545
18	1.1961	1.4282	1.7024	2.0258	2.4066	2.8543	3.3799	3.9960	4.7171	5.5599
19	1.2081	1.4568	1.7535	2.1068	2.5270	3.0256	3.6165	4.3157	5.1417	6.1159
20	1.2202	1.4859	1.8061	2.1911	2.6533	3.2071	3.8697	4.6610	5.6044	6.7275
21	1.2324	1.5157	1.8603	2.2788	2.7860	3.3996	4.1406	5.0338	6.1088	7.4002
22	1.2447	1.5460	1.9161	2.3699	2.9253	3.6035	4.4304	5.4365	6.6586	8.1403
23	1.2572	1.5769	1.9736	2.4647	3.0715	3.8197	4.7405	5.8715	7.2579	8.9543
24	1.2697	1.6084	2.0328	2.5633	3.2251	4.0489	5.0724	6.3412	7.9111	9.8497
25	1.2824	1.6406	2.0938	2.6658	3.3864	4.2919	5.4274	6.8485	8.6231	10.8347
26	1.2953	1.6734	2.1566	2.7725	3.5557	4.5494	5.8074	7.3964	9.3992	11.9182
27	1.3082	1.7069	2.2213	2.8834	3.7335	4.8223	6.2139	7.9881	10.2451	13.1100
28	1.3213	1.7410	2.2879	2.9987	3.9201	5.1117	6.6488	8.6271	11.1671	14.4210
29	1.3345	1.7758	2.3566	3.1187	4.1161	5.4184	7.1143	9.3173	12.1722	15.8631
30	1.3478	1.8114	2.4273	3.2434	4.3219	5.7435	7.6123	10.0627	13.2677	17.4494
35	1.4166	1.9999	2.8139	3.9461	5.5160	7.6861	10.6766	14.7853	20.4140	28.1024
40	1.4889	2.2080	3.2620	4.8010	7.0400	10.2857	14.9745	21.7245	31.4094	45.2593
45	1.5648	2.4379	3.7816	5.8412	8.9850	13.7646	21.0025	31.9204	48.3273	72.8905
50	1.6446	2.6916	4.3839	7.1067	11.4674	18.4202	29.4570	46.9016	74.3575	117.3909

11%	12%	13%	14%	15%	16%	17%	18%	19%	20%
1.1100	1.1200	1.1300	1.1400	1.1500	1.1600	1.1700	1.1800	1.1900	1.2000
1.2321	1.2544	1.2769	1.2996	1.3225	1.3456	1.3689	1.3924	1.4161	1.4400
1.3676	1.4049	1.4429	1.4815	1.5209	1.5609	1.6016	1.6430	1.6852	1.7280
1.5181	1.5735	1.6305	1.6890	1.7490	1.8106	1.8739	1.9388	2.0053	2.0736
1.6851	1.7623	1.8424	1.9254	2.0114	2.1003	2.1924	2.2878	2.3864	2.4883
1.8704	1.9738	2.0820	2.1950	2.3131	2.4364	2.5652	2.6996	2.8398	2.9860
2.0762	2.2107	2.3526	2.5023	2.6600	2.8262	3.0012	3.1855	3.3793	3.5832
2.3045	2.4760	2.6584	2.8526	3.0590	3.2784	3.5115	3.7589	4.0214	4.2998
2.5580	2.7731	3.0040	3.2519	3.5179	3.8030	4.1084	4.4355	4.7854	5.1598
2.8394	3.1058	3.3946	3.7072	4.0456	4.4114	4.8068	5.2338	5.6947	6.1917
3.1518	3.4785	3.8359	4.2262	4.6524	5.1173	5.6240	6.1759	6.7767	7.4301
3.4985	3.8960	4.3345	4.8179	5.3503	5.9360	6.5801	7.2876	8.0642	8.9161
3.8833	4.3635	4.8980	5.4924	6.1528	6.8858	7.6987	8.5994	9.5964	10.6993
4.3104	4.8871	5.5348	6.2613	7.0757	7.9875	9.0075	10.1472	11.4198	12.8392
4.7846	5.4736	6.2543	7.1379	8.1371	9.2655	10.5387	11.9737	13.5895	15.4070
5.3109	6.1304	7.0673	8.1372	9.3576	10.7480	12.3303	14.1290	16.1715	18.4884
5.8951	6.8660	7.9861	9.2765	10.7613	12.4677	14.4265	16.6722	19.2441	22.1861
6.5436	7.6900	9.0243	10.5752	12.3755	14.4625	16.8790	19.6733	22.9005	26.6233
7.2633	8.6128	10.1974	12.0557	14.2318	16.7765	19.7484	23.2144	27.2516	31.9480
8.0623	9.6463	11.5231	13.7435	16.3665	19.4608	23.1056	27.3930	32.4294	38.3376
8.9492	10.8038	13.0211	15.6676	18.8215	22.5745	27.0336	32.3238	38.5910	46.0051
9.9336	12.1003	14.7138	17.8610	21.6447	26.1864	31.6293	38.1421	45.9233	55.2061
11.0263	13.5523	16.6266	20.3616	24.8915	30.3762	37.0062	45.0076	54.6487	66.2474
12.2392	15.1786	18.7881	23.2122	28.6252	35.2364	43.2973	53.1090	65.0320	79.4968
13.5855	17.0001	21.2305	26.4619	32.9190	40.8742	50.6578	62.6686	77.3881	95.3962
15.0799	19.0401	23.9905	30.1666	37.8568	47.4141	59.2697	73.9490	92.0918	114.4755
16.7386	21.3249	27 1093	34.3899	43.5353	55.0004	69.3455	87.2598	109.5893	137.3706
18.5799	23.8839	30.6335	39.2045	50.0656	63.8004	81.1342	102.9666	130.4112	164.8447
20.6237	26.7499	34.6158	44.6931	57.5755	74.0085	94.9271	121.5005	155.1893	197.8136
22.8923	29.9599	39.1159	50.9502	66.2118	85.8499	111.0647	143.3706	184.6753	237.3763
38.5749	52.7996	72.0685	98.1002	133.1755	180.3141	243.5035	327.9973	440.7006	590.6682
65.0009	93.0510	132.7816	188.8835	267.8635	378.7212	533.8687	750.3783	1051.6675	1469.7716
109.5302	163.9876	244.6414	363.6791	538.7693	795.4438	1170.4794	1716.6839	2509.6506	3657.2620
184.5648	289.0022	450.7359	700.2330	1083.6574	1670.7038	2566.2153	3927.3569	5988.9139	9100.4382

TABLE A–2

Present Value of $1

$$\text{Present Value} = \frac{1}{(1+i)^n}$$

Number of Periods	1%	2%	3%	4%	5%	6%	7%	8%	9%
1	0.9901	0.9804	0.9709	0.9615	0.9524	0.9434	0.9346	0.9259	0.9174
2	0.9803	0.9612	0.9426	0.9246	0.9070	0.8900	0.8734	0.8573	0.8417
3	0.9706	0.9423	0.9151	0.8890	0.8638	0.8396	0.8163	0.7938	0.7722
4	0.9610	0.9238	0.8885	0.8548	0.8227	0.7921	0.7629	0.7350	0.7084
5	0.9515	0.9057	0.8626	0.8219	0.7835	0.7473	0.7130	0.6806	0.6499
6	0.9420	0.8880	0.8375	0.7903	0.7462	0.7050	0.6663	0.6302	0.5963
7	0.9327	0.8706	0.8131	0.7599	0.7107	0.6651	0.6227	0.5835	0.5470
8	0.9235	0.8535	0.7894	0.7307	0.6768	0.6274	0.5820	0.5403	0.5019
9	0.9143	0.8368	0.7664	0.7026	0.6446	0.5919	0.5439	0.5002	0.4604
10	0.9053	0.8203	0.7441	0.6756	0.6139	0.5584	0.5083	0.4632	0.4224
11	0.8963	0.8043	0.7224	0.6496	0.5847	0.5268	0.4751	0.4289	0.3875
12	0.8874	0.7885	0.7014	0.6246	0.5568	0.4970	0.4440	0.3971	0.3555
13	0.8787	0.7730	0.6810	0.6006	0.5303	0.4688	0.4150	0.3677	0.3262
14	0.8700	0.7579	0.6611	0.5775	0.5051	0.4423	0.3878	0.3405	0.2992
15	0.8613	0.7430	0.6419	0.5553	0.4810	0.4173	0.3624	0.3152	0.2745
16	0.8528	0.7284	0.6232	0.5339	0.4581	0.3936	0.3387	0.2919	0.2519
17	0.8444	0.7142	0.6050	0.5134	0.4363	0.3714	0.3166	0.2703	0.2311
18	0.8360	0.7002	0.5874	0.4936	0.4155	0.3503	0.2959	0.2502	0.2120
19	0.8277	0.6864	0.5703	0.4746	0.3957	0.3305	0.2765	0.2317	0.1945
20	0.8195	0.6730	0.5537	0.4564	0.3769	0.3118	0.2584	0.2145	0.1784
21	0.8114	0.6598	0.5375	0.4388	0.3589	0.2942	0.2415	0.1987	0.1637
22	0.8034	0.6468	0.5219	0.4220	0.3418	0.2775	0.2257	0.1839	0.1502
23	0.7954	0.6342	0.5067	0.4057	0.3256	0.2618	0.2109	0.1703	0.1378
24	0.7876	0.6217	0.4919	0.3901	0.3101	0.2470	0.1971	0.1577	0.1264
25	0.7798	0.6095	0.4776	0.3751	0.2953	0.2330	0.1842	0.1460	0.1160
26	0.7720	0.5976	0.4637	0.3607	0.2812	0.2198	0.1722	0.1352	0.1064
27	0.7644	0.5859	0.4502	0.3468	0.2678	0.2074	0.1609	0.1252	0.0976
28	0.7568	0.5744	0.4371	0.3335	0.2551	0.1956	0.1504	0.1159	0.0895
29	0.7493	0.5631	0.4243	0.3207	0.2429	0.1846	0.1406	0.1073	0.0822
30	0.7419	0.5521	0.4120	0.3083	0.2314	0.1741	0.1314	0.0994	0.0754
35	0.7059	0.5000	0.3554	0.2534	0.1813	0.1301	0.0937	0.0676	0.0490
40	0.6717	0.4529	0.3066	0.2083	0.1420	0.0972	0.0668	0.0460	0.0318
45	0.6391	0.4102	0.2644	0.1712	0.1113	0.0727	0.0476	0.0313	0.0207
50	0.6080	0.3715	0.2281	0.1407	0.0872	0.0543	0.0339	0.0213	0.0134

10%	11%	12%	13%	14%	15%	16%	17%	18%	19%	20%
0.9091	0.9009	0.8929	0.8850	0.8772	0.8696	0.8621	0.8547	0.8475	0.8403	0.8333
0.8264	0.8116	0.7972	0.7831	0.7695	0.7561	0.7432	0.7305	0.7182	0.7062	0.6944
0.7513	0.7312	0.7118	0.6931	0.6750	0.6575	0.6407	0.6244	0.6086	0.5934	0.5787
0.6830	0.6587	0.6355	0.6133	0.5921	0.5718	0.5523	0.5337	0.5158	0.4987	0.4823
0.6209	0.5935	0.5674	0.5428	0.5194	0.4972	0.4761	0.4561	0.4371	0.4190	0.4019
0.5645	0.5346	0.5066	0.4803	0.4556	0.4323	0.4104	0.3898	0.3704	0.3521	0.3349
0.5132	0.4817	0.4523	0.4251	0.3996	0.3759	0.3538	0.3332	0.3139	0.2959	0.2791
0.4665	0.4339	0.4039	0.3762	0.3506	0.3269	0.3050	0.2848	0.2660	0.2487	0.2326
0.4241	0.3909	0.3606	0.3329	0.3075	0.2843	0.2630	0.2434	0 2255	0.2090	0.1938
0.3855	0.3522	0.3220	0.2946	0.2697	0.2472	0.2267	0.2080	0.1911	0.1756	0.1615
0.3505	0.3173	0.2875	0.2607	0.2366	0.2149	0.1954	0.1778	0.1619	0.1476	0.1346
0.3186	0.2858	0.2567	0.2307	0.2076	0.1869	0.1685	0.1520	0.1372	0.1240	0.1122
0.2897	0.2575	0.2292	0.2042	0.1821	0.1625	0.1452	0.1299	0.1163	0.1042	0.0935
0.2633	0.2320	0.2046	0.1807	0.1597	0.1413	0.1252	0.1110	0.0985	0.0876	0.0779
0.2394	0.2090	0.1827	0.1599	0.1401	0.1229	0.1079	0.0949	0.0835	0.0736	0.0649
0.2176	0.1883	0.1631	0.1415	0.1229	0.1069	0.0930	0.0811	0.0708	0.0618	0.0541
0.1978	0.1696	0.1456	0.1252	0.1078	0.0929	0.0802	0.0693	0.0600	0.0520	0.0451
0.1799	0.1528	0.1300	0.1108	0.0946	0.0808	0.0691	0.0592	0.0508	0.0437	0.0376
0.1635	0.1377	0.1161	0.0981	0.0829	0.0703	0.0596	0.0506	0.0431	0.0367	0.0313
0.1486	0.1240	0.1037	0.0868	0.0728	0.0611	0.0514	0.0433	0.0365	0.0308	0.0261
0.1351	0.1117	0.0926	0.0768	0.0638	0.0531	0.0443	0.0370	0.0309	0.0259	0.0217
0.1228	0.1007	0.0826	0.0680	0.0560	0.0462	0.0382	0.0316	0.0262	0.0218	0.0181
0.1117	0.0907	0.0738	0.0601	0.0491	0.0402	0.0329	0.0270	0.0222	0.0183	0.0151
0.1015	0.0817	0.0659	0.0532	0.0431	0.0349	0.0284	0.0231	0.0188	0.0154	0.0126
0.0923	0.0736	0.0588	0.0471	0.0378	0.0304	0.0245	0.0197	0.0160	0.0129	0.0105
0.0839	0.0663	0.0525	0.0417	0.0331	0.0264	0.0211	0.0169	0.0135	0.0109	0.0087
0.0763	0.0597	0.0469	0.0369	0.0291	0.0230	0.0182	0.0144	0.0115	0.0091	0.0073
0.0693	0.0538	0.0419	0.0326	0.0255	0.0200	0.0157	0.0123	0.0097	0.0077	0.0061
0.0630	0.0485	0.0374	0.0289	0.0224	0.0174	0.0135	0.0105	0.0082	0.0064	0.0051
0.0573	0.0437	0.0334	0.0256	0.0196	0.0151	0.0116	0.0090	0.0070	0.0054	0.0042
0.0356	0.0259	0.0189	0.0139	0.0102	0.0075	0.0055	0.0041	0.0030	0.0023	0.0017
0.0221	0.0154	0.0107	0.0075	0.0053	0.0037	0.0026	0.0019	0.0013	0.0010	0.0007
0.0137	0.0091	0.0061	0.0041	0.0027	0.0019	0.0013	0.0009	0.0006	0.0004	0.0003
0.0085	0.0054	0.0035	0.0022	0.0014	0.0009	0.0006	0.0004	0.0003	0.0002	0.0001

TABLE A–3

Future Value of an Ordinary Annuity Payment of $1 for n Periods

$$FVA = \frac{(1 + i)^n - 1}{i}$$

Number of Periods	1%	2%	3%	4%	5%	6%	7%	8%	9%	10%
1	1.0000	1.0000	1.0000	1.0000	1.0000	1.0000	1.0000	1.0000	1.0000	1.0000
2	2.0100	2.0200	2.0300	2.0400	2.0500	2.0600	2.0700	2.0800	2.0900	2.1000
3	3.0301	3.0604	0.0909	3.1216	3.1525	3.1836	3.2149	3.2464	3.2781	3.3100
4	4.0604	4.1216	4.1836	4.2465	4.3101	4.3746	4.4399	4.5061	4.5731	4.6410
5	5.1010	5.2040	5.3091	5.4163	5.5256	5.6371	5.7507	5.8666	5.9847	6.1051
6	6.1520	6.3081	6.4684	6.6330	6.8019	6.9753	7.1533	7.3359	7.5233	7.7156
7	7.2135	7.4343	7.6625	7.8983	8.1420	8.3938	8.6540	8.9228	9.2004	9.4872
8	8.2857	8.5830	8.8923	9.2142	9.5491	9.8975	10.2598	10.6366	11.0285	11.4359
9	9.3685	9.7546	10.1591	10.5828	11.0266	11.4913	11.9780	12.4876	13.0210	13.5795
10	10.4622	10.9497	11.4639	12.0061	12.5779	13.1808	13.8164	14.4866	15.1929	15.9374
11	11.5668	12.1687	12.8078	13.4864	14.2068	14.9716	15.7836	16.6455	17.5603	18.5312
12	12.6825	13.4121	14.1920	15.0258	15.9171	16.8699	17.8855	18.9771	20.1407	21.3843
13	13.8093	14.6803	15.6178	16.6268	17.7130	18.8821	20.1406	21.4953	22.9534	24.5227
14	14.9474	15.9739	17.0863	18.2919	19.5986	21.0151	22.5505	24.2149	26.0192	27.9750
15	16.0969	17.2934	18.5989	20.0236	21.5786	23.2760	25.1290	27.1521	29.3609	31.7725
16	17.2579	18.6393	20.1569	21.8245	23.6575	25.6725	27.8881	30.3243	33.0034	35.9497
17	18.4304	20.0121	21.7616	23.6975	25.8404	28.2129	30.8402	33.7502	36.9737	40.5447
18	19.6147	21.4123	23.4144	25.6454	28.1324	30.9057	33.9990	37.4502	41.3013	45.5992
19	10.8109	22.8406	25.1169	27.6712	30.5390	33.7600	37.3790	41.4463	46.0185	51.1591
20	22.0190	24.2974	26.8704	29.7781	33.0660	36.7856	40.9955	45.7620	51.1601	57.2750
21	23.2392	25.7833	28.6765	31.9692	35.7193	39.9927	44.8652	50.4229	56.7645	64.0025
22	24.4716	27.2990	30.5368	34.2480	38.5052	43.3923	49.0057	55.4568	62.8733	71.4027
23	25.7163	28.8450	32.4529	36.6179	41.4305	46.9958	53.4361	60.8933	69.5319	79.5430
24	26.9735	30.4219	34.4265	39.0826	44.5020	50.8156	58.1767	66.7648	76.7898	88.4973
25	28.2432	32.0303	36.4593	41.6459	47.7271	54.8645	63.2490	73.1059	84.7009	98.3471
26	29.5256	33.6709	38.5530	44.3117	51.1135	59.1564	68.6765	79.9544	93.3240	109.1818
27	30.8209	35.3443	40.7096	47.0842	54.6691	63.7058	74.4838	87.3508	102.7231	121.0999
28	32.1291	37.0512	42.9309	49.9676	58.4026	68.5281	80.6977	95.3388	112.9682	134.2099
29	33.4504	38.7922	45.2189	52.9663	62.3227	73.6398	87.3465	103.9659	124.1354	148.6309
30	34.7849	40.5681	47.5754	56.0849	66.4388	79.0582	94.4608	113.2832	136.3075	164.4940
35	41.6603	49.9945	60.4621	73.6522	90.3203	111.4348	138.2369	172.3168	215.7108	271.0244
40	48.8864	60.4020	75.4013	95.0255	120.7998	154.7620	199.6351	259.0565	337.8824	442.5926
45	56.4811	71.8927	92.7199	121.0294	159.7002	212.7435	285.7493	386.5056	525.8587	718.9048
50	64.4632	84.5794	112.7969	152.6671	209.3480	290.3359	406.5289	573.7702	815.0836	1163.9085

11%	12%	13%	14%	15%	16%	17%	18%	19%	20%
1.0000	1.0000	1.0000	1.0000	1.0000	1.0000	1.0000	1.0000	1.0000	1.0000
2.1100	2.1200	2.1300	2.1400	2.1500	2.1600	2.1700	2.1800	2.1900	2.2000
3.3421	3.3744	3.4069	3.4396	3.4725	3.5056	3.5389	3.5724	3.6061	3.6400
4.7097	4.7793	4.8498	4.9211	4.9934	5.0665	5.1405	5.2154	5.2913	5.3680
6.2278	6.3528	6.4803	6.6101	6.7424	6.8771	7.0144	7.1542	7.2966	7.4416
7.9129	8.1152	8.3227	8.5355	8.7537	8.9775	9.2068	9.4420	9.6830	9.9299
9.7833	10.0890	10.4047	10.7305	11.0668	11.4139	11.7720	12.1415	12.5227	12.9159
11.8594	12.2997	12.7573	13.2328	13.7268	14.2401	14.7733	15.3270	15.9020	16.4991
14.1640	14.7757	15.4157	16.0853	16.7858	17.5185	18.2847	19.0859	19.9234	20.7989
16.7220	17.5487	18.4197	19.3373	20.3037	21.3215	22.3931	23.5213	24.7089	25.9587
19.5614	20.6546	21.8143	23.0445	24.3493	25.7329	27.1999	28.7551	30.4035	32.1504
22.7132	24.1331	25.6502	27.2707	29.0017	30.8502	32.8239	34.9311	37.1802	39.5805
26.2116	28.0291	29.9847	32.0887	34.3519	36.7862	39.4040	42.2187	45.2445	48.4966
30.0949	32.3926	34.8827	37.5811	40.5047	43.6720	47.1027	50.8180	54.8409	59.1959
34.4054	37.2797	40.4175	43.8424	47.5804	51.6595	56.1101	60.9653	66.2607	72.0351
39.1899	42.7533	46.6717	50.9804	55.7175	60.9250	66.6488	72.9390	79.8502	87.4421
44.5008	48.8837	53.7391	59.1176	65.0751	71.6730	78.9792	87.0680	96.0218	105.9306
50.3959	55.7497	61.7251	68.3941	75.8364	84.1407	93.4056	103.7403	115.2659	128.1167
56.9395	63.4397	70.7494	78.9692	88.2118	98.6032	110.2846	123.4135	138.1664	154.7400
64.2028	72.0524	80.9468	91.0249	102.4436	115.3797	130.0329	146.6280	165.4180	186.6880
72.2651	81.6987	92.4699	104.7684	118.8101	134.8405	153.1385	174.0210	197.8474	225.0256
81.2143	92.5026	105.4910	120.4360	137.6316	157.4150	180.1721	206.3448	236.4385	271.0307
91.1479	104.6029	120.2048	138.2970	159.2764	183.6014	211.8013	244.4868	282.3618	326.2369
102.1742	118.1552	136.8315	158.6586	184.1678	213.9776	248.8076	289.4945	337.0105	392.4842
114.4133	133.3339	155.6196	181.8708	212.7930	249.2140	292.1049	342.6035	402.0425	471.9811
127.9988	150.3339	176.8501	208.3327	245.7120	290.0883	342.7627	405.2721	479.4306	567.3773
143.0786	169.3740	200.8406	238.4993	283.5688	337.5024	402.0323	479.2211	571.5224	681.8528
159.8173	190.6989	227.9499	272.8892	327.1041	392.5028	471.3778	566.4809	681.1116	819.2233
178.3972	214.5828	258.5834	312.0937	377.1697	456.3032	552.5121	669.4475	811.5228	984.0680
199.0209	241.3327	293.1992	356.7868	434.7451	530.3117	647.4391	790.9480	966.7122	1181.8816
341.5896	431.6635	546.6808	693.5727	881.1702	1120.7130	1426.4910	1816.6516	2314.2137	2948.3411
581.8261	767.0914	1013.7042	1342.0251	1779.0903	2360.7572	3134.5218	4163.2130	5529.8290	7343.8578
986.6386	1358.2300	1874.1646	2590.5648	3585.1285	4965.2739	6879.2907	9531.5771	13203.4242	18281.3099
1668.7712	2400.0182	3459.5071	4994.5213	7217.7163	10435.6488	15089.5017	21813.0937	31515.3363	45497.1908

TABLE A–4

Present Value of an Ordinary Annuity Payment of $1 for n Periods

$$PVA = \frac{1 - \frac{1}{(1+i)^n}}{i}$$

Number of Periods	1%	2%	3%	4%	5%	6%	7%	8%	9%
1	0.9901	0.9804	0.9709	0.9615	0.9524	0.9434	0.9346	0.9259	0.9174
2	1.9704	1.9416	1.9135	1.8861	1.8594	1.8334	1.8080	1.7833	1.7591
3	2.9410	2.8839	2.8286	2.7751	2.7232	2.6730	2.6243	2.5771	2.5313
4	3.9020	3.8077	3.7171	3.6299	3.5460	3.4651	3.3872	3.3121	3.2397
5	4.8534	4.7135	4.5797	4.4518	4.3295	5.2124	4.1002	3.9927	3.8897
6	5.7955	5.6014	5.4172	5.2421	5.0757	4.9173	4.7665	4.6229	4.4859
7	6.7282	6.4720	6.2303	6.0021	5.7864	5.5824	5.3893	5.2064	5.0330
8	7.6517	7.3255	7.0197	6.7327	6.4632	6.2098	5.9713	5.7466	5.5348
9	8.5660	8.1622	7.7861	7.4353	7.1078	6.8017	6.5152	6.2469	5.9952
10	9.4713	8.9826	8.5302	8.1109	7.7217	7.3601	7.0236	6.7101	6.4177
11	10.3676	9.7868	9.2526	8.7605	8.3064	7.8869	7.4987	7 1390	6.8052
12	11.2551	10.5753	9.9540	9.3851	8.8633	8.3838	7.9427	7.5361	7.1607
13	12.1337	11.3484	10.6350	9.9856	9.3936	8.8527	8.3577	7.9038	7.4869
14	13.0037	12.1062	11.2961	10.5631	9.8986	9.2950	8.7455	8.2442	7.7862
15	13.8651	12.8493	11.9379	11.1184	10.3797	9.7122	9.1079	8.5595	8.0607
16	14.7179	13.5777	12.5611	11.6523	10.8378	10.1059	9.4466	8.8514	8.3126
17	15.5623	14.2929	13.1661	12.1657	11.2741	10.4773	9.7632	9.1216	8.5436
18	16.3983	14.9920	13.7535	12.6593	11.6896	10.8276	10.0591	9.3719	8.7556
19	17.2260	15.6785	14.3238	13.1339	12.0853	11.1581	10.3356	9.6036	8.9501
20	18.0456	16.3514	14.8775	13.5903	12.4622	11.4699	10.5940	9.8181	9.1285
21	18.8570	17.0112	15.4150	14.0292	12.8212	11.7641	10.8355	10.0168	9.2922
22	19.6604	17.6580	15.9369	14.4511	13.1630	12.0416	11.0612	10.2007	9.4424
23	20.4558	18.2922	16.4436	14.8568	13.4886	12.3034	11.2722	10.3711	9.5802
24	21.2434	18.9139	16.9355	15.2470	13.7986	12.5504	11.4693	10.5288	9.7066
25	22.0232	19.5235	17.4131	15.6221	14.0939	12.7834	11.6536	10.6748	9.8226
26	22.7952	20.1210	17.8768	15.9828	14.3752	13.0032	11.8258	10.8100	9.9290
27	23.5596	20.7069	18.3270	16.3296	14.6430	13.2105	11.9867	10.9352	10.0266
28	24.3164	21.2813	18.7641	16.6631	14.8981	13.4062	12.1371	11.0511	10.1161
29	25.0658	21.8444	19.1885	16.9837	15.1411	13.5907	12.2777	11.1584	10.1983
30	25.8077	22.3965	19.6004	17.2920	15.3725	13.7648	12.4090	11.2578	10.2737
35	29.4086	24.9986	21.4872	18.6646	16.3742	14.4982	12.9477	11.6546	10.5668
40	32.8347	27.3555	23.1148	19.7928	17.1591	15.0463	13.3317	11.9246	10.7574
45	36.0945	29.4902	24.5187	20.7200	17.7741	15.4558	13.6055	12.1084	10.8812
50	39.1961	31.4236	25.7298	21.4822	18.2559	15.7619	13.8007	12.2335	10.9617

10%	11%	12%	13%	14%	15%	16%	17%	18%	19%	20%
0.9091	0.9009	0.8929	0.8850	0.8772	0.8696	0.8621	0.8547	0.8475	0.8403	0.8333
1.7355	1.7125	1.6901	1.6681	1.6467	1.6257	1.6052	1.5852	1.5656	1.5465	1.5278
2.4869	2.4437	2.4018	2.3612	2.3216	2.2832	2.2459	2.2096	2.1743	2.1399	2.1065
3.1699	3.1024	3.0373	2.9745	2.9137	2.8550	2.7982	2.7432	2.6901	2.6386	2.5887
3.7908	3.6959	3.6048	3.5172	3.4331	3.3522	3.2743	3.1993	3.1272	3.0576	2.9906
4.3553	4.2305	4.1114	3.9975	3.8887	3.7845	3.6847	3.5892	3.4976	3.4098	3.3255
4.8684	4.7122	4.5638	4.4226	4.2883	4.1604	4.0386	3.9224	3.8115	3.7057	3.6046
5.3349	5.1461	4.9676	4.7988	4.6389	4.4873	4.3436	4.2072	4.0776	3.9544	3.8372
5.7590	5.5370	5.3282	5.1317	4.9464	4.7716	4.6065	4.4506	4.3030	4.1633	4.0310
6.1446	5.8892	5.6502	5.4262	5.2161	5.0188	4.8332	4.6586	4.4941	4.3389	4.1925
6.4951	6.2065	5.9377	5.6869	5.4527	5.2337	5.0286	4.8364	4.6560	4.4865	4.3271
6.8137	6.4924	6.1944	5.9176	5.6603	5.4206	5.1971	4.9884	4.7932	4.6105	4.4392
7.1034	6.7499	6.4235	6.1218	5.8424	5.5831	5.3423	5.1183	4.9095	4.7147	4.5327
7.3667	6.9819	6.6282	6.3025	6.0021	5.7245	5.4675	5.2293	5.0081	4.8023	4.6106
7.6061	7.1909	6.8109	6.4624	6.1422	5.8474	5.5755	5.3242	5.0916	4.8759	4.6755
7.8237	7.3792	6.9740	6.6039	6.2651	5.9542	5.6685	5.4053	5.1624	4.9377	4.7296
8.0216	7.5488	7.1196	6.7291	6.3729	6.0472	5.7487	5.4746	5.2223	4.9897	4.7746
8.2014	7.7016	7.2497	6.8399	6.4674	6.1280	5.8178	5.5339	5.2732	5.0333	4.8122
8.3649	7.8393	7.3658	6.9380	6.5504	6.1982	5.8775	5.5845	5.3162	5.0700	4.8435
8.5136	7.9633	7.4694	7.0248	6.6231	6.2593	5.9288	5.6278	5.3527	5.1009	4.8696
8.6487	8.0751	7.5620	7.1016	6.6870	6.3125	5.9731	5.6648	5.3837	5.1268	4.8913
8.7715	8.1757	7.6446	7.1695	6.7429	6.3587	6.0113	5.6964	5.4099	5.1486	4.9094
8.8832	8.2664	7.7184	7.2297	6.7921	6.3988	6.0442	5.7234	5.4321	5.1668	4.9245
8.9847	8.3481	7.7843	7.2829	6.8351	6.4338	6.0726	5.7465	5.4509	5.1822	4.9371
9.0770	8.4217	7.8431	7.3300	6.8729	6.4641	6.0971	5.7662	5.4669	5.1951	4.9476
9.1609	8.4881	7.8957	7.3717	6.9061	6.4906	6.1182	5.7831	5.4804	5.2060	4.9563
9.2372	8.5478	7.9426	7.4086	6.9352	6.5135	6.1364	5.7975	5.4919	5.2151	4.9636
9.3066	8.6016	7.9844	7.4412	6.9607	6.5335	6.1520	5.8099	5.5016	5.2228	4.9697
9.3696	8.6501	8.0218	7.4701	6.9830	6.5509	6.1656	5.8204	5.5098	5.2292	4.9747
9.4269	8.6938	8.0552	7.4957	7.0027	6.5660	6.1772	5.8294	5.5168	5.2347	4.9789
9.6442	8.8552	8.1755	7.5856	7.0700	6.6166	6.2153	5.8582	5.5386	5.2512	4.9915
9.7791	8.9511	8.2438	7.6344	7.1050	6.6418	6.2335	5.8713	5.5482	5.2582	4.9966
9.8628	9.0079	8.2825	7.6609	7.1232	6.6543	6.2421	5.8773	5.5523	5.2611	4.9986
9.9148	9.0417	8.3045	7.6752	7.1327	6.6605	6.2463	5.8801	5.5541	5.2623	4.9995

INDEX